# The Economics of
# HACCP
## Costs and Benefits

**Edited by**
**Laurian J. Unnevehr**
Department of Agricultural and Consumer Economics
University of Illinois, Urbana

St. Paul, Minnesota, U.S.A.

This book has been reproduced directly from
computer-generated copy submitted in final form
to Eagan Press by the editor of the volume. No editing
or proofreading has been done by the Press.

Library of Congress Catalog Card Number: 99-69124
International Standard Book Number: 1-891127-16-0

©2000 by the American Association of Cereal Chemists, Inc.

All rights reserved.
No portion of this book may be reproduced in any form, including photocopy,
microfilm, information storage and retrieval system, computer database
or software, or by any means, including electronic or mechanical, without
written permission from the publisher.

Copyright is not claimed in any portion of this work written by United States
government employees as a part of their official duties.

Reference in this publication to a trademark, proprietary product,
or company name by personnel of the U.S. Department of Agriculture
or anyone else is intended for explicit description only and does not imply
approval or recommendation of the product to the exclusion of others
that may be suitable.

Printed in the United States of America on acid-free paper

American Association of Cereal Chemists, Inc.
3340 Pilot Knob Road
St. Paul, MN 55121-2097, USA

# Preface

The Hazard Analysis Critical Control Point (HACCP) system is widely used in the food industry to prevent food safety hazards or to ensure product quality. During the last few years, it has been mandated by U.S. federal regulation for seafood, meat, poultry, and has been proposed for fruit juice, in an effort to reduce risks from food borne microbial pathogens. Many other industrialized nations have mandated HACCP for part or all of their food industries. Yet little economic research has been conducted to assess how cost effective HACCP may be for improving food safety or how it may affect food markets, consumers, and industry. Both private industry and public agencies need more information about how to integrate risk assessment, cost-benefit analysis, and economic analysis into the HACCP framework.

The NE-165 Regional Research Project on Private Strategies, Public Policies and Food System Performance, with support from the Universities of Illinois, Connecticut, and Massachusetts, and from the Farm Foundation, organized a conference in June 1998 on "The Economics of HACCP." The purpose was to provide current information on the economics of HACCP, and to foster communication among universities, public agencies and industry. Over 130 participants, from 13 countries and a wide variety of institutions, attended the conference. Dr. Catherine Woteki, USDA Undersecretary for Food Safety, gave the opening address.

This volume contains the conference papers, which covered several emerging issues for economics and agribusiness research related to the use of HACCP in the food industry:

- *How can we best evaluate HACCP as a public policy tool to improve food safety?* Several papers debated the best approaches to evaluating the public costs and benefits of mandating HACCP; other studies presented ex-post evidence regarding costs and benefits following regulation. The ex-post estimates often differ from the ex-ante impact assessments by public agencies, and provide lessons for future policy analysis.
- *To what extent are there private incentives to adopt HACCP?* Many studies documented the private benefits and incentives that are motivating firms to adopt HACCP or to improve food safety through other means. These were very apparent in the studies of food retailing, where customer contact provides immediacy to such concerns. Other studies emphasized how such incentives are being transmitted through the food chain from consumers to food processors, handlers, and producers.

- *How can HACCP costs best be modeled as part of the production process?* Several papers explored different ways of modeling HACCP costs, choosing among alternative hazard preventions, or incorporating HACCP into a production function framework. Finding appropriate models is important for future economic research.
- *How can the benefits of HACCP be compared to the costs?* Several studies attempted to quantify private and public benefits from HACCP and to link those benefits to costs. The purpose of such comparisons is to find the most cost-effective means of improving food safety. This will be continue to be a challenge for HACCP research.

The papers in this volume also provide new evidence supporting several generalizations about HACCP that have been reported earlier in the economics or agribusiness literature:

- There are economies of scale in HACCP implementation. The large costs of developing and implementing a plan are not scale neutral and will be lower on a per unit basis for larger food processors.
- The costs of HACCP include a substantial "human capital" component in plan development, training of personnel, and on-going monitoring activities. HACCP costs also may include investments in specific processes that require new physical capital and result in additional operating expenses.
- Many food industry firms have private market incentives to adopt HACCP or to improve food safety, but these are often difficult to quantify.
- HACCP is of growing importance in many countries, in food retailing, and in international food product trade.

The organization of the conference and publication of this volume were supported by the efforts of the conference steering committee: Walt Armbruster, Farm Foundation; Julie Caswell, University of Massachusetts; Neal Hooker, Texas A&M University; Helen Jensen, Iowa State University; Tanya Roberts, USDA/ERS; Cathy Wessells, University of Rhode Island; and Richard Williams, U.S. Food and Drug Administration. Special thanks are due to Alesia Strawn, Communications Specialist, University of Illinois, for production of the print version.

We hope that this volume may be of use to those concerned with food safety in both the public and private sectors.

*Laurian Unnevehr*
Conference Steering Committee Chair
University of Illinois

# Contents

**Part I: HACCP and Food Safety as Public Policy Issues in the United States** ............................................................................... 1

1. USDA Research Needs in Food Safety. *Catherine Woteki* ...................... 3

**Analyzing the Impact of HACCP Regulations in the U.S.**

2. HACCP Adoption in the U.S. Food Industry. *Sheila A. Martin and Donald W. Anderson* ........................................................................... 15

3. HACCP in Pork Processing: Costs and Benefits. *Helen H. Jensen and Laurian J. Unnevehr* ................................................. 29

4. The Cost of HACCP Implementation in the Seafood Industry: A Case Study of Breaded Fish. *Corinna Colatore and Julie A. Caswell* .............................................................................. 45

5. Measuring the Costs and Benefits of Interventions at Different Points in the Production Process: Lessons, Questions, and Comments. *Neal H. Hooker* ............................................................... 69

6. The Cost of Quality in the Meat Industry: Implications for HACCP Regulation. *John M. Antle* ...................................................... 81

7. HACCP Principles for Regulatory Analysis. *Richard B. Belzer* ............ 97

8. Benefit-Cost Analysis of Reducing Salmonella Enteritidis: Regulating Shell Eggs Refrigeration. *Hyder Lakhani* .............................. 125

**Measuring the Distributional Impact of U.S. HACCP Regulations**

9. The Distributional Effects of Food Safety Regulation in the Egg Industry. *Christiana E. Hilmer, Walter N. Thurman and Roberta A. Morales* ............................................................................. 133

10. The Costs, Benefits and Distributional Consequences of Improvements in Food Safety: The Case of HACCP. *Elise H. Golan, Katherine L. Ralston, Paul D. Frenzen and Stephen J. Vogel* .................................................................. 149

## Market Incentives for Quality and Safety Improvement

11. Market Influences on Sanitation and Process Control Deficiencies in Selected U.S. Slaughter Industries. *Michael Ollinger* ...................................................................... 171

12. The Cost of an Outbreak in the Fresh Strawberry Market. *Thomas W. Worth* ...................................................................... 187

13. Cost-Effective Hazard Control in Food Handling. *John A. Fox and David A. Hennessy* ............................................................. 199

14. A Real Option Approach to Valuing Food Safety Risks. *Victoria Salin* ............................................................................. 225

15. Economic Efficiency Analysis of HACCP in the U.S. Red Meat Industry. *William E. Nganje and Michael A. Mazzocco* .................. 241

## Part II: International Perspectives on HACCP Costs and Benefits ............................................................................ 267

## Implications of HACCP for Vertical Coordination

16. HACCP, Vertical Coordination and Competitiveness in the Food Industry. *Gerrit W. Ziggers* ..................................... 269

17. The Vertical Organization of Food Chains and Health and Safety Efforts. *Frank H. J. Bunte* ......................................... 285

## HACCP in Food Retailing

18. Applying HACCP to Small Retailers and Caterers: A Cost Benefit Approach. *Matthew P. Mortlock, Adrian C. Peters and C. J. Griffith* ...................................................................... 301

19. The 1996 *E. coli* O157 Outbreak and the Introduction of HACCP in Japan. *Atsushi Maruyama, Shinichi Kurihara and Tomoyoshi Matsuda* ............................................................ 315

20. Analysis of Implementation and Costs of HACCP System in Foodservices Industries in the County of Campinas, Brazil. *Marcia R. D. Buchweitz and Elisabete Salay* ............................................. 335

## HACCP Costs and Benefits

21. Costs and Benefits of Implementing HACCP in the UK Dairy Processing Sector. *Spencer Henson, Georgina Holt and James Northen* ................................................................................. 347

22. HACCP and the Dairy Industry: An Overview of International and U.S. Experiences. *Brian W. Gould, Marianne Smukowski and J. Russell Bishop* ............................................................................. 365

23. Costs to Upgrade the Bangladesh Frozen Shrimp Processing Sector to Adequate Technical and Sanitary Standards and to Maintain a HACCP Program. *James C. Cato and Carlos A. Lima dos Santos* ................................................................. 385

24. The Economics of HACCP Application in Argentine Fish Products. *Aurora Zugarramurdi, M. A. Parin, L. Gadaleta and Hector M. Lupin* ............................................................................. 403

# PART I
# HACCP and Food Safety as Public Policy Issues in the United States

Chapter 1

# USDA Research Needs in Food Safety

Catherine Woteki[1]

## Introduction

It's a pleasure to be here today and to open this very important conference on the economics of HACCP. HACCP is playing a prominent and growing role in establishing public policy for food safety within the U.S. Department of Agriculture (USDA) and the Food and Drug Administration (FDA). Today, I would like to provide you with an overview of our research needs in the food safety strategy for meat, poultry, and egg products -- the foods regulated by the USDA.

USDA's overall public health goal, as it relates to food safety, is to reduce the incidence of foodborne illness associated with meat, poultry, and egg products. Several years ago, following an outbreak of foodborne illness on the West Coast, USDA began a strategy for change to meet this public health goal.

Since then, several major initiatives have been implemented, including safe handling labels on products, expanded testing for microbial pathogens, performance standards for pathogen reduction, and mandatory Hazard Analysis and Critical Control Points (HACCP) systems to prevent and control contamination. Our strategy addresses the entire farm-to-table continuum because a multi-faceted approach is necessary to solve today's complex food safety problems.

We have embarked on this strategy and implemented these changes with full recognition that there are many gaps in our knowledge about the hazards in food, the risks they pose, the interventions needed to address them, and the benefits and costs of interventions. But we needed to make regulatory and policy decisions based on the best information available today, recognizing that adjustments in those policies might be needed as new information becomes available.

This is not a new situation for public health or regulatory agencies. Looking back at history, we can find many examples where action on a public health problem has been taken before all of the answers are in. In 1849, a large outbreak of cholera occurred in London, and John Snow, a founding member of the London Epidemiological Society, set out to find the source. By studying the distribution of cases, he was able to conclude the problem was coming from a well in one part of town. To control the epidemic, he simply took the handle off of the pump. It was not until some 34 years later, in 1883, that the cholera vibrio was identified, but he was able to make the association between the disease and the source -- and end the epidemic -- without knowing all of the details.

## Risk Assessment Applications

In the almost 150 years since that cholera epidemic, public health researchers have developed a variety of methods to help decision-makers understand and manage health hazards. Risk assessment is one method that is playing a major role in helping us to further improve food safety and in planning future initiatives. It has the potential to improve our ability to reduce foodborne illness in a number of ways.

Risk assessment provides structured information that allows risk managers to identify interventions that can lead to public health improvement and provides a basis for them to use in weighing the options. These options include regulatory action when necessary, but also include a broad range of options such as voluntary activities, educational initiatives, and research to close critical data gaps. In a few moments I will talk more about how we are using a microbial risk assessment for *Salmonella* Enteritidis in eggs and egg products in making risk management decisions.

Risk assessment also can be used to identify data gaps and target research that should have the greatest value in terms of public health impact. We have already begun the process of identifying research needs based on public health impact, but risk assessments will allow us to continue this work in a more quantitative manner. I'll talk more about this approach to research priority-setting as well.

Additionally, risk assessments will help industry to develop more effective HACCP plans. We are in the process of implementing a HACCP requirement in all plants that slaughter and process meat and poultry. For the future, risk assessments will help plants to scientifically develop HACCP plans.

For instance, plants can use a risk assessment to help identify hazards that are reasonably likely to occur. The best information plants may have now is qualitative -- whether a hazard presents a high, medium, or low risk. The real benefit to this change is that a hazard would be defined in terms of the risk of an adverse human health consequence, rather than in terms of contamination of a carcass or product.

Risk assessments also play an important role in international trade by ensuring that countries establish food safety requirements that are scientifically sound and by providing a means for determining equivalent levels of public health protection between countries. Without a systematic assessment of risk, countries may set import requirements that are not related to food safety and could create artificial barriers to trade. Recognizing the importance of this science-based approach to fair trade, the World Trade Organization requires each country's food safety measures to be based on risk assessment. The Codex Alimentarius Commission, which establishes international food safety standards, is now developing principles for using risk assessment in establishing such standards.

**Challenges of Microbial Risk Assessments**

While there are many potential applications for risk assessment, we are realistic about the challenges of conducting formal, quantitative risk assessments. Risk assessment has its roots in toxicology and carcinogenity studies, and its application to microbial pathogens poses some significant challenges. One challenge, of course, relates to the fact that unlike chemical, environmental, or toxicological contaminants, bacteria can multiply and produce toxins as conditions change as food moves through the farm-to-table continuum. Fortunately, researchers are making progress in developing the predictive models and other tools that will meet the technical requirements for quantifying estimates of risk.

In addition to this technical difficulty, we have many data gaps that limit the precision we can achieve through risk assessments. For instance, we have little information to accurately estimate the relationship between the quantity of a biological agent and the frequency and magnitude of adverse human health effects, particularly as this might relate to susceptible sub-populations. We also have limited information on exposure assessment -- the foods consumed by populations and the probability of contamination.

**Current Risk Assessment Activities**

Despite these challenges, I believe we are making good progress, and I would like to describe some of the risk assessment activities in which USDA is involved.

First, we have created a structure within USDA for addressing risk assessments. The Food Safety and Inspection Service (FSIS), the USDA agency with responsibility for the safety of meat, poultry, and egg products, has undergone a reorganization that establishes a new Division of Epidemiology and Risk Assessment within the Office of Public Health and Science. This division will ensure that the risk assessment paradigm is incorporated into the spectrum of work throughout the agency. The reorganization also created an Office of Policy, Program Development and Evaluation that houses our food safety risk managers. This separation of risk assessment from risk management activities will ensure the scientific integrity of the process and the products of agency-conducted risk assessments.

Also within USDA is the Office of Risk Assessment and Cost-Benefit Analysis headed by Dr. Nell Ahl, which is responsible for ensuring that major regulations proposed by USDA are based on sound scientific and economic analysis. For regulations having an annual economic impact of at least $100 million in 1994 dollars, USDA must conduct a thorough analysis that makes clear the nature of the risk, alternative ways of reducing it, the reasoning that justifies the proposed rule, and a comparison of the likely costs and benefits of reducing the risk.

USDA's Economic Research Service (ERS) is playing an important role in conducting economic research on the costs of foodborne illness and in analyzing the costs and benefits to perform cost-benefit analyses of interventions to control pathogens.

Second, in addition to establishing a structure within USDA to address risk assessment, we also are encouraging research to develop predictive models and other tools that will meet the technical requirements for conducting risk assessments.

This work is being done through the Interagency Risk Assessment Consortium, which was established as part of President Clinton's Food Safety Initiative. The Consortium links all federal agencies with risk-management responsibilities for food safety to collectively advance the science of microbial risk assessment.

Third, we are working in conjunction with other government agencies and academia to conduct risk assessments addressing meat, poultry, and egg products.

This summer, we plan to begin a risk assessment for *E. coli* O157:H7 in hamburger. We have already done some important developmental work in designing a structure for conducting such an assessment. In addition, we recently completed in conjunction with USDA's Agricultural Research Service a study to determine the prevalence of premature browning of hamburger, and the data from this study will feed into the larger risk assessment for *E. coli* O157:H7.

In addition to this risk assessment for *E. coli* O157:H7 in hamburger, we have entered into a cooperative agreement with Harvard University's School of Public Health for a risk analysis of Bovine Spongiform Encephalopathy (BSE). No cases of BSE have been identified in the United States, but we believe it would be beneficial to analyze and evaluate USDA's current measures to prevent it and identify any additional measures that may be warranted to ultimately protect the public health.

**Risk Assessment for *Salmonella* Enteritidis**

I would like to take a few minutes to discuss the risk assessment recently completed on *Salmonella* Enteritidis (SE) in eggs and egg products. This is our first quantitative, farm-to-table microbial risk assessment, and I expect it to serve as a prototype for future risk assessments. The final document is available on the FSIS web site at www.fsis.usda.gov. The scientists who conducted the risk assessment formally presented the results at a USDA-sponsored Risk Forum in Washington, DC in June 1998.

The risk assessment was conducted by a multi-disciplinary team with members from government and academia. The team was led by Dr. Roberta Morales, who was at that time on contract to USDA, and Dr. Richard Whiting, who was then with USDA's Agricultural Research Service. A number of USDA agencies had members on the team -- the Agricultural Research Service, the Animal and Plant Health Inspection Service, the Economic Research Service, the Agricultural Marketing Service, and the Office of Risk Assessment and Cost-Benefit Analysis. From the Department of Health and Human Services, we had representation from the Food and Drug Administration and the Centers for Disease Control and Prevention. Two academic institutions, North Carolina State University and Delaware Valley College, were also involved.

We began the SE risk assessment in December 1996 in response to an increasing number of human illnesses attributed to the consumption of contaminated eggs. Data from the Centers for Disease Control and Prevention indicate that SE is one of the most commonly reported causes of bacterial foodborne illness in the United States and has been increasing since 1976. This increase has occurred despite a variety of initiatives we, and the industry, have had underway to address the SE problem.

The SE risk assessment has several objectives. First, it is intended to characterize, using the data available, the adverse public health effects associated with consuming shell eggs and egg products contaminated with SE. A second goal is to identify data needs and prioritize future data collection efforts to reduce the areas of uncertainty with the assessment. Third, the risk assessment is designed to identify and evaluate potential risk reduction strategies, using a farm-to-table model.

The risk assessment extends from production to consumption of shell eggs and processed egg products. This reflects our belief that to appropriately address the problem of SE, a comprehensive strategy with multiple interventions is needed. Thus, the risk assessment has several modules reflecting each stage of the farm-to-table continuum.

The first module, shell egg production, simulates the daily SE-positive egg infection frequency for U.S. commercial flocks. The second module, shell egg processing and distribution, follows the shell eggs from collection on the farm through processing, transportation, and storage. The third module, egg products processing and distribution, tracks the change in numbers of SE organisms in egg processing plants from receiving through pasteurization. The fourth module, preparation and consumption, describes exposure from the consumption of eggs and egg-containing foods that are contaminated with SE. The fifth module, public health outcomes, links exposure to eggs and egg products containing SE with the public health outcomes of morbidity and mortality.

It would take more time than I have available today to present the findings of this risk assessment, so I will briefly review some of the results and some of the important principles we learned from the process. I urge you to take a closer look at the entire document.

First, we now know more about the incidence of illness attributed to SE in shell eggs and egg products. Out of an average of 46.8 billion shell eggs produced per year in the United States, we estimate that, on average, 2.3 million eggs contain SE. While this translates into a very small percentage of contaminated eggs, the implications for human health are quite significant -- consumption of these eggs results in an average of 661,633 human illnesses per year. We predict that 94 percent of these individuals recover without seeking medical care.

These numbers, however, are associated with a degree of uncertainty, largely because of the data gaps encountered. This relates to the second thing we learned from the risk assessment -- that there is much more we need to know about shell eggs and egg products. We encountered many data gaps during the risk assessment process, which required us to make a number of assumptions.

For example, we need more information on the association between environmental factors and management practices with SE-positive flocks on the farm. We need cooling curves in order to predict the temperature of an egg at a specified time during the processing of shell eggs, given the initial temperature of the egg, ambient air temperature, and packaging characteristics. We also need more information about industry practices regarding the storage of eggs, and consumer practices regarding the preparation and consumption of eggs. These are just a few of the many data gaps identified.

Another major result from the risk assessment is that we now have a farm-to-table model computer program we can use to determine the effects of specific interventions on the incidence of illness. In fact, as part of the risk assessment, the team evaluated a number of possible interventions on the expected number of human illnesses. They included shell egg cooling, diverting eggs from flocks with a high prevalence of SE-positive hens to breaker plants for pasteurization, and reducing the prevalence of SE-positive flocks. The potential uses for such a model in determining appropriate risk reduction strategies are many.

From this exercise, we also learned a very important principle: a broadly based policy is more likely to be effective than a policy directed solely at one area of the egg production to consumption chain. That is why our strategy to address the problem of SE in eggs will be a multi-faceted strategy that addresses the entire farm-to-table chain.

**Risk Management**

Now that we have completed this risk assessment, the results will be available to risk managers as they develop a comprehensive strategy to address the problem of SE in eggs and egg products.

On May 19, we published in conjunction with the Food and Drug Administration (FDA), an Advance Notice of Proposed Rulemaking (ANPR) in the *Federal Register* to initiate a comprehensive and coordinated process of addressing the SE problem in shell eggs and to solicit input from the public on strategies. This ANPR was published jointly by USDA and FDA because we share the responsibility for egg safety.

A number of other initiatives are already underway by USDA and FDA to address the safety of both shell eggs and egg products. For instance, FSIS is in the process of developing a proposal to mandate HACCP in egg processing plants similar to the requirement that now exists for meat and poultry plants.

The risk assessment will be very useful as we proceed with this process of risk management analysis. While the process of risk management analysis is not new in FSIS, what *is* new is that we now have a farm-to-table model with which we can predict the effects of possible interventions and determine which provide the best return in terms of public health protection. Our risk managers will evaluate all of the options available and select appropriate ones based on factors such as the level of risk, feasibility, and economic costs.

While we have issued the risk assessment as a final report, we must be careful not to consider the risk assessment itself as final. It is a fluid entity that will continually be updated as new information becomes available, making the risk assessment even more valuable as a management tool.

**Risk Communication**

I have not yet touched upon the third component of risk analysis -- risk communication. Our experiences with the SE risk assessment reinforced some principles of risk communication, and I would like to share those with you.

Good communication is important from the very beginning of the risk assessment. Everyone -- the affected industry, the public health and scientific communities, consumers, and regulators -- must have a clear understanding of the project and the opportunity to provide input. In particular, risk managers must be brought in at the beginning stages to ensure that there is agreement on the goals of the project, and communication between the risk assessors and managers should continue throughout the process.

Stakeholder input is imperative throughout the risk assessment process. We held several meetings to solicit input from stakeholders during our risk assessment, and we found it very important to the integrity of the project. We intend to continue this open process as risk management decisions are made.

Communication with the public will also be important as we determine what educational strategies are necessary to reduce the risk of foodborne illness from SE in shell eggs and egg products. It will be extremely important that we communicate risks to the public in a proper context and in a helpful manner with a well-thought-out and consistent message.

We are getting some experience with risk communication right now, in fact. We are using a study conducted by the Agricultural Research Service on the premature browning of hamburgers to determine what should be the appropriate message to consumers on how to determine the doneness of hamburgers. We must balance the science with practicality because a science-based recommendation is of little use unless consumers will follow it.

## Research

Research is very important to the success of these regulatory and non-regulatory food safety initiatives. In order to effectively address the safety of meat, poultry, and egg products, we need to know more about the hazards in these foods and their relation to illness.

Our programs must be responsive to new information and new data. Because our food safety strategy has broadened to cover the farm-to-table continuum, our research agenda must also address the entire continuum.

On a very basic level, we must encourage fundamental research on the ecology of human pathogens in animals. We need to know, for example, how these pathogens grow, develop and colonize in livestock; how they acquire the ability to produce toxins; and how they acquire antibiotic resistance. This fundamental science will set the stage for potential breakthroughs in vaccines, competitive exclusion, and other preventive approaches.

We also need research to help us make regulatory decisions that are based on the most current science. HACCP provides an important framework for improving food safety within plants, but it must be combined with science-based performance standards that HACCP systems are designed to achieve. Our ultimate goal is to base performance standards on quantitative risk assessments. We need extensive, new data to conduct these risk assessments.

**Current Estimate**

Where will this data come from? Agricultural research is now a shared responsibility of both the public and private sectors (see Table 1). The federal government has played a major role in supporting agricultural research for over a century. It supports intramural research through the Agricultural Research Service, the Forest Service, and the Economic Research Service. It supports the collection of valuable statistical data through the National Agricultural Statistics Service. It also funds extramural research at state institutions, which is administered by the Cooperative State Research, Education, and Extension Service. This strong, federally-funded research base is needed because there are research problems and issues of national importance that may receive too little attention from individual states or regions. This research also serves the needs of the regulatory and action agencies.

Despite the extensive investments made by the federal government, the lack of growth in public agricultural research expenditures -- federal and state -- constrains the ability of the public agricultural research system to respond to new demands. Federal expenditures have not really grown in real terms since the mid-1970s. The recently enacted research title reauthorization will help to turn around this dismal situation. It provided $120 million per year in competitive grants for the next five years from mandatory funds, and food safety is one of the priority areas.

At the same time, we know that over the past 30 years, the importance of the private sector in both funding and conducting agricultural research is growing.

Agricultural research will continue to require involvement by the federal government in areas where private incentives are weak, and many aspects of food safety research fall into this area. But I believe the growing importance of food safety and the impact it can have on business are providing a growing incentive for private industry to support fundamental, as well as applied, food safety research. I believe this mutual interest in food safety research provides opportunities for partnerships in the future.

**TABLE 1**
**Research, Education, and Economics Agencies**

|  | Staff Years (ceiling) | FY '97 Budget (millions) |
|---|---|---|
| ARS | 7,901 | $800 |
| CSREES | 399 | $912 |
| ERS | 620 | $53 |
| NASS | 1,130 | $101 |
| Total -- REE | 9,950 | $1,848 |

**Food Safety Research Agenda**

FSIS is not a research agency, but it needs to identify research that is needed to improve food safety. That means reaching out to the research agencies within the federal government, and to scientists in academia and the private sector. For that reason, in 1997, the Agency developed a Food Safety Research Agenda as one means of communicating with those outside the Agency about its priorities in food safety research. As a first step, FSIS established a Food Safety Research Working Group, which was composed of government scientists representing a broad base of expertise.

The working group used human health effects as the basis for determining FSIS research needs. Using the criteria, the group reached a consensus on the major research questions that needed to be answered. They identified general research questions as well as research needs that are unique to the following pathogens: *Salmonella*, *Campylobacter*, *Listeria*, and *Enterohemorrhagic E. coli*, including *E. coli* O157:H7.

**Criteria Used to Develop the Research Agenda**
1. Incidence of Adverse Health Outcome
2. Causes of Adverse Health Outcome
   a. Chemical
   b. Physical
   c. Biological
3. Linkage (etiological/vehicle linked to food)
4. Outcomes
   a. Sequela
   b. Deaths
   c. Distribution (demographics/populations)
   d. Costs
      *Medical
      *Productivity Losses
   e. Public Sensitivity/Perceptions

These are the questions from the research agenda that were developed for *E. coli*.

***E. coli* Research Questions**

1. What is the incidence of EHEC and *E. coli* O157:H7 disease/infection in humans and animals in the United States? What is the relative incidence among different subpopulations?

2. What are the virulence factors associated with EHEC? Are all Shiga toxin-producing E.coli (STEC) pathogenic for humans, i.e., are all STEC also considered EHEC? Which virulence factors are associated with bloody diarrhea, hemolytic uremic syndrome, or other sequelae?

3. How do EHEC colonize both animals and humans?

4. What is the infectious dose and the dose-response relationship of EHEC and *E. coli* O157:H7 for humans and animals? Does a threshold exist below which illness does not occur? Is a zero tolerance standard supportable by scientific evidence? Can dose response data calculated for Shigella sp. or S. dysenteriae type 1 be used for EHEC and *E. coli* O157:H7?

5. What is known about the microbial ecology of EHEC and *E. coli* O157:H7? What are the environmental reservoirs for EHEC and *E. coli* O157:H7 along the farm-to-table continuum? What are survival and growth characteristics before and after cooking?

6. Should we be screening *E. coli* from human disease, and/or from food, for toxin production, or for the presence of stx, eae, hyl, EHEC plasmid, adhesins, etc.? Should we be screening human fecal specimens and/or foods for the presence of Shiga toxins?

The questions that the working group developed are important for several reasons. First, because the groups used public health criteria as a means of determining research priorities, we can consider the pathogens identified by the group to be the major pathogens of concern for future research.

Second, the research agenda reflects the direction taken by the President's Food Safety Initiative. The initiative supports the use of risk assessment as a means of characterizing risks to human health associated with foodborne hazards and assisting regulators in making decisions about where in the food production and marketing chain to allocate resources to control those hazards.

The challenge for the future will be to integrate all of the research needs stated in the FSIS Research Agenda and the President's Food Safety Initiative into an operational plan that reflects the emphasis on cooperation and partnerships.

To assist in this process, an interagency working group convened by the White House Office of Science and Technology Policy has been working to coordinate federal research priorities and planning. The goal of this working group will be to develop a coordinated federal food safety research plan that will extend to our research partners in states, industry, and academia. We expect the group to report this summer.

## Closing

In closing, I hope I have provided a useful update on USDA's food safety initiatives and described the challenges for the future. True progress will require that all of us work together to meet mutual goals. I look forward to working with you to see these goals become reality.

**Notes**

[1] Dr. Woteki is Under Secretary for Food Safety, Office of the Under Secretary for Food Safety, U.S. Department of Agriculture.

Chapter 2

# HACCP Adoption in the U.S. Food Industry

Sheila A. Martin and Donald W. Anderson[1]

## Introduction

As the Food and Drug Administration (FDA) considers broadening Hazard Analysis and Critical Control Points (HACCP) requirements to regulate additional segments of the food industry, it needs to know how these requirements will affect each industry segment. FDA needs this information not only to satisfy its legal requirements under Executive Order 12866 and the Regulatory Flexibility Act, but also to ensure that it can structure the regulation in the most cost-effective manner. Armed with information about the cost of compliance to each industry segment under alternative regulatory scenarios, FDA can consider these costs in combination with the expected benefits of HACCP implementation in these industries to provide the greatest level of protection for the nation's food supply while minimizing the costs of compliance.

Costs to industry of complying with new HACCP regulations depend not only on the specific requirements of the regulation, but also on the status of food safety practices in the regulated plants at the time the regulation takes affect. As HACCP and HACCP-related food safety practices diffuse throughout the food industry, the regulatory costs of new HACCP regulations will fall as fewer plants must make significant process changes in order to come into compliance. On the other hand, the benefits of HACCP regulation will fall as well because regulators cannot take credit for benefits that occur as a result of market forces that lead industry to adopt HACCP practices on its own.

HACCP is quickly becoming standard practice in a number of segments of the food industry, and HACCP regulations have played an important part in this diffusion. The FDA has already implemented a HACCP rule for the seafood industry and is in the process of implementing a HACCP regulation for juice. USDA passed the meat and poultry HACCP rule in 1996, and all establishments are required to comply by the year 2000. As these regulations have been announced and enacted, HACCP practices

have diffused not only among plants in the regulated sector but also among plants that:
- supply the regulated sector,
- are owned by a company that also owns plants in the regulated sector, or
- compete with companies that are implementing HACCP for either of the above reasons.

Because the cost of HACCP to industry depends in part on the current status of regulated plants regarding HACCP and other food safety practices, several important questions are relevant to the issue of HACCP costs:
- How widespread are HACCP and other food safety practices among food processing plants?
- Is the food industry already following the prerequisite programs essential to effective HACCP?
- Will plants have to make significant changes in their manufacturing processes to implement HACCP?

This paper reports on the results of a survey that was conducted to determine the current and future status of the food safety practices of plants under FDA's jurisdiction. Until now, very little information was available on the current food safety practices of plants. Although other studies have estimated costs of HACCP regulations (Anderson et al. 1997; Nganje 1997; USDA 1996; Knutson et al. 1995; Martin et al. 1993), most focus on the cost of HACCP planning and not on the potentially significant changes in HACCP costs that may be required of many plants complying with regulatory HACCP. In conjunction with information on the cost of implementing each of the possibly required process changes, this information about the incidence of specific food safety processes will allow FDA to consider the cost of HACCP regulations of alternative designs.

The first section of this paper describes how the survey results inform a broader study of the cost of HACCP to industry and our approach to estimating these costs. The second section describes the survey instrument, sample design, and survey implementation procedures. In the third section, we describe some preliminary results from the survey. Finally, we provide a few observations about the implications of the survey results.

## Approach

This survey collected data to support a study of the costs of HACCP under a number of regulatory scenarios. Our approach to estimating the cost of HACCP consists of two analytical steps:

Step 1: Determine the baseline rate of HACCP compliance.

Step 2: Determine the cost of moving from baseline to compliance.

As shown in Figure 1, the baseline can change over time. As more plants implement HACCP, the cost of a future HACCP regulation falls. We define regulatory costs as those costs that occur as a direct result of activities undertaken to comply with the regulation. By this definition, the costs to plants that implement HACCP in the absence of a regulation should not be included in regulatory costs. This definition may be somewhat controversial. One could argue that the threat of a regulation may encourage some plants to adopt HACCP prior to the enactment of a regulation and that these costs should be attributed to the regulation as well. Nevertheless, in this study, HACCP regulatory costs include only the costs of moving from baseline practices to regulatory compliance.

Determining a baseline measurement of HACCP compliance posed a particularly complex problem: how should we define the specific practices that a plant would have to follow in order to comply with a HACCP regulation that does not yet exist? This was particularly difficult because:

- we are examining the entire FDA-regulated food industry (except seafood);
- a HACCP regulation may differ significantly for different industries;
- the HACCP regulation may allow plants significant flexibility in implementing HACCP, which would imply that a variety of practices might be used by different plants;
- we wanted to capture not only the costs of HACCP training and HACCP plan development, but also the process changes needed to implement the HACCP plan; and
- we wanted to provide FDA with an analysis that would be valid under a variety of regulatory scenarios.

To meet these significant challenges, we worked with FDA's Office of

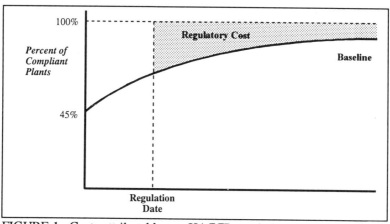

FIGURE 1. Costs attributable to a HACCP regulation.

the Strategic Manager for HACCP and a number of food safety experts and industry associations to develop a list of process changes that could possibly be required under an FDA HACCP rule for a variety of industries. These processes became the core of the survey. Using the survey data and information about the cost of each potentially required process change, we will develop a spreadsheet model that allows FDA to calculate the costs of a HACCP regulation that includes any subset of these potentially required processes.

## Survey Procedures

### Survey Instrument Design

The survey instrument was divided into three sections:
- HACCP development and training,
- food safety processes, and
- sanitation programs.

The section about HACCP development and training asked respondents specific questions about their progress toward HACCP plan development and implementation and about the extent to which their employees had received HACCP training. The second section was designed to capture information about specific food safety practices that may be necessary to implement HACCP. This section was divided into 15 different generic food processes, shown in Table 1. For each of these generic processes, we included questions about several food safety procedures. In total, the survey included questions about 105 food safety procedures. However, because most plants do not conduct all of the 15 generic food processes,

**TABLE 1**
**Generic Food Processes**

| | |
|---|---|
| ✓ Receiving | ✓ Heat treatment |
| ✓ Storing (refrigerated and dry) | ✓ Pasteurization |
| ✓ Controlling product/temperature during processing | ✓ Packaging |
| ✓ Batching | ✓ Inspecting container seal integrity |
| ✓ Culturing and fermenting | ✓ Metal detection |
| ✓ Drying | ✓ Testing finished product |
| ✓ Low water activity processing | ✓ Refrigerated distribution |
| ✓ Baking | |

not all of the food safety procedures applied to each plant. This structure allowed us to customize the survey for each plant. For example, a typical bakery might answer questions about receiving, dry and refrigerated storage, batching, baking, packaging, and testing the finished product. Gateway questions introducing each section (e.g., "Does your plant bake any product or any of its ingredients in this facility?") allowed us to determine which segments of the survey each plant should complete.

The final section of the survey asked respondents specifically about their sanitation procedures. Although these procedures are already required by Good Management Practices (GMPs), many plants are not fully complying with these GMPs. HACCP implementation will require that plants implement prerequisite programs such as sanitation programs before HACCP can be fully operational. Thus, FDA considers the costs of catching up to current GMPs part of the cost of HACCP compliance and wants to know the extent to which currently required sanitation practices are not being followed.

**Sample Design**

FDA's Official Establishment Inventory (OEI) database provided the sampling frame for the survey. The OEI contains information on all plants under FDA's jurisdiction. We defined the survey universe as the 13,060 plants in the OEI that manufacture or repackage food products in the U.S. for human consumption, with the exception of the seafood industry, for which a HACCP cost analysis has already been conducted. Food importers and foreign processors are not part of the sampling frame, except for food importers that also either manufacture or repack foods.

To facilitate the sample design and selection, we supplemented the OEI database with additional variables from American Business Lists (ABL). We matched names and addresses of plants from the OEI database with plant and corporate parent company names and addresses from ABL and added the primary SIC code, number of employees, and the annual revenue of the plant. We used the SIC codes to identify and exclude establishments that, although they are in the OEI database as manufacturers or repackers of a relevant product, are primarily retail establishments that will be exempt from HACCP. We used the employment and revenue data from ABL to identify small and large businesses according to Small Business Administration size standards for each four-digit SIC code.

The primary purpose of stratification is to ensure that substantive differences in key outcomes between population subdivisions are detected with acceptable statistical power. In this case, subdivisions of the population of particular interest are *company size* and *industry* because these characteristics will be important factors influencing the cost of HACCP. We stratified the universe of 13,060 establishments into 47 food industries, or substrata, and grouped them into six broader food industry categories, or superstrata. These superstrata represent groupings of similar food industries identified by their primary four-digit SIC codes as follows:

1. Premarket Services and Other Products
2. Animal Protein Products
3. Preserved/Additive/Supplement Products
4. Cereals/Grains/Baked Products
5. Beverages/Sugared Products
6. Products for Sensitive Consumers

Each of the six superstrata were stratified further into two size categories, plants owned by small companies and plants owned by large companies, resulting in 12 sampling strata. Because FDA is particularly interested in superstratum 6 -- Products for Sensitive Consumers -- we surveyed all of the plants in this superstratum.

Because this study may contribute to regulatory impact analyses, we based our size strata on the definitions used by the Small Business Administration (SBA). The SBA defines size based on the size of the company, rather than the size of the plant. Thus, the small category includes plants that are owned by small companies, while the large category includes plants owned by large companies. The definitions of *small* and *large* vary by SIC code; however, companies with more than 500 employees are considered large in most cases.

Table 2 shows the number of establishments in the sampling universe, divided into the 12 sampling strata, and the sample allocation. Our targeted number of respondents was 1,231. Our sample allocation was designed to yield 240 respondents from the first five superstrata (recall that the sixth superstrata was censused). We developed the sample

**TABLE 2**

**Survey Universe and Final Sample**

| | Survey Universe | | Sample | | Completed Interviews | |
|---|---|---|---|---|---|---|
| | Large | Small | Large | Small | Large | Small |
| **Premarket services and other products** | 697 | 3,565 | 182 | 371 | 50 | 63 |
| **Animal protein products** | 309 | 1,193 | 133 | 448 | 40 | 90 |
| **Preserved/additive/ supplement products** | 420 | 1,914 | 147 | 427 | 45 | 88 |
| **Cereals/grains/baked products** | 445 | 1,972 | 154 | 420 | 32 | 74 |
| **Beverages/sugared products** | 519 | 1,980 | 161 | 406 | 24 | 75 |
| **Products for sensitive consumers** | 25 | 21 | 25 | 21 | 8 | 6 |
| **Subtotal** | 2,415 | 10,645 | 802 | 2,093 | 199 | 396 |
| **Total** | | 13,060 | | 2,895 | | 595 |

allocation by estimating best- and worst-case scenarios for contact and response rates and selected a much larger sample, which was divided into ten waves in each superstratum. Thus, if the actual sample yield was between our best and worst assumptions, we could assign as many of the ten waves as needed to obtain the desired number of completed interviews.

As we discuss later, more than one-third of the sampled plants were ineligible for the survey, and the contact and response rates for the eligible plants were close to our pre-survey worst assumptions. These factors caused the number of completed interviews to be considerably less than initially planned but still sufficient for most of the planned tabulations and analyses.

**Survey Procedures**

We conducted the survey using the following three-step procedure. First, establishments in the sample were called and screened for eligibility. Establishments had to meet the following criteria to be eligible for the survey:
- the plant must manufacture or repack food product for human consumption, and
- the plant's primary industry classification must be one of the 47 industries in the study.

If eligible, we identified an appropriate contact person (e.g., plant manager, HACCP team leader, or QA manager) and asked him or her to participate in the survey. If the contact person agreed to participate, we scheduled a date and time for the follow-up telephone interview. Next, we mailed an introductory packet to those establishments recruited. The introductory packet included a cover letter, an information sheet, cue sheets to refer to during the telephone interview, and a list of definitions. Finally, we called respondents back at the scheduled time to conduct the follow-up telephone interview. On average, the follow-up telephone interview lasted about 30 minutes.

**Survey Response**

We completed a total of 595 interviews -- 199 with large plants and 396 with small plants. Table 2 shows the number of completed interviews by sampling strata.

A number of plants were ineligible because they did not have a phone or were out of business. The eligibility rate was 64 percent. The overall response rate for all plants was 32 percent.

Although we did not complete our targeted number of interviews, the number of completes is sufficient for most of the planned tabulations and analyses. The shortfall in the number of respondents was caused by a combination of lower-than-expected eligibility rates and lower-than-expected response rates. Our eligibility rates were lower than anticipated because of the age of the sampling frame due to both the delay in starting the full-scale interviews and inaccuracies in the initial OEI database. Many plants were either out of business or no longer manufacture or repack food products.

## Preliminary Results

### HACCP Training and Knowledge

Table 3 presents the distribution of the respondents' answers to several questions about HACCP training and knowledge. First, we asked the respondents, "Have you or any staff at your plant received training, such as workshops, seminars, courses, or independent study, regarding HACCP?" Then we asked them to specify the source of the training they received. Finally, we asked them to rank their familiarity with HACCP on a five-point Likert scale, with one being "unfamiliar" and five being "very familiar."

The results show that 63 percent of plants have staff that have received some form of HACCP training. Although almost 60 percent of plants owned by small firms have received some form of HACCP training, only 12 percent would rank themselves as very familiar with HACCP. The most common form of training for both large and small companies is industry trade groups. Ten percent of plants owned by large companies and 22

TABLE 3
HACCP Training and Knowledge

|  | Small | Large | Significance[a] | Overall |
|---|---|---|---|---|
| **Percent of plants with HACCP training** | 59% | 79% | * | 63% |
| **Most common source of training[b]** | Trade group | Trade group | N/A | Trade group |
| **Percent of plants rating themselves "very familiar" with HACCP[c]** | 12% | 21% | * | 14% |
| **Percent of plants rating themselves as "unfamiliar" with HACCP** | 22% | 10% | * | 19% |

[a] An asterisk in this column indicates that there is a statistically significant difference between the percentage of small and large plants at the 95 percent confidence level; N/A indicates that the significance does not apply to this question.

[b] Choices for source of training included (1) trade group; (2) government agency; (3) college or university (e.g., extension service); (4) consultant or consulting firm; (5) independent study; (6) other.

[c] Respondents chose a number from one to five, with one being unfamiliar and five being very familiar.

percent of plants owned by small companies still consider themselves unfamiliar with HACCP.

**HACCP Implementation**

Table 4 shows the percentage of plants that have conducted a hazard analysis, written a HACCP plan, and implemented a HACCP plan for at least one product. Overall, about 58 percent of plants have conducted a hazard analysis, but only 45 percent have written or implemented a HACCP plan. Small companies lag behind large companies with respect to HACCP implementation. The industries leading the implementation of HACCP are Animal Protein Products, Cereals/Grains/Baked Products, and Products for Sensitive Consumers.

**HACCP Diffusion**

Figure 2 plots the results of answers to two questions. First, we asked plants that have implemented HACCP the year in which they first implemented HACCP. Then we asked plants that have not implemented HACCP about their plans regarding HACCP in the absence of an FDA regulation. The exact wording of the question was, "Which of the following describes your plans regarding HACCP, if the government does not issue HACCP regulations? Do you: (1) intend to implement a HACCP

TABLE 4
Percentage of Plants that Have Implemented HACCP

|  | Small | Large | Significance[a] | Overall |
|---|---|---|---|---|
| Conducted a hazard analysis for at least one product | 52% | 78% | * | 58% |
| Written a HACCP plan for at least one product | 39% | 69% | * | 45% |
| Implemented a HACCP plan for at least one product | 39% | 67% | * | 45% |
| Lead industries (implementation)[b] | Animal protein products (59%) | Cereals/ grains/ baked products (87%) | N/A | Products for sensitive consumers (74%) |

[a]An asterisk in this column indicates that there is a statistically significant difference between the percentage of small and large plants at the 95 percent confidence level; NA indicates that the significance does not apply to this question.

[b]This row shows the industry stratum with the greatest percentage of plants that have implemented a HACCP plan for at least one product. The percentage for that stratum appears in parentheses.

plan within the next year, (2) intend to implement a HACCP plan within the next five years, or (3) have no plans to implement a HACCP system?"

Relative to its speed during the 1980s, the diffusion of HACCP has been relatively rapid since 1990. Although some companies plan to implement HACCP within the next few years, about 30 percent of plants owned by small companies and 12 percent of plants owned by large companies have no plans to implement HACCP if FDA does not require it. This implies that these companies do not face market forces that provide the incentive to implement HACCP in the absence of regulation.

**Process Complexity**

The complexity of a plant's process contributes to its HACCP implementation cost for two reasons:
1. the HACCP plan may include a greater number of critical control points, increasing the cost of developing and implementing the HACCP plan; or
2. the plant may need to implement a greater number of process changes as part of a HACCP plan.

We define process complexity as the number of different generic food processes taking place at the plant. Recall that we defined 15 generic food processes, which are shown in Table 1. We calculated the distribution of the plants by the value of the variable *complexity*. For all plants, the value of this variable ranged from 1 to 12. Large plants appear to have slightly more complex processes; 58 percent of large plants have six or more processes, while only 50 percent of small plants have six or more

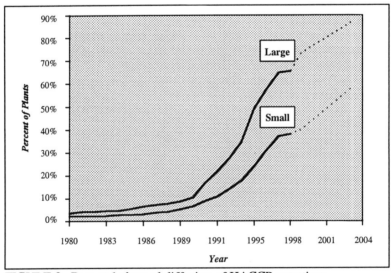

FIGURE 2. Past and planned diffusion of HACCP over time.

processes. However, it appears that the differences in complexity between large and small plants are not great. This observation agrees with a similar finding in RTI's study of HACCP needs for small meat and poultry plants. RTI found that small plants actually had more complex processes than large plants (Anderson 1997) and would therefore require HACCP plans that are at least as complex as those of large plants.

### Sanitation Processes

Table 5 shows some of the key results regarding sanitation processes. Despite current GMPs, only 76 percent of plants currently have a written sanitation program, and only about 78 percent keep records to verify that sanitation inspections have taken place. Small plants generally lag behind large plants in this area. If these plants are regulated under a HACCP rule, part of their costs will be catching up so that they are compliant with GMPs.

### Process Changes

Table 6 presents the percentage of plants following several food safety practices. The practices shown in the table are representative of the 105 different safety practices included in the survey. As explained above, not all processes were relevant to each plant. For example, we only asked about time temperature devices on ovens if a plant conducted baking. Thus, the percentages reflect the number of plants following each procedure as a percentage of those plants in which the practice is relevant.

The table shows that, in general, large plants lead small plants in applying these food safety processes. However, the differences between large and small companies are statistically significant for only a few questions. The standard errors for some procedures were larger than others in part because a fewer number of respondents answered these questions. For example, only plants that bake product answered questions referring to ovens. Where the differences are significant, a greater percentage of small plants will have to make process changes in order to comply with a HACCP rule that requires these processes. For example, suppose FDA

**TABLE 5**

**Percentage of Plants Meeting Selected Sanitation Requirements**

|  | Small | Large | Significance[a] | Overall |
|---|---|---|---|---|
| **Plant has a written** |  |  |  |  |
| sanitation program | 71% | 91% | * | 76% |
| **Plant keeps records to** |  |  |  |  |
| **verify sanitation** |  |  |  |  |
| inspections | 73% | 87% | * | 78% |

[a]An asterisk in this column indicates that there was a statistically significant difference between the percentage of small and large plants at the 95 percent confidence level.

determined that the practice of obtaining vendor guarantees of the safety of ingredients and materials would spread the responsibility for food safety among all segments of the food industry, from farm to table. If FDA decided to require that all plants acquire written vendor guarantees that raw materials meet specific requirements for biological, chemical, or physical contamination, 31 percent of regulated plants owned by small firms and 13 percent of the regulated plants owned by large firms would need to put this process in place to comply with the regulation.

## Observations

These preliminary data indicate that the costs associated with a HACCP regulation will fall as companies continue to adopt HACCP on their own in response to market forces. However, because the benefits of HACCP

### TABLE 6
### Percentage of Plants Practicing Food Safety Procedures

| Procedure | Small | Large | Significance[a] | Overall |
|---|---|---|---|---|
| **Requiring written vendor guarantees** | 69 | 87 | * | 73 |
| **Performing biological testing of ingredients** | 50 | 64 | * | 53 |
| **Monitoring time and temperature of raw materials** | 45 | 53 | | 47 |
| **Calibrating thermometers to traceable standards** | 52 | 72 | * | 57 |
| **Weighing food additives** | 91 | 82 | | 89 |
| **Monitoring water activity during drying** | 45 | 46 | | 45 |
| **Using time/temperature-recording devices on ovens** | 51 | 60 | | 53 |
| **Verifying the pasteurization process** | 87 | 96 | | 90 |
| **Using a tamper-evident seal** | 54 | 54 | | 54 |
| **Performing destructive testing of seals** | 49 | 61 | | 52 |
| **Monitoring temperature during refrigerated distribution** | 67 | 68 | | 67 |

[a]An asterisk in this column indicates that there was a statistically significant difference between the percentage of small and large plants at the 95 percent confidence level.

regulations will also fall, it is unclear whether the cost effectiveness of a HACCP regulation will rise or fall over time. We suspect that the industry segments that are least likely to adopt HACCP in the absence of an FDA regulation are those in which the market incentives to do so are not sufficiently strong to justify the costs. These could include industry segments in which food safety problems are rare, there is little brand recognition or loyalty, or it is difficult to trace the source of a particular batch of product. In these markets, companies may have little to gain from improving their food safety record.

HACCP is clearly spreading faster in large firms than in small firms, and many food safety practices are also more common in large firms. While this is not surprising, it does have implications for the expected burden of a HACCP regulation. Several factors increase the cost burden of HACCP on small firms:

- They have fewer food safety practices in place;
- they are less likely to be familiar with HACCP;
- their processes are just as complex as larger firms;
- they have fewer company resources available to help them develop and implement a HACCP plan; and
- fixed costs are spread over a smaller volume, increasing the per-unit costs.

We plan to conduct additional analyses of these data to develop a better understanding of their implications for the cost burden of HACCP regulations. Ultimately, we will construct a tool that FDA can use to consider the cost implications of alternative regulatory designs. FDA must consider these costs in relation to the public benefits of alternative regulatory scenarios. For example, FDA may consider developing a HACCP regulation for a specific industry that exempts specific industry segments from certain requirements or allows for a longer implementation period. The information collected in this survey combined with information on the potential benefits of these policies will provide FDA with the data necessary to consider the costs and benefits of these kinds of regulatory alternatives.

**Notes**

[1] Sheila A. Martin is Senior Economist at the Center for Economics Research, Research Triangle Institute. Donald W. Anderson is Senior Economist at the Center for Economics Research, Research Triangle Institute. This research was supported by the Food and Drug Administration under Contract Number 223-96-2290 and Contract Number 223-91-2238. We are indebted to Dr. David J. Zorn for his assistance throughout this project.

## References

Anderson, D.W., J.L. Teague, and D.B. Phillips. 1997. Hazard Analysis and Critical Control Point (HACCP) Needs Assessment for Small Business. Prepared for the U.S. Department of Agriculture, Food Safety and Inspection Service, June.

Knutson, R.D., H.R. Cross, G.R. Acuff, L.H. Russell, F.O. Boadu, J.P. Nichols, S. Wang, L.J. Ringer, A. B. Childers, Jr. And J.W. Savell. 1995. Reforming Meat and Poultry Inspection: Impacts of Policy Option. Unpublished manuscript, Institute for Food Science and Engineering (IFSE), Agricultural and Food Policy Center, Center for Food Safety, Texas A&M University, April.

Martin, S.A., B.J. Bowland, B. Calingaert, N. Dean and D. Ward. 1993. Economic Analysis of HACCP Procedures for the Seafood Industry. Volumes 1 and 2. Prepared for the U.S. Food and Drug Administration, Center for Food Safety and Applied Nutrition, November.

Nganje, William E. 1997. HACCP Costs of Small Meat Packing Plants: Summary of Survey Results. Working Paper 970527, Food and Agribusiness Management program, University of Illinois, Urbana, IL, May.

U.S. Department of Agriculture, Food Safety and Inspection Service. 1996. Pathogen Reduction; Hazard Analysis and Critical Control Point (HACCP) systems; Final Rule. Federal Register, Part II, 651, no. 144(25 July 1996): 38805-38989.

Chapter 3

# HACCP in Pork Processing: Costs and Benefits

Helen H. Jensen and Laurian J. Unnevehr[1]

## Introduction

Food safety regulations issued in July 1996 mark a new approach to ensuring the safety of meat and poultry products. The U.S. Department of Agriculture's (USDA) Food Safety and Inspection Service (FSIS) moved from a system of carcass-by-carcass inspection to an approach that relies on science-based risk assessment and prevention through the use of Hazard Analysis and Critical Control Point (HACCP) systems. Under the new regulations, the government requires meat processors to put a HACCP plan in place, to conduct periodic tests for microbial pathogens, and to reduce the incidence of pathogens. The new regulations shift greater responsibility for deciding *how* to improve food safety in the processing sector to processors themselves. Thus, the intent of this regulation was to promote more efficient resource allocation in food safety improvement (reducing inputs in control and/or improving food safety outcomes).

In addition to the need to improve the safety of food products to meet new federal standards, firms also have private incentives to improve both food safety and the shelf life of meat products. Currently, these private incentives are most apparent in growing export markets for meat products, but also occur through contracting of final product from large purchasers, such as ground beef for fast food restaurants (Seward). Thus, industry has both market and regulatory incentives to improve food safety, and to do so in the most cost-effective manner.

The demand for improved food safety has induced changes in methods used in meat processing for pathogen control. New technologies for pathogen control include both specific interventions or actions in the production process as well as new methods of managing process control (i.e., HACCP). The adoption of the new technologies allows the processing firm to achieve a safer food product through reduced pathogen levels. The challenge for the industry is to evaluate which set of interventions is the most cost-effective for achieving pathogen control.

In this paper we investigate the production of food safety in meat processing in order to better understand the costs and benefits of changing food safety levels. The motivation for doing this is to provide better information for the marginal benefit/cost analysis of food safety interventions. This is the type of information that is needed to assess the cost-effectiveness of the new food safety regulation, as discussed in Unnevehr and Jensen and in MacDonald and Crutchfield. The FSIS impact assessment of the rule on food safety (USDA/FSIS 1996a) was limited by lack of information on the marginal costs of food safety production.

Here, we specifically address: a) the structure of costs incurred by the firm in applying interventions to control food safety in meat processing; b) new data on the cost and effectiveness of selected food safety interventions in pork processing; and c) an economic framework for choosing optimal sets of interventions. The intent is to provide basic information on the cost frontier and, hence, marginal costs associated with improved pathogen control at the plant level.

The paper is organized as follows: in the next section we provide an overview of the HACCP-based pathogen reduction regulation and previous estimates of the total cost of regulation. Next, we discuss the structure of costs and benefits for food safety improvement in pork processing. Then, we propose a model for evaluating adoption of selected technologies available to pork firms for pathogen control. The model results highlight the tradeoffs between private and public objectives for pork processing firms and reveal how steeply marginal costs rise as pathogen standards are tightened. In the final section we offer some implications with respect to the overall costs of achieving greater food safety.

## HACCP Regulation and Industry Costs of Improving Food Safety

Government intervention can take many forms, including direct regulation. How the regulation is specified has an effect on both the allocation decisions of the firm as well as the firm's costs and profits under the regulation (Helfand). The new FSIS rule regarding pathogen reduction combines both a process standard by requiring the adoption of a HACCP system and performance standard in setting allowable levels for *salmonella* and generic *E.coli* in products (Unnevehr and Jensen). According to Helfand, this type of combined standard theoretically encourages high levels of production but tends to reduce profits more than a simple performance standard.

In the case of microbial pathogens, performance standards are costly to monitor and enforce for many different pathogens. Thus, the combined performance/process standard represents an attempt to improve overall food safety without undue testing costs. Although there is no single indicator pathogen that can be used to evaluate the safety of products, testing for *salmonella* (by FSIS) on raw meat products is used to verify that standards for this microbial pathogen are being met; testing for generic *E.coli* (by the firms) on carcasses is used to verify the process control for fecal contamination (Crutchfield et al. 1997). HACCP systems that reduce these two pathogen may be assumed to result in overall improvements in food safety.

The use of HACCP as the basis of pathogen control in plants has basically two components, as previous studies have recognized. The first component is the pure process control aspect of training, monitoring, record keeping, and testing, which has been the focus of previous estimates of the costs of the regulation to industry (Roberts, Buzby, and Ollinger). The second component is the cost of specific interventions to reduce pathogens. Plants incur these costs in order to meet pathogen reduction goals; hence, these costs need to be considered as costs of the pathogen reduction regulation (MacDonald and Crutchfield). Relatively little is known about the second set of costs, in part because there is uncertainty regarding how much new technology will be needed to meet specific pathogen reduction targets. Earlier forms of the FSIS regulation mandated that each firm would have to introduce at least one antimicrobial technique in the production process, but this requirement was abandoned in favor of allowing firms greater flexibility in meeting performance standards.

Roberts, Buzby, and Ollinger provide a summary of the costs for the meat and poultry industries estimated by the FSIS (both preliminary and revised) and by the Institute for Food Science and Engineering (IFSE) at Texas A&M. The annual costs of process control under HACCP consisted of planning and training, record keeping, and testing. The revised FSIS regulation cost estimates for these recurring process control efforts was $75 million; IFSE estimated these costs at $953 million. One source of the difference in the estimates was a very high estimate of testing costs from IFSE. They assumed that industry would have to incur costs over and above the required tests for *E.coli*, simply to monitor performance of their HACCP systems. The wide variation in estimated costs of implementing HACCP shows the inherent uncertainties and wide range of possible assumptions (e.g., the number of critical control points).

Regarding the second major component, process modification costs, FSIS reported an estimated range of $5.5 to $20 million (Roberts, Buzby, and Ollinger). The modification cost estimates, however, are very uncertain because the extent of necessary modifications to meet performance standards is unknown. The original FSIS and the IFSE cost estimates did not include these costs explicitly. The later, revised, FSIS estimates include explicitly costs for out-of-compliance beef and pork plants to adopt steam vacuum systems and for poultry plants to adopt antimicrobial rinses (Roberts, Buzby, and Ollinger). However, the steam vacuum technology is only one of several potential interventions in beef and pork (Jensen, Unnevehr, and Gomez).

Thus, none of the past cost estimates provides much information to support the choice of any particular performance standard based on marginal cost/benefit analysis. Furthermore, there is little available information to guide choices faced by meat processing firms in adopting different technologies for pathogen control. Therefore, we explore sources of new cost information below, but first we review the issues facing pork processing firms in evaluating pathogen reduction alternatives.

## Issues in Evaluating Costs and Benefits of Pathogen Reduction in Pork

Benefits of pathogen reduction or control include both private and public benefits. Crutchfield et al. provide one estimate of public benefits that includes cost of illness, lost productivity, and loss of life. They estimate that foodborne illnesses attributed to meat and poultry alone from six microbial pathogens cost the U.S. economy $2.0 to $6.7 billion annually for 1995. Table 1 compares their total foodborne cost of illness estimates for selected pathogens with the prevalence of the same pathogens on pork carcasses in the 1995-96 USDA Microbiological Baseline. Although prevalence for most individual pathogens is low, pork is a potential source of four economically important pathogens: *staphyloccocus aureus, clostridium perfringens, campylobacter jejuni/coli*, and *salmonella*. It should be noted that traceback to individual food sources is difficult.[2] Nevertheless, there are potential public benefits from reducing the incidence of these pathogens on pork carcasses, which may

### TABLE 1
Prevalence of Selected Microorganisms on Pork Carcasses and Their Costs of Foodborne Illness from Meat and Poultry Alone

| Microorganism | Percent of Samples Positive on Pork Carcasses | Costs of Foodborne Illness (Low Estimates) (Billion $) |
|---|---|---|
| Total Coliforms | 45.4 | NA |
| E. coli (biotype I) | 31.0 | NA |
| Clostridium perfringens | 10.4 | 0.1-0.3 |
| Staphyloccocus aureus | 16.0 | 0.6-1.7 |
| Listeria monocytogenes | 7.4 | 0.1-0.7 |
| Campylobacter jejuni/coli | 31.5 | 0.5-0.9 |
| E. coli O157:H7 | 0.0 | 0.2-0.7 |
| Salmonella | 8.7 | 0.5-2.4 |

Sources: USDA/FSIS Microbiological Baseline Data Collection, April 1995-March 1996; and USDA Economic Research Service (Crutchfield et al.).

presumably reduce their prevalence later in the food chain.

Possible private benefits from reducing prevalence of pathogens include improvements in shelf life, access to new markets such as export markets, retention of customers, decreased scrap or reworking of product, and reduced product liability. These benefits are clear to many pork processing firms, but it is difficult to assign a specific dollar value to any of them. Access to export markets may be an important motivation. Shipping to those markets in Asia requires both extended shelf life and assuring buyers of the highest possible level of food safety.

We turn now to consideration of firm level issues in controlling foodborne pathogens. The major stages of the production process for pork include: incoming animals, pre-evisceration, post-evisceration, chilling, fabrication, and packing. Each stage can have monitored CCPs and/or some microbial control interventions. If firms want to reduce pathogens on pork carcasses they must consider two issues: a) how to control multiple pathogen targets; and b) where in the process to intervene.

Microbial pathogen control in the slaughter and processing environment involves control of hazards of various types. Some hazards are brought into the plant with the animals (many pathogens such as *salmonella* live in the enteric systems of animals); other hazards contaminate product through worker or other environmental contamination (such as *staphyloccocus aureus* or *listeria*). Some hazards grow (multiply) on product; others do not multiply. Thus specific HACCP controls may or may not control more than one pathogen. For example, a recent plant study by Saide-Albornoz et al. found prevalence levels for four pathogens, including *Salmonella spp.*, declined as carcasses passed through processing stages from singeing to chilling, but prevalence of *S. aureus* increased. *S. aureus* is usually carried by humans and the increased levels observed at later stages of processing were probably due to the increased human handling.

We have focused below on control of pathogens from the enteric systems of animals, which is the current focus of regulatory activity. In this case, methods applied to control of one pathogen often affect or control other pathogens as well, but perhaps not to the same degree. Generic *E. coli* is associated with fecal contamination of product, and its presence is likely to be an indicator of other associated contamination (or the potential for contamination), e.g., from *salmonella*. Total aerobic bacteria affects shelf life, and may be controlled by similar interventions. Thus certain kinds of safety and quality can be jointly produced through particular production processes (e.g., chilling carcasses or acid rinses).

The HACCP framework provides guidance on the issue of where to intervene in the process. During slaughter, evisceration, and chilling carcasses, the process provides opportunity for carcass contamination and cross-contamination. Presence or growth of pathogens can be affected by temperature, environment (e.g., acidity), physical pressures (e.g., washing), and time of year or day during which processing occurs. Thus, a HACCP system recognizes the need for control and monitoring throughout the production process and helps plants identify where to intervene. Pathogen reduction efforts at different intervention points, often at Critical Control Points (CCPs), affect the level of pathogens at that point in the process, but they can also reduce subsequent hazards. A simple example would be whether a hot water carcass rinse is applied before or after evisceration, or at both times.

Pork processing firms may adopt HACCP systems which monitor and verify the control from existing procedures inherent to the process of slaughter, evisceration, chilling, and fabrication. Examples of such procedures include scalding, singeing, chilling, or knife rinsing between carcasses. Presumably HACCP would make these procedures more effective through increased employee awareness and reduced variability in implementation. But firms may also find that existing procedures do not accomplish desired pathogen reductions. Thus, as private and public demand for food safety grows, firms may seek additional interventions focused on pathogen reduction. During the rest of the paper, we consider the cost-effectiveness of such interventions in pork processing.

## Cost-Effectiveness of Different Technologies for Pathogen Control in Pork

In the past few years, several new and existing technologies have been more widely adopted and adapted for pathogen reduction in the pork packing industry. Interventions available for pork include carcass wash, sanitizing sprays, steam vacuum, and carcass (hot water) pasteurizer. The carcass wash is a cabinet that provides a hot water rinse to the carcass, and has been in use for over 25 years. Washes can be applied either pre- or post- evisceration, and can be applied at different temperatures, with different pressures, and, for sanitizing sprays, at different levels of acidity. Sanitizing sprays, usually acetic acid, are most often used post-evisceration, in combination with hot water rinses. These sprays are a relatively new control technology that has been adopted during only the last 5 to 10 years. Steam vacuums, also relatively new, are used to remove contamination from specific parts of the carcass, and may be utilized at different points in the process. Hot water pasteurizers have been developed in Canada for hog carcasses, but have not yet been adopted in the United States. The adoption of a new technology in processing must be approved by FSIS before its use.

Costs of technologies used to increase food safety in product include both fixed (equipment) and variable costs. Data regarding costs of equipment and inputs required for operation were obtained directly from input suppliers of new technologies.[3] Comparable operating and depreciation costs were constructed for all technologies with representative prices for energy, water, labor, and capital. These cost estimates are representative of large plants, i.e., pork packing plants processing 800 to 1,200 carcasses per hour, which account for over 85 percent of total pork supply. Fixed costs are highest for the newest technology, pasteurizers, and much lower for other interventions (Table 2). Energy and water are the principal components of variable costs. Variable costs are also highest for pasteurizers, due to their high energy costs. Total costs per carcass are thus highest for carcass pasteurizers, followed by sanitizing spray systems, steam vacuum, and hog carcass wash. Total costs range from $0.05 per carcass for washes at 55°C to

## TABLE 2
### Fixed, Variable, and Total Costs ($/Carcass) of Different Technologies: Pork

|  | Hog Carcass Wash[a] | Sanitizing Spray System[a] | Carcass Pasteurizer[b] | Steam Vacuum[c] |
|---|---|---|---|---|
| **Fixed Costs** | | | | |
| Nominal cost equipment | 10,900 | 32,900 | 200,000 | 12,500 |
| Installation | 12,000 | | | |
| Freight | 7,000 | 7,000 | | |
| Spare parts | 2,281 | | | |
| Total | 32,181 | 39,900 | 200,000 | 12,500 |
| Medium term fixed costs per carcass[d] | 0.00655 | 0.00812 | 0.04069 | 0.00254 |
| **Variable Costs** | | | | |
| Water | 0.00140 | 0.00008 | 0.00021 | 0.00003 |
| Electric | 0.00052 | 0.00557 | 0.00174 | 0.00063 |
| Effluent | 0.00141 | 0.00008 | 0.00021 | 0.00024 |
| Natural gas | 0.04201 | 0.08402 | 0.11004 | 0.00000 |
| Labor | 0.00271 | 0.00271 | 0.00271 | 0.01300 |
| Solution | NA | 0.00500 | NA | NA |
| Total variable cost ($/carcass)[d] | 0.04804 | 0.09746 | 0.11491 | 0.01390 |
| **Total Costs** | | | | |
| Total costs per carcass | 0.05459 | 0.10557 | 0.15559 | 0.08220 |

[a]Sources: CHAD Co. and Birko Co. These costs are for a 55°C rinse. Costs for a 25°C and a 65°C rinse would be $0.02659 and $0.08260 per carcass, respectively.
[b]Stanfos Inc.
[c]Jarvis Co. Total cost based on use of five vacuums.
[d]Based on plant processing 1,200 carcasses per hour, 16 hours a day, 260 days a year. Medium term fixed costs use a ten-year depreciation period and a 10 percent annual interest rate.

nearly $0.16 per carcass for hot water pasteurizers, and can be up to $0.20 for high temperature washes of 65°C. The newer technologies have higher total costs than the older technology of low temperature carcass washing.

Interventions are often used in combination for pathogen control, and such combinations can result in pathogen reduction that is non-additive. Thus, evaluation of alternative interventions would ideally include evaluation of combinations of interventions or use of interventions at different points in the process. However, those types of studies are unusual, and furthermore, the literature on pathogen reduction technologies for pork is much smaller than that for beef (see review article by Siragusa).

Data regarding pathogen reductions are drawn from two published studies by meat scientists.[4] Dickson reports reductions in total aerobic

### TABLE 3
#### Mean Pathogen Reduction of Different Technologies in Hog Carcasses ($\log_{10}$ Counts)

| Type of Microorganism | (7) Carcass Pasteur.[a] | (1) Water Rinse (25°C)[b] | (2) Water Rinse (25°C) and Sanit. Sp.[b] | (3) Water Rinse (55°C)[b] | (4) Water Rinse (55°C) and Sanit. Sp.[b] | (5) Water Rinse (65°C)[b] | (6) Water Rinse (65°C) and Sanit. Sp.[b] |
|---|---|---|---|---|---|---|---|
| **Total Aerobic Bacteria (TAB)** | | | | | | | |
| Before treatment | 2.38 | 4.5 | 4.5 | 4.5 | 4.5 | 4.5 | 4.5 |
| After treatment | 0.39 | 3.49 | 2.25 | 2.64 | 2.25 | 2.06 | 1.76 |
| % reduction | 83.61 | 22.44 | 50.00 | 41.33 | 50.00 | 54.22 | 60.89 |
| **Total Enterics (TE)** | | | | | | | |
| Before treatment | 2.7 | 4.1 | 4.1 | 4.1 | 4.1 | 4.1 | 4.1 |
| After treatment | 0.0 | 2.71 | 1.13 | 1.41 | 1.48 | 1.68 | 0.0 |
| % reduction | 100.00 | 34.15 | 72.44 | 65.61 | 63.90 | 59.02 | 100.00 |
| **Cost** | | | | | | | |
| 1,200 carc/h ($/carcass) | 0.15559 | 0.02659 | 0.14057 | 0.05459 | 0.16857 | 0.08260 | 0.19658 |

[a]Source: Gill, Bedard, and Jones (1997). The samples were taken from parts other than the anal area of the carcass. The samples were taken during the plant operation, and were not contaminated intentionally.

[b]Source: Dickson (1997). In this experiment the carcasses were intentionall contaminated.

bacteria and total enterics for water rinses at different temperatures and with or without sanitizing sprays; data regarding the carcass pasteurizer are available from Gill, Bedard, and Jones (Table 3). In the Dickson study, carcasses were innoculated with relatively high levels of pathogens, whereas they were not in the Gill, Bedard, and Jones study. The Dickson study shows that higher reductions occur as water temperature increases and as rinses are combined with sanitizing sprays, and that reductions are generally to one-half of the initial levels. The Gill, Bedard, and Jones study shows that the carcass pasteurizer virtually eliminates the lower levels observed during processing.[5]

Table 3 also shows that costs increase as more energy is used to heat water ($0.03 to $0.08 per carcass) and as sanitizing sprays are added ($0.14 to $0.20 per carcass). Greater pathogen reductions in pork are associated with higher costs. The use of the sanitizing spray with the highest water temperature (6) provides the greatest pathogen reduction at more than double the cost over the use of highest temperature water rinses alone.[6]

## A Model for Minimizing the Costs of Pathogen Reduction

Firms can use one or more interventions to reduce pathogen levels on carcasses at the end of the kill floor process. Interventions may result in different levels of reduction for different pathogens. We assume that each intervention reduces pathogen levels by some percentage amount from the initial level on the carcass. The economic problem is to choose the most cost-effective set of interventions to meet a set of pathogen standards. This can be formulated as:

$$\min_{X_i} \sum_{i=1}^{N} C_i X_i$$

s.t.

$$I_j * \prod_{i=1}^{N} (1 - P_{ij} X_i) \leq S_j \quad j = 1, 2, \ldots, J \text{ (nonlinear constraints)}$$

$X_i$ : binary variable, $\forall X_i$; $i = 1, 2, \ldots, N$.

where:

N: Number of activities
J: Number of pathogen varieties to monitor by HACCP
$X_i$: Activity i, binary variable
$C_i$: cost of activity i
$P_{ij}$: percentage of pathogen j reduced by performing activity i

$S_j$: maximum number of pathogen j allowed by regulation
$I_j$: Initial level of pathogen j

This model chooses the least cost set of N possible interventions to achieve standards, Sj, for j pathogens, given the initial levels of pathogens (Ij), the effectiveness of interventions (Pij), and their costs (Ci). In this formulation, the order of the interventions does not matter (as it might in practice). Each intervention can only be used once, which accords with how interventions have been adopted by stages in plants.

This model was implemented for the set of rinse and spray interventions at different temperatures from Table 3. Table 4 reports the optimal costs and sets of interventions for initial pathogen levels equal to those in the Dickson study. The model chooses ten different optimal combinations of activities as pathogen standards are tightened, with corresponding costs increasing from $0.03 to $0.47 per carcass. For example, reduction of aerobic bacteria to 1.25 CFU and of enterics to 0.75 CFU requires use of two hot water rinses at 55°C and 65°C, at a cost of $0.137 cents per carcass. Costs increase steeply as desired pathogen levels approach zero. For aerobic bacteria counts of 0.25 CFU and no detectable enterics, costs are $0.47 per carcass, and four different interventions are used. Sanitizing sprays enter the optimal set of interventions only in the last four combinations for standards lower than 0.75 and 0.25 CFU respectively for TAB and TE. It is interesting that among the least cost combinations of technologies selected, the one that was required for the greatest relative pathogen control is comparable to the recommended

### TABLE 4
**Least Cost Combinations of Washes and Sanitizing Sprays to Achieve Different Pathogen Standards**

| Pathogen Standard | | Cost | Activity | | | | Final Level | |
|---|---|---|---|---|---|---|---|---|
| TAB | TE | | | | | | TAB | TE |
| 4.50 | 4.25 | 0 | 0 | 0 | 0 | 0 | 4.50 | 4.10 |
| 3.50 | 2.75 | 0.0266 | 0 | 0 | 0 | 1 | 3.49 | 2.70 |
| 2.75 | 1.50 | 0.0546 | 0 | 0 | 0 | 3 | 2.64 | 1.41 |
| 2.25 | 1.00 | 0.0812 | 0 | 0 | 1 | 3 | 2.05 | 0.93 |
| 1.75 | 1.25 | 0.1092 | 0 | 0 | 1 | 5 | 1.60 | 1.11 |
| 1.25 | 0.75 | 0.1372 | 0 | 0 | 3 | 5 | 1.21 | 0.58 |
| 1.00 | 0.50 | 0.1638 | 0 | 1 | 3 | 5 | 0.94 | 0.38 |
| 0.75 | 0.25 | 0.2778 | 0 | 2 | 3 | 5 | 0.60 | 0.16 |
| 0.50 | 0.25 | 0.3043 | 1 | 2 | 3 | 5 | 0.47 | 0.10 |
| 0.50 | 0.00 | 0.3338 | 0 | 3 | 5 | 6 | 0.47 | 0.00 |
| 0.25 | 0.00 | 0.4743 | 2 | 3 | 5 | 6 | 0.24 | 0.00 |

Note: Initial level of TAB is 4.5 and TE is 4.1.

complete set of equipment sold by Chad Co.

Figure 1 shows the three dimensional cost surface, which combines the step cost functions for reduction of TAB and TE. The figure demonstrates that costs rise more steeply for near elimination of TAB than for TE. Thus, these data show that costs are higher for improving shelf life than they are for improving food safety.

We also performed a preliminary analysis of the hot water carcass pasteurizer. Although the data in Table 3 are not directly comparable to those for hot water rinses and sprays, it seemed useful to explore conditions under which the pasteurizer might be adopted. The pasteurizer reduces the costs of achieving very low levels of pathogens (Table 5). For TAB levels of 0.25 CFU and non-detectable TE, the cost is $0.29 per carcass. This is achieved through combining rinses at 55°C and 65°C with the hot water pasteurizer. Thus the carcass pasteurizer may be adopted as the desirable level of control increases.

## Comparison with Overall Processing and HACCP Costs

Costs of intervention per carcass are small in comparison to total costs of processing in large plants. For pork, Melton and Huffman estimate the value-added packing costs to be $0.10 per pound for 1988; in current dollars, this would be $30 per carcass. In comparison, Hayenga estimates that large hog packing plants today have variable costs of $22 per carcass,

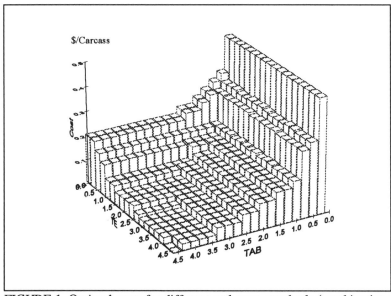

FIGURE 1. Optimal costs for different pathogen standards (combinations of carcass wash and sanitizing sprays).

and fixed costs of $6 per carcass, for a single shift, large plant. He estimates total costs to be $28 for a single shift and $23 for a double shift operation.

The additional costs of $0.20 for hot water rinses and sanitizing sprays represent an increase of less than 1 percent (0.7 - 0.9 percent). Our highest cost optimal combination for pathogen reduction would be $0.47, which would be only 1 to 2 percent of total processing costs. Thus, these new technologies for large plants represent a relatively small potential increase relative to other determinants of cost variation in the industry, such as scale or number of shifts. In a competitive industry, however, achieving efficiency in meeting the new regulation represents a significant challenge to firms.

Our overall estimate of HACCP costs is somewhat higher than the final FSIS cost estimates of $0.00003/lb. (or $.0056 per carcass) for large hog firms (Crutchfield et al.). These costs represent all of the costs of implementing HACCP, of which process modifications were only assumed to be a small part. FSIS assumed that half of pork and beef plants would adopt steam vacuums to achieve additional pathogen reductions, and that these would cost about $0.08 per carcass (similar to our estimates in Table 2). FSIS did not consider the costs of any other potential interventions. Thus, if more plants adopt the other technologies considered above the costs of pathogen reduction could be higher.

Current premiums for quality through carcass value pricing cause variations of plus/minus $5 per carcass (Lee). Thus incentives for improving food safety to hog producers are likely to be very small relative to incentives for delivering high quality animals with desired weight and back fat. While our data are only preliminary, they point to the relatively

**TABLE 5**

**Least Cost Combinations of Washes, Sanitizing Sprays, and Hot Water Pasteurizer to Achieve Different Pathogen Standards**

| Pathogen Standard | | Cost | Activity | | | | | Final Level | |
|---|---|---|---|---|---|---|---|---|---|
| TAB | TE | | | | | | | TAB | TE |
| 4.50 | 4.25 | 0 | 0 | 0 | 0 | 0 | 0 | 4.50 | 4.10 |
| 3.50 | 2.75 | 0.0266 | 0 | 0 | 0 | 0 | 1 | 3.49 | 2.70 |
| 2.75 | 1.50 | 0.0546 | 0 | 0 | 0 | 0 | 3 | 2.64 | 1.41 |
| 2.25 | 1.00 | 0.0812 | 0 | 0 | 0 | 1 | 3 | 2.05 | 0.93 |
| 1.75 | 1.25 | 0.1092 | 0 | 0 | 0 | 1 | 5 | 1.60 | 1.11 |
| 1.25 | 0.75 | 0.1372 | 0 | 0 | 0 | 3 | 5 | 1.21 | 0.58 |
| 0.75 | 0.00 | 0.1556 | 0 | 0 | 0 | 0 | 7 | 0.74 | 0.00 |
| 0.50 | 0.00 | 0.2102 | 0 | 0 | 0 | 3 | 7 | 0.43 | 0.00 |
| 0.25 | 0.00 | 0.2928 | 0 | 0 | 3 | 5 | 7 | 0.20 | 0.00 |

Note: Initial level of TAB is 4.5 and TE is 4.1.

small costs of post-evisceration control.

Another technology for reducing risk of foodborne illness from meats is irradiation. The federal government is currently evaluating changes in regulation to allow its use for red meat. Irradiation cost estimates for ground beef product are between $0.02 and $0.05 per pound (McCafferty; Morrison, Buzby, and Lin). Comparable costs would be expected for pork products of similar nature (e.g., thickness). Hence, irradiation offers an alternative technology, although a relatively high cost one. It is likely that given the relative costs, when used, it would be in combination with other technologies.

## Conclusions

Demand for safer meat products and new food safety regulations have led to the development of new technologies for pathogen control. To date, there has been relatively little information on the cost-effectiveness of various technologies for improved food safety. Our data indicate costs of individual technologies are in the range of $0.03 to $0.20 per carcass for hogs, and that optimal combinations of technologies may cost as much as $0.47 per carcass.

The cost estimates for specific interventions show that power and water are important to achieve greater pathogen reductions. Thus operating costs for interventions are highly dependent on water and power rates, which are likely key variables in current decisions to locate and operate plants. In addition, labor costs (including training and turnover costs) are likely to become more important to holding down costs of monitoring and control (about half of estimated total costs of HACCP). The cost issues surrounding food safety represent a subset of on-going cost issues facing the pork processing industry. Food safety performance will be tied to other performance issues in the meat industry.

As public and private demand for food safety grows, firms need to be able to evaluate the optimal (least-cost) combinations throughout the process to produce desired levels of safety for final products. Our simple optimization model shows how economic and microbiological information might be combined to better understand the trade-offs in achieving multiple pathogen reduction targets. This model can be extended to include desired reductions in variability which might be achieved through HACCP.

It is clear that the cost function is upward sloping for microbial pathogen reduction in the pork industry. Greater pathogen reductions can only be achieved at higher cost, and thus regulation will be least burdensome if it allows firms to choose cost-effective interventions. Good information will be required to choose the most cost effective technologies to manage risks that may be highly variable. As the desired level of pathogens is reduced, more technologies will be adopted and costs will increase. However, our preliminary data also show that new technologies such as the hot water carcass pasteurizer may reduce costs in comparison to combinations of existing technologies.

Our estimated costs of pathogen reduction measures represent less than 2 percent of packing costs, although we caution that the total costs of HACCP must also include monitoring and testing costs. These estimates are considerably larger than initial FSIS estimates of HACCP costs to industry, but improvements in food safety may be achieved through relatively modest investments in large plants.

We caution that these results are preliminary in several senses -- more studies of pathogen reduction under plant conditions are needed; new technologies are emerging to control pathogens; and they represent only part of the costs of a full HACCP system that includes monitoring and verification. Some interventions or combinations of interventions appear to dominate and will be more cost-effective. But, their effectiveness in real world situations is still unclear. Plants may obtain their own information about cost-effectiveness based on internal review; however, that information is only available post-adoption. Therefore, much experimentation will be necessary; industry should evaluate new options carefully and may want to foster more public research to compare and fine-tune technologies.

**Notes**

[1]H. Jensen is Professor of Economics at Iowa State University and L. Unnevehr is Professor in the Department of Agricultural and Consumer Economics at the University of Illinois at Urbana-Champaign. Senior authorship is shared. Thanks for valuable research assistance go to Patricia Batres-Marquez, Miguel I. Gomez, and Chang-chou Chiang. This research was supported in part by USDA/CSREES No. 96-35400-3750 and No. 97-34211-3956, and National Pork Producers Council.

[2]The prevalence of microbial pathogens is lower in general on beef carcasses than on pork carcasses, but much higher in ground beef and in broiler rinse fluid (USDA Microbiologial Baseline Data). Whether these prevalence levels translate into illnesses depends upon handling and preparation.

[3]We are grateful to the following companies for sharing information with us: CHAD Co., Stanfos Inc., Jarvis Co., and Birko Co.

[4]Two issues confound comparability among pathogen reduction studies. First, some studies observe pathogen levels in plants, which are generally low, and therefore observed reductions are also small. Other studies innoculate carcasses with high levels of pathogens in order to observe measurable and significant reductions following interventions. Second, few studies consider all possible combinations of interventions that a plant might consider, including the use of interventions at different points in processing.

⁵Other studies of rinses and sprays (Fu et al.; Frederick et al.; Epling et al.) show comparable reductions for *salmonella* and *campylobacter*, so the Dickson results may be taken as representative of these kinds of controlled laboratory experiments. As the pasteurizer is a very new technology, we have only the Gill, Bedard, and Jones study as evidence. We do not have data regarding the steam vacuum.

⁶See Jensen, Unnevehr, and Gomez for a similar analysis of pathogen reduction technologies in beef.

**References**

Birko Company. Personal communication with Gene Chase, company representative, Henderson, CO, November 1997.

Buzby, J.C., T.Roberts, C.-T. Jordan Lin, and J.M. MacDonald. "Bacterial Foodborne Disease: Medical Costs and Productivity Losses." Economic Research Service, United States Department of Agriculture, *Agricultural Economic Report* No. 741, August 1996.

CHAD Company. Personal communication with Mike Gangel, President, Lexena, KS, September 1997.

Crutchfield, S. R., J. C.Buzby, T. Roberts, M. Ollinger, and C.-T. J. Lin. "An Economic Assessment of Food Safety Regulations: The New Approach to Meat and Poultry Inspection." Economic Research Service, United States Department of Agriculture, *Agricultural Economic Report* No. 755, July 1997.

Dickson, J. S. "Hot Water Rinses as a Bacteriological Intervention Strategy on Swine Carcasses." Proceedings of The Food Safety Consortium Annual Meeting, Kansas City, MO, October 1997.

Epling, L.K., J.A. Carpenter, and L.C. Blankenship. "Prevalence of Campybacter spp. And Salmonella spp. on Pork Carcasses and the Reduction Effected by Spraying with Lactic Acid." Journal of Food Protection, Vol. 56(6), 1993: 536-537.

Frederick, T.L., M.F. Miller, L.D. Thompson, and C.B. Ramsey. "Microbiological Properties of Pork Cheek Meat as Affected by Acetic Acid and Temperature." *Journal of Food Science*, Vol. 59(2), 1994: 300-302.

Fun, A.-H., J.G. Sebranek, and E.A. Murano. "Microbial and Quality Characteristics of Pork Cuts from Carcasses Treated with Sanitizing Sprays." *Journal of Food Science*, Vol. 59(2), 1994: 306-309.

Gill, C. O., D. Bedard, and T. Jones. "The Decontaminating Performance of a Commercial Apparatus for Pasteurizing Polished Pig Carcasses". *Food Microbiology* 14, 1997:71-79.

Hayenga, Marvin L. "Cost Structures of Pork Slaughter and Processing Firms: Behavioral and Performance Implications." Department of Economics, Iowa State University Staff Paper No. 287. Ames, IA: Iowa State University. November 13, 1997.

Helfand, Gloria E. "Standards Versus Standards: The Effects of Different Pollution Restrictions." *American Economic Review* 81, (June 1991):622-34.

Jarvis Company. Personal communication with Tommy Fulgham, Manager, Meat Machinery Division, Middletown, CT, October 1997.

Jensen, H.H., L.J. Unnevehr, and M.I. Gomez. "Costs of Improving Food Safety in the Meat Sector" *Journal of Agricultural and Applied Economics*, Vol. 20, No. 1, Spring 1998, in press.

Lee, Kyu-Hee. "Evaluation of Packers' Pricing Mechanisms in the Hog Industry: A Simulation Approach." Master's thesis, Department of Agricultural Economics, University of Illinois at Urbana-Champaign, 1995.

MacDonald, James M. and Stephen Crutchfield. "Modeling the Costs of Food Safety Regulation." *American Journal of Agricultural Economics* 78 (December 1996),1285-90.

McCafferty, Dennis. "Get ready for irradiated meat." *USA Weekend*. January 23-25, 1998.

Melton, Bryan E. and Wallace E. Huffman. "Beef and Pork Packing Costs and Input Demands: Effects of Unionization and Technology." *American Journal of Agricultural Economics* 77 (August 1995):471-85.

Morrison, Rosanna, Jean C. Buzby, and C.-T. Jordan Lin. "Irradiating Ground Beef to Enhance Food Safety." *Food Review,* Jan.-April 1997:33-37.

Roberts, Tanya, Jean C. Buzby, and Michael Ollinger. "Using Benefit and Cost Information to Evaluate a Food Safety Regulation: HACCP for Meat and Poultry." *American Journal of Agricultural Economics* 78 (December 1996):1297-1301.

Saide-Albornoz, Jaime J., C. Lynn Knipe, Elsa A. Murano, and George W. Beran. "Contamination of Pork Carcasses during Slaughter, Fabrication, and Chilled Storage." *Journal of Food Protection* 58 (September 1995):993-997.

Seward, Skip. "Integrating Data for Risk Management: Comments." in *Tracking Foodborne Pathogens from Farm to Table: Data Needs to Evaluate Control Options*, ed. Roberts et al., USDA Economic Research Service Miscellaneous Publication No. 1532, December 1995.

Siragusa, G.R. "The Effectiveness of Carcass Decontamination Systems for Controlling the Presence of Pathogens on the Surfaces of Meat and Animal Carcasses." *Journal of Food Safety* 15, 1995: 229-238.

Stanfos, Inc. Personal communication with Lang Jameson, company representative, Edmonton, Alberta, Canada, September 1997.

U.S. Department of Agriculture, Food Safety and Inspection Service (USDA/FSIS), *Pathogen Reduction: Hazard Analysis and Critical Control Point (HACCP) Systems, Final Rule*, Washington, D.C., July 1996a.

U.S. Department of Agriculture. Food Safety Inspection Service (USDA/FSIS), Supplement: Final Regulatory Assessment for Docket No. 93-016F *Pathogen Reduction; Hazard Analysis and Critical Control Point (HACCP) Systems*. Washington, D.C., July 1996b.

U.S. Department of Agriculture. Food Safety Inspection Service (USDA/FSIS), "Nationwide Pork Microbiological Baseline Data Collection Program: Market Hogs," June 1996.

Unnevehr, L. J. and H. H.Jensen. "HACCP as a Regulatory Innovation to Improve Food Safety in the Meat Industry." *American Journal of Agricultural Economics* 78 (August 1996):764-69.

Chapter 4

# The Cost of HACCP Implementation in the Seafood Industry: A Case Study of Breaded Fish

Corinna Colatore and Julie A. Caswell[1]

## Introduction

Governments across the world are increasingly mandating the use of Hazard Analysis and Critical Control Points (HACCP) approaches to assuring food safety. In the United States, HACCP was first mandated in 1995 for the seafood industry, with full implementation to take place by December 1997. The adoption of HACCP as a regulatory approach in the United States is based in part on an estimation of the approach's benefits and costs. However, accurately estimating benefits and costs prior to implementation is difficult. As implementation occurs, better estimates should be possible based on actual experience.

The U.S. Food and Drug Administration (FDA) made a considerable effort to forecast the costs of HACCP implementation related to its seafood regulation (Williams and Zorn 1994, 1995). The Regulatory Impact Analyses associated with the proposed and final rules were based on available data, which were incomplete, imprecise, or both. With the regulation in effect, it is now possible to measure the costs of HACCP adoption based on actual industry experience (Colatore 1998).

## HACCP: A New Approach to Seafood Safety

The final rule of the new FDA seafood regulation was published in the December 18, 1995 *Federal Register*, in title 21 of the Code of Federal Regulations (CFR), parts 123 and 1240. The regulation introduces a requirement that seafood processors have a HACCP plan in place that incorporates seven basic principles: conduct a hazard analysis; identify the Critical Control Points (CCPs) of each specific production process; establish Critical Limits (CLs) for preventive measures associated with each identified CCP; establish CCP monitoring requirements; establish corrective actions to be taken when monitoring shows that a CL has been exceeded; establish effective record keeping procedures that document the entire HACCP system; and establish supervising procedures to verify that the HACCP system is working correctly.

Under the new regulation, domestic processors and importers of fishery products are required to write and implement a HACCP plan specific to each establishment. The regulation also requires that some tasks be performed by trained individuals and suggests sanitation procedures. This preventive approach is intended to assure that all the hazards posed to human health by seafood products are known, considered, and systematically avoided. The HACCP approach offers the advantage of making manufacturers more responsible for the safety of their products. The regulation also is expected to improve the efficiency of inspection for both imported and domestic products. The examination of HACCP records will enable inspectors to see how the processing facility (even if it is in another country) operates over time and to determine whether problems have occurred and how they were addressed. This has the effect of increasing the frequency of inspections without increasing the costs.

The benefits of this regulation in terms of food safety must be compared to the costs it introduces. Forecasting the costs of HACCP is difficult because it is a very flexible system; it allows each processor to choose a different way to comply with the regulation. Different choices have different costs, and the uncertainties associated with the first undermines a precise evaluation of the second. Using a combination of different data sources, FDA developed a Final Regulatory Impact Analysis and estimated a global cost of compliance with the Seafood HACCP regulation ranging from $677 to $1,400 million (FRIA 1995). The limit of the FRIA is intrinsically the reliability of the data used, which was limited at the time of the forecasts. This study quantifies the real costs faced by a group of seafood companies in HACCP adoption and compares these costs to the FDA predictions. It is particularly important to quantify HACCP costs in the seafood industry because in the United States it was the first food sector to face mandatory requirements.

## Defining the "Cost of HACCP"

Two problems arise in defining the term the "cost of HACCP." The first is that the HACCP plans implemented by seafood companies may be more complex and expensive than the FDA regulation requires. Some costs, though not directly required, may be considered useful or necessary by a company to make the HACCP plan work. Given this problem, it is necessary to distinguish between the *Total Cost of HACCP Adoption* incurred by a company and its *Cost of Implementing the Minimum FDA Requirements.* A second problem arises because not every seafood company implemented HACCP because it was required by the FDA regulation. Some companies may have implemented HACCP independently for their own business purposes, not because it became an FDA requirement. Consequently, not all the cost of meeting the minimum FDA requirements can be regarded as resulting from the FDA rule. Considering this problem, it is necessary to distinguish a third measure that is the *Incremental Cost of HACCP Adoption Attributable to the FDA Regulation*.

The three related but distinct estimates of HACCP costs are defined as:

1. ***Total cost of HACCP adoption***: cost of HACCP plans as adopted by the companies (e.g., they may include more CCPs than required by the FDA regulation). The cost is counted regardless of whether the plan was adopted in response to the FDA regulation. This therefore includes voluntary and mandatory adoption.
2. ***Cost of implementing the minimum FDA requirements***: cost for the companies of adopting a HACCP plan that met FDA minimum requirements (i.e., had the minimum number of CCPs to meet mandated requirements). The cost is counted regardless of whether the plan was adopted in response to the FDA regulation. It therefore includes voluntary and mandatory adoption.
3. ***Incremental cost of HACCP adoption attributable to the FDA regulation***: cost for the companies of adopting HACCP plans that meet the minimum FDA requirements net of voluntary HACCP adoption that did or would have occurred without the regulation.

These separate estimates of the costs of HACCP adoption have not been clearly delineated in previous cost studies.

## An Application: The Cost of HACCP Adoption in the Breaded Fish Industry

We chose the breaded fish industry for an in-depth case study measuring the costs of HACCP adoption. The "breaded fish" segment includes a group of products with these common characteristics:

1. The raw material used is either cod, haddock, pollack, flounder, sole, perch, or whiting.
2. The finfish meat is subjected to heading, eviscerating, filleting, skinning, and freezing before undergoing a process of breading or battering.
3. The finished product is commercially sold frozen, in the shape of portions, sticks, and fillets.

The breaded fish production process has a wide variety of potential hazards associated with both human health (safety-related hazards) and the quality of the finished product (economic fraud hazards). The FDA Seafood HACCP regulation requires control over only safety-related hazards. For breaded fish processing, these are (FDA 1994):

1. *Microbiological growth in batter:* the temperature of the batter must be carefully monitored because dangerous microorganisms such as *Staphylococcus aureus* can grow in a batter stored at room temperature. *Staphylococcus aureus* enterotoxin is heat stable and can cause disease even if the product is consumed after being fully cooked. For this reason, the temperature of the batter mix should not exceed 55°F (12.8°C) for more than twelve hours, cumulatively, or 70°F (21.1°C) for more than four hours, cumulatively.
2. *Metal inclusion:* metal-to-metal contact, such as during the cutting operation, can potentially introduce metal fragments into the product. Such fragments are a hazard to the consumer when they are 0.125" (3 mm) or larger in any dimension. A metal detection device can detect metal fragments lodged within the product.

To comply with the FDA regulation, a breaded fish processing company must have implemented a HACCP plan by December 18, 1997. Following the FDA requirements, the plan could be limited to as few as two CCPs in order to control the two safety-related hazards described above. Though not forced to, companies may also control non-safety related hazards as part of what they define as their HACCP plans. A typical breaded fish processing company generally implements five or six CCPs. The most common CCPs used in breaded fish processing are shown in Table 1.

**Survey Method and Instruments**

A sample of eight companies processing breaded fish in Massachusetts

**TABLE 1**
**Major CCPs for Breaded Finfish Production**

|     | CCP | Hazards | Control |
| --- | --- | --- | --- |
| I   | Receipt of raw materials | *Chemical contamination* | Certification |
|     |     | *Filth* | Visual check |
|     |     | *Decomposition* | Temp. check |
|     |     | *Species substitution* | Visual check |
| II  | Raw materials examination | *Chemical contamination* | Lab analysis |
|     |     | *Filth* | Visual check |
|     |     | *Decomposition* | Lab analysis |
|     |     | *Misuse of food additives* | Lab analysis |
| III | Batter mix storage-recirculation | *Microbiological growth* | Temp. check |
| IV  | Finished product examination | *Overbreading* | Weight check |
|     |     | *Decomposition* | Sensory analysis |
| V   | Packing | *Short weight* | Weight check |
| VI  | Metal detection | *Metal inclusion* | Metal detection |

was interviewed on their costs of implementing a HACCP plan, costs of complying with the FDA's requirements, and views of the HACCP regulation itself. The cost data were collected in personal interviews with quality control personnel using detailed interview protocols and survey instruments modeled after procedures developed at the Research Triangle Institute (Martin et al. 1993). The complete instruments are available in Colatore and Caswell (1998). Information gathered included:

1. When and why did the company implement HACCP? This information is important to separating companies that implemented HACCP only to be in compliance with FDA requirements from those that would have adopted in any case or did so much before the regulation was issued.
2. What are the plant's production, size, sales, costs, number of employees, etc.? This information allows an understanding of the company's size and baseline costs.
3. What were the design and development costs for the HACCP plan? For example, how much time and money was spent in writing the plan, for training, and for hiring new employees?
4. What are the plant's processes and CCPs?
5. For each CCP, what, if any, changes were introduced in the control procedure with HACCP adoption and what were their estimated costs? For example, the cost of an additional 30 minutes of control from a Quality Assurance Manager who earns $25 per hour is $12.50. In general, when an additional duty, A (e.g., calibrating additional thermometers) requiring time, $t_A$, is performed, its cost can be calculated in different ways:
   a. If a new employee is hired to do the job, the rest of the employees work exactly as before, and the production is not affected by A, then the cost of A is the additional salary paid to the new employee.
   b. If no employee was hired, A will be performed by employee E who was already working for the company. Assuming that the all employees are working efficiently, E can:
      I. Stop doing his/her usual duties to perform A. In this case, the production speed will be affected and the amount of product produced in a day will be reduced. The cost associated with this production loss (C) is calculated as: $C = R * (V_f - V_i)$ where R is the loss of product, $V_f$ is the value of the finished product, and $V_i$ is the value of the raw material.
      II. Perform his/her usual duties, and perform A in overtime hours. The production would not be affected in this case and the cost for the company is the extra time taken multiplied by the wage rate.

Assuming the wage system is efficient, these methods should give the same result with the company indifferent between them. The wage system efficiency assumption states that each employee is paid by the company a wage of $X/hr which is equal to the value of the additional product he/she gives the company. If he/she did not work, the company would produce Y less product. The value of this lost product will be: $Y(V_f - V_i)$, and this monetary value will be exactly $X/hr. In this study, the cost of additional duties was calculated using method (a) when a new employee was hired; method (b.i.) when it was possible to quantify the production loss; and method (b.ii.) in all the other cases.

Final cost data reported here are net of cost decreases. In other words, if a new procedure increases some costs and decreases others, the net change is reported.

6. For each sanitation procedure, what, if any, changes were introduced in the control procedure with HACCP adoption and what were their estimated costs? Reported cost data are again net of cost decreases.

## Cost of HACCP in the Breaded Fish Industry

The data collected in the interviews were used to quantify the three HACCP cost estimates, as previously defined. The costs were broken into eight categories, briefly described in Table 2. All reported costs are for the first year of HACCP adoption.

### I. Total Cost of HACCP Adoption

This section reports the first year costs of HACCP adoption for the interviewed companies. The average for the eight companies is shown in the first column of Table 3.

*Complexity of the HACCP plan.* Though the required number of CCPs under the FDA regulation is only two or three, the majority of companies implemented six CCPs and the average was 5.25. Real HACCP plans, as defined by the companies, are more complex than the FDA regulation requires because the interviewed companies included both safety and non-safety related CCPs. The quality of the final product is very important. For example, although not dangerous to human health, decomposition of the raw fish is easily detected by consumers through a smell of ammonium. For this reason, companies frequently perform an organoleptic examination on raw material and finished product. This control is usually very expensive, involving destructive sampling. Other CCPs usually included are checking the weight and label of the package, and checking the percentage of fish flesh in the finished product (i.e., checking for over-breading).

*Time and cost of plan design.* The time needed to develop a HACCP plan varied very much among the companies, though the complexity of the process is the same among them. There was a large range in the time

**TABLE 2**

**Description of Variables**

| Variable | Description |
|---|---|
| *Complexity of the HACCP plan* | Number of Critical Control Points in the HACCP plan. A HACCP plan designed to comply with the FDA regulation is simple and covers only safety-related CCPs. Real HACCP plans are often more complex and also cover quality related CCPs. |
| *Time and Cost of Plan Design* | Time needed and associated costs of designing the HACCP plan. It includes research, writing, rewriting if the plan does not work, and implementing. An additional cost variable, *Changes of Previous Plan Due to the Regulation*, was introduced for cases where a working HACCP plan had to be changed due to the regulation. |
| *Cost of Training* | It includes any employee of the company who was sent to be trained in HACCP when the company decided to implement HACCP. It includes, for every trained employee, the cost of the course, travel and lodging expenses, and productivity loss. In addition, it includes the productivity loss associated with internal training (i.e., the training of ALL employees to the concept of HACCP). |
| *Cost of Control and Record-Keeping* | Includes: |
| •*Additional Monitoring Cost* | •Cost associated with additional time spent in monitoring CCPs. |
| •*Additional Lab Analysis Cost* | • Additional cost associated with external lab work. This can include microbiological or chemical tests on representative samples of raw material or finished product. |
| •*Annualized Equipment Cost* | •Cost of new equipment purchased. The start-up costs associated with the purchase of capital goods, such as new equipment, were calculated as an annualized cost, using the formula (Price Gittinger, 1973): $A = V_0 * \{I / [1-(1+I)^n]\}$, where A is the annualized cost, $V_0$ is the cost of the equipment, n is years of life, and I is the discount rate (value set at 8%). |
| •*Corrective Action Cost* | •Cost associated with the occurrence of a critical limit (CL) deviation. This cost includes labor, product destroyed or reworked, analytical tests of potentially hazardous product, etc. |
| •*Cost of New Employees Hired for Monitoring* | •Cost of eventual employees hired in order to cover the increase in time needed for control procedures. |

needed to write a HACCP plan, from a few days to several months. The average number of person hours needed for plan design was 615 and the average cost was $19,320. This cost is explained if we consider that the actual writing of the plan is only a stage in the process of development which includes reading, writing, implementing, revising, and writing it again if it does not work. The process is one of trial and error rather than a simple writing exercise.

*Cost of training.* The final version of the HACCP regulation does not require a trained individual to be present full time in a plant, but its requirement that records be signed by a trained individual makes it logical for a company to have such a person full time. In order to implement HACCP efficiently, companies may want to send more than one employee to be trained. In this study, the average was 6-7 trained people per company, and only one company had a trained individual prior to implementing HACCP. The average expenditure for external training was $5,841. The cost for each trained employee was $865, due to the cost of a three-day course, the cost of lost productivity (the wages for three lost days of work), and travel expenses (hotel, food, travel).

### TABLE 2 (continued)
### Description of Variables

| Variable | Description |
| --- | --- |
| *Review Costs* | Cost of daily reviewing the records and of reviewing the whole HACCP plan. This review is generally done twice a year. |
| *Sanitation Costs* | The FDA regulation does not require any changes with respect to an already effective GMP regulation. Though not required to, some companies may have faced some increased sanitation costs. Examples of sanitation costs are: Writing and implementing a Sanitation Standard Operating Procedure (SSOP) plan; microbiological analysis, such as swiping the machinery; new cleaning equipment; hiring new sanitation personnel, etc. |
| *Validation Costs* | Change to companies' costs of certifying their product. |
| *Total HACCP Cost* | Sum of the above 7 cost categories. |
| *% [Total HACCP Costs/ATC]* | % [Total HACCP Costs/ATC] is the total cost of implementing HACCP as a percentage of Annual Total Costs for the interviewed companies. ATC were derived from the total sales of the interviewed companies using the following formula: Total Costs = (1 - RA) * Total Sales, where RA is the Median value of the Return on Assets. Its value for the frozen fish industry was 6.0, in 1997 (Dun & Bradstreet 1997). |

## TABLE 3
### Average First Year Costs of HACCP Adoption for Breaded Fish Companies, Survey Results and FDA Forecasts

| Cost | Total Cost | Survey Results | | FDA Forecasts | |
| --- | --- | --- | --- | --- | --- |
| | | Cost of Minimum FDA Requirements | Cost Attributable to FDA Reg. | Adjusted NMFS Model | FDA Model |
| **Complexity of the HACCP plan** | | | | | |
| Number of CCPs | 5.25 | 2 | 0.75 | 2 | 2 |
| **Time and Cost of Plan Design** | | | | | |
| HACCP Plan Design Time (person hours) | 615 | 299 | 97 | 24 | 48 |
| HACCP Plan Design Cost | $19,320 | $8,725 | $3,678 | $542 | $533 |
| Changes of Previous Plan Due to Regulation | ---- | ---- | $385 | ---- | ---- |
| **Cost of Training** | | | | | |
| Number of Employees Trained | 6.75 | 0.9 | 0.25 | 1 | 1 |
| Training Cost | $16,393 | $4,885 | $1,491 | $1,344 | $760 |
| **Cost of Control and Record-keeping** | $92,958 | $17,358 | $16,740 | $15,384 | $3,513 |
| Additional Monitoring Time (person hours) | 710 | 178 | 192 | 625 | 350 |
| Additional Monitoring Cost | $7,135 | $1,424 | $1,624 | $8,160 | $3,513 |
| Additional Lab. Analysis Cost | $43,975 | $0 | $0 | $1,200 | $0 |
| Annualized Equipment Cost | $7,615 | $7,372 | $6,528 | $4,824 | $0 |
| Corrective Action Cost | $24,726 | $8,562 | $8,588 | $1,200 | $0 |
| Number of Employees Hired for Monitoring | 0.4 | 0 | 0 | 0 | 0 |
| Cost of New Employees Hired for Monitoring | $9,507 | $0 | $0 | $0 | $0 |
| **Review Cost** | | | | | |
| Record Reviewing Cost | $7,676 | $2,924 | $1,597 | $900 | $693 |
| Review of the Whole HACCP Plan and Verification of its Effectiveness | $1,130 | $430 | $102 | $300 | $353 |
| **Sanitation Cost** | $31,523 | $0 | $0 | $7,464 | $1,126 |
| **Validation Cost** | -$55,495 | $0 | $0 | $0 | $0 |
| **Total Cost** | $113,505 | $34,323 | $23,993 | $25,934 | $6,978 |
| % [Total Cost/ATC] | 0.17% | 0.05% | 0.03% | ---- | ---- |

The external training was not the most expensive part of a company's training program. Several companies scheduled meetings where they stopped production, showed slides and talked about the concept of hazard prevention with their employees. The cost of this kind of training is high. On average, the interviewed companies spent $10,551 in the process. Though not required by FDA, the training of employees is of fundamental importance for the success of HACCP implementation. It is a simplification of HACCP and an underestimation of its power to think that it is only necessary to write a plan. The changes in the working attitude are so deep that it takes a considerable amount of time and effort for all employees to understand the concept of HACCP. Overall, the average training expenditure was $16,393.

*Cost of control and record-keeping.* The average time increase in monitoring and record-keeping for the interviewed companies was 710 hours per year, which represented an average increase in costs of $7,135. In general, companies were performing these controls even before introducing HACCP but had to increase the frequency or spend more time in the record-keeping process. A few companies had to perform new lab analysis, such as finished product testing, and consequently there was an average cost increase of $43,975. In addition, the majority of the interviewed companies had to buy new equipment, most of which was safety-related. The most frequent type of new equipment was thermometers. A few companies had to buy metal detectors or batter cooling systems, which were very expensive. The average annualized cost of new equipment was $7,615.

Corrective action costs are expensive in terms of managerial labor, analytical tests of potentially hazardous products, hazardous products destroyed or reworked, and lost production time. In the surveyed firms, it was very difficult to predict the frequency with which a deviation occurs and quantify its cost. Several deviations seem related only to the initial period of HACCP implementation. With time, HACCP should facilitate a company's prevention of deviations. As a result, the cost of deviations from critical limits should largely be considered as implementation costs. The first year average cost for corrective actions was $24,726.

A last expenditure source was new employees. The HACCP regulation does not require that companies hire any new employees, but for three companies the increased tasks were too many to be performed by the same employees. In addition to the annual wage paid by the company, the cost of new employees also includes training and was, on average, $9,507. Considering the monitoring, lab analysis, equipment, corrective actions, and new employees, the total average cost of controlling CCPs was $92,958.

*Review cost.* The regulation requires records to be reviewed and signed by a trained individual within two weeks of production. The average cost of reviewing records is estimated to be $7,676. The process takes an average of 347 person hours per year, and in a few cases required the purchase or updating of a computer. The HACCP regulation also requires that the plan be reviewed regularly. In general, companies review the plan every three months or twice a year. The average cost of reviewing, verifying, and eventually rewriting a HACCP plan is $1,130 per year.

*Sanitation cost.* Though the FDA regulation does not require breaded fish companies to make any changes with respect to an already effective Good Manufacturing Practice (GMP) regulation, all the interviewed companies faced some increased sanitation costs. Though not required to, all companies implemented a written Sanitation Standard Operating Procedures (SSOP) plan, the writing and implementing of which was generally as expensive as the parallel HACCP plan. In addition, some companies increased microbiological analysis, bought new cleaning equipment, or hired new sanitation personnel. The average expenditure for additional sanitation procedures was $31,523.

*Validation cost.* Although the FDA is responsible for mandatory seafood inspection, the U.S. Department of Commerce (USDC) has for many years offered a voluntary inspection system under the National Marine Fishery Service (NMFS). Under this program companies pay a full-time NMFS inspector to be on-site during each shift, which allows them to place a "packed under federal inspection" seal on their product. This system is quite expensive for processors (around $80,000 per year) but offers the advantage of certifying the quality of the product. This type of certification is particularly important for companies that sell to customers who require that they demonstrate that their product reaches a standard quality level.

Several years ago, NMFS started to develop a more efficient seafood surveillance system called the "HACCP-Based Inspection Program." Under it, companies implement a HACCP and SSOP plan, which has to be verified and validated by NMFS. Once the plan is implemented, NMFS only inspects the company periodically. The frequency of inspections depends on company performance. Because it is based on a quality certification approach, the newer NMFS inspection system requires a company to have a HACCP plan which covers CCPs for food safety as well as for economic fraud hazards. Implementing such a complex HACCP plan requires a large investment. On the other hand, this certification system, costing around $15,000-30,000 per year, represents a substantial savings for companies compared to continuous on-site inspection. Hence, by participating in the HACCP-based program, a company would save around $50,000 per year in validation procedures.

In this study, all eight interviewed companies were USDC certified; the majority had discontinued the NMFS voluntary inspection program and joined the HACCP-based one. Hence, they saved money in the validation process by implementing HACCP. This explains why validation costs are negative on average. For companies that maintained the NMFS voluntary inspection program, validation costs did not change with HACCP implementation. Instead, companies that shifted from the NMFS voluntary inspection program to the USDC's HACCP program had high savings due to HACCP implementation. On average, the validation costs were -$55,495.

*Total cost.* Not considering decreases in validation costs, the average first year cost for the interviewed companies to adopt HACCP was $169,000. However, considering the decrease in validation costs, the net first year cost was $113,505, which represents an increase of 0.17 percent in their annual total costs.

## II. Cost of Implementing the Minimum FDA Requirements

This section reports the costs of implementing HACCP if the interviewed companies only implemented the minimum FDA requirements. The average data for the interviewed firms is shown in column two of Table 3.

*Complexity of the HACCP plan.* The HACCP plan required by FDA is very simple. It is limited to only safety related CCPs, which in the breaded fish case are at minimum two: batter temperature control and metal detection. Since none of the interviewed companies implemented only the minimum FDA requirements, some assumptions, explained below, were used to calculate the costs discussed in the following paragraphs.

*Time and cost of plan design.* The time needed to develop a HACCP plan is assumed to be proportional to the number of CCPs in it. Hence the development time and cost for a HACCP plan with two CCPs can be calculated as a fraction of the time actually needed by the interviewed companies. For example, if a company developed a HACCP plan with six CCPs in 120 person hours, it is assumed that it would have spent 40 person hours in developing one with two CCPs. Following this assumption, the average time and cost of designing a HACCP plan incorporating the minimum FDA requirements are 299 person hours and $8,725. These figures probably underestimate the development time and cost of the minimum plan because the steps of reading the requirements and documenting the hazard analysis do not increase proportionally with the number of CCPs.

*Cost of training.* The least cost solution to be in compliance with the FDA requirements in terms of training is for a company to send one employee to be trained externally. No requirements are made regarding internal training of employees. We used the proportionality assumption to calculate the cost of this task. The average cost of complying with the minimum FDA training requirements was $4,885. It was calculated as the sum of two factors: the cost of external training for one employee and a proportion of the cost of internal training.

*Cost of control and record-keeping.* Since data were available for each CCP separately, the costs of monitoring and record keeping for the minimum FDA requirements could be directly calculated. The average time increase in monitoring and record-keeping for safety related CCPs was 178 person hours, and the cost was $1,424. No additional lab analysis cost was introduced. The annualized cost of the equipment bought for safety related purposes was $7,372. Almost all the equipment the interviewed companies had to buy was safety related (e.g., thermometers, metal detectors, batter cooling systems). The corrective action costs were in general high. A few companies had problems with maintaining the required batter temperature level and had high costs due to critical limit deviations. The average corrective action costs for safety related CCPs was $8,562, but, as mentioned before, this cost should be considered as an implementation cost more than an ongoing one. Considering the monitoring, equipment, corrective actions, and lab analysis, the total cost of monitoring CCPs was $17,358.

*Review cost.* The average cost of reviewing safety related records was $2,924. The average cost of reviewing the HACCP plan, verifying it, and eventually rewriting it was $430. These figures were calculated using the proportionality assumption.

*Sanitation cost.* Since the HACCP regulation does not require breaded fish companies to make any changes with respect to the GMP regulation, the cost of sanitation was not considered in calculating the costs of the minimum FDA requirements.

*Validation cost.* The savings in the validation program due to the implementation of HACCP are not considered in calculating the costs of the minimum FDA requirements. If a company followed only the minimum requirements, it could not shift from the NMFS voluntary inspection program to the USDC's HACCP program.

*Total cost.* The average first year cost the interviewed companies would have had to introduce a HACCP plan that only met the minimum FDA requirements was $34,323, an increase of 0.05 percent in their annual total costs. This value is about a third of the net total cost of HACCP implementation found in the previous section. By implementing a HACCP plan more complex than the FDA minimum, the average company spent three times more than required.

### III. Incremental Cost of HACCP Adoption Attributable to the FDA Regulation

This section reports the incremental costs of HACCP adoption attributable to the FDA regulation. The data for this estimate are shown in column three of Table 3. Though the FDA regulation requires companies to implement HACCP, not all the costs of implementing HACCP are attributable to the FDA regulation. The majority of the interviewed companies implemented HACCP for reasons other than that it was required by FDA. These included an interest in improving efficiency; the benefits in terms of product differentiation; the savings in validation with USDC's HACCP program; and the requirement of quality standards from customers. The regulation should not be considered responsible for the costs of implementing HACCP in companies that did or would have adopted HACCP anyway.

To calculate the costs attributable to the FDA regulation, we assumed that for companies that implemented HACCP for economic reasons independent of the regulation, the costs of only minor changes, if any, required by the regulation are attributable to FDA. For companies that implemented HACCP because FDA required it, we assumed the costs of implementing the minimum requirements are attributable to the regulation. Following these assumptions, the majority of the interviewed companies had zero costs attributable to the FDA regulation.

*Complexity of the HACCP plan.* The average number of CCPs implemented due to the FDA regulation is 0.75. The reason for this low value is that the majority of the interviewed companies had a HACCP plan already in place when the FDA regulation came into effect and they did not need to change it.

*Time and cost of plan design.* For the same reason, the average time and cost of plan design due to the regulation were 97 person hours and $3,678. Some companies had to modify the already existing HACCP plan. The cost of these modifications was $385.

*Cost of training.* An average of 0.25 people received training in order to comply with the FDA regulation. The average cost of training was $1,491.

*Cost of control and record-keeping.* The average time increase in monitoring and record-keeping was 192 person hours and the cost was $1,624. No additional lab analysis cost was introduced. The annualized cost of the equipment bought to comply with the FDA regulation was $6,528. The average corrective action cost was $8,588. The total cost of monitoring, equipment, corrective actions, and lab analysis attributable to the FDA regulation was $16,740.

*Review cost.* The average cost of reviewing records was $1,597. The average cost of reviewing the HACCP plan, verifying it, and eventually rewriting it was $102.

*Sanitation cost.* Since the HACCP regulation does not require breaded fish companies to make any changes with respect to the GMP regulation, the FDA is not responsible for any change in the sanitation procedures.

*Validation cost.* The savings in the validation program due to the implementation of HACCP are not considered in calculating the costs attributable to the FDA regulation. If a company followed only the minimum requirements, it could not shift from the NMFS voluntary inspection program to the USDC's HACCP program.

*Total cost*. The average cost the interviewed companies introduced in order to comply with the FDA regulation was $23,993, an increase of 0.04 percent in their annual total costs. Only about 20 percent of the total cost of implementing HACCP among these companies is attributable to the FDA regulation.

## Comparing Actual to Predicted HACCP Costs

In this section, the cost results from the survey are compared with cost estimates developed by FDA in its Regulatory Impact Analysis. In order to forecast the cost of HACCP, the FDA used several data sources and developed two models to predict the costs borne by average seafood companies in complying with the regulation. The most detailed FDA model is the "Adjusted NMFS Model" (Preliminary Regulatory Impact Analysis (PRIA), Williams and Zorn 1994; Final Regulatory Impact Analysis (FRIA), Williams and Zorn 1995). The main data sources used to develop this model are two reports of a study conducted under contract with the National Marine Fishery Service (NMFS). The study (Kearney 1989, 1991) was conducted by:

- Randomly selecting 130 processing plants not operating under HACCP;
- For each plant evaluating the deficiencies and modeling a correction procedure. This step consists of a thorough *in situ* observation of all aspects of a plant's activities, a resulting list of suggested steps to meet the requirements, and a detailed accounting of the costs of implementing these steps. The result is an estimation of the costs for each plant.

In the Adjusted NMFS model, FDA extrapolated the results of the Kearney study in order to forecast the costs for the whole seafood industry. This model has two weaknesses. First, the data source is based essentially on *suggested procedures*. Though very accurate, these procedures may be quite different from those that processors would have implemented. Second, in extrapolating the results of the Kearney study, the FDA implicitly assumed that (1) every company in the U.S. would implement HACCP as a result of the introduction of the new regulation and (2) they would implement only the minimum requirements. The validity of these two assumptions can be argued. As reported in the previous sections, our survey results show that the majority of the interviewed companies implemented HACCP for reasons other than that it was required and implemented more than was required.

In the Preliminary Regulatory Impact Analysis for the seafood HACCP rule, the costs of HACCP adoption were reported for specific segments of the seafood industry, including for the breaded fish industry. In the Final Regulatory Impact Analysis (FRIA), predicted costs were not reported for individual industry segments. Through consultation with FDA, we were able to construct the predicted costs of HACCP for the breaded fish industry using the assumptions of the FRIA. These Adjusted NMFS Model forecasts are shown in column four of Table 3. By comparing these data with the survey results, it is possible to assess the accuracy of the Adjusted NMFS Model as a forecasting method. These comparisons are:

*Complexity of the HACCP plan.* In the Adjusted NMFS Model, it was assumed that companies would implement only the minimum required number of CCPs, which is two in the breaded fish case. The same number of CCPs were used in the calculations for the cost of minimum FDA requirements from the survey results (column two of Table 3) because that estimate was based on the same assumption. This assumption is not confirmed by the actual experience of the interviewed companies (column one of Table 3). The majority of the interviewed companies implemented six CCPs and the average was 5.25. The companies implemented more CCPs because they included CCPs for non-safety related hazards in their HACCP plans. They did so because they generally did not find it effective to implement a minimum HACCP plan. Considering only the CCPs attributable to the FDA regulation (column three of Table 3), the average number of CCPs implemented only because it was required was 0.75.

*Time and cost of plan design.* In the Adjusted NMFS Model, FDA estimated that developing a HACCP plan for breaded fish would require 24 hours of managerial labor. The cost of developing the plan was estimated as $542. From this study, the average time spent in designing a HACCP plan as defined by the companies was much higher at 615 person hours and the cost was $19,320. If the interviewed companies only met the minimum FDA requirements, the time and expenditure for designing an HACCP plan would have been 299 person hours and $8,725. If considering only the time and expenditure attributable to the FDA regulation, they would have been 97 person hours and $3,678.

It is not surprising that the survey-based costs are higher than the FDA predictions. The experience of the people involved in the process is the most important variable influencing the time needed. The time estimate developed by FDA would be realistic if all companies had perfect information about HACCP prior to implementing it, which was not the case.

In the Adjusted NMFS Model, FDA assumed that companies would use an employee to develop the HACCP plan. This assumption is in general correct; only one company hired a consultant. In addition, no company used the help of an association, which in FDA's predictions would have reduced the costs.

*Cost of training.* The FDA assumed that small companies would send one employee to external training, while large companies had at least one trained individual before implementing HACCP. The estimated cost for external training is $1,344 in the adjusted NMFS study, where it is assumed that only a minority of the companies would have travel expenses. The cost of internal training was not considered in FDA's forecast, causing an underestimation of the cost of training.

The actual cost of training as derived from the survey was higher than the FDA forecasts. In general, the interviewed companies had to train more than one employee in order to be successful in the implementation of HACCP. The average was 6-7 people per company receiving external training and only one company had a trained individual prior to implementing HACCP. The average cost of training each employee among the surveyed firms was only $865, due to the fact that only three companies had travel expenses. In addition, the cost of internal training of employees was $10,552. The total cost of training was, on average, $16,393.

If considering only the minimum FDA requirements, all companies sent an employee for external training, except the company that had a trained individual already in place. As a result, the average number of employees sent for external training was 0.9. The cost of internal and external training was $4,885. The employees trained only because of the FDA regulation were 0.25 because the majority of the interviewed companies trained their employees before the regulation was introduced. The cost of training was $1,491.

*Cost of control and record-keeping.* In the NMFS model, an average seafood company is expected to spend 625 additional hours per year in monitoring and record keeping with an associated cost of $8,160. These figures are probably an underestimate, since they are derived from companies that have implemented a "negative record-keeping" system called NUOCA (Notations of Unusual Occurrences and Corrective Actions). Under this system, records are made only when a deviation occurs. FDA instead requires that records be kept regularly regardless of the occurrence of a deviation.

These figures cannot easily be compared with the survey estimates: while the adjusted NMFS model calculates the time and cost of monitoring assuming that no controls were in place prior to the implementation of HACCP, the survey shows that the majority of the interviewed companies had some kind of quality control system even before HACCP. Hence, the time and cost increase was mainly due to record-keeping. The average time and cost increase in monitoring and record keeping for the interviewed companies was 710 hours per year and $7,135. These figures are, coincidentally, similar to the adjusted NMFS forecast, which is both an underestimate (because they consider only two CCPs and because they are based on the NUOCA record-keeping system) and an overestimate (because they assumed that no monitoring was in place before the introduction of HACCP).

The overestimation problem in the FDA forecast becomes obvious if it is compared with the survey based estimate of the cost of the minimum FDA requirements (column two of Table 3) and the incremental cost attributable to the FDA regulation (column three of Table 3). By assuming that no control was in place prior to HACCP, FDA's estimates are higher than the survey results. In practice, every breaded fish company had some kind of quality control in order to be on the market. The increase in monitoring time is mainly due to record keeping, while the monitoring *per se* did not increase much in the majority of the interviewed companies.

Lab analysis costs were $1,200 in the adjusted NMFS model. The average expense of the interviewed companies was much higher ($43,975), but it was limited to non-safety related CCPs. There were no lab analysis costs relative to minimum requirements or attributable to the FDA regulation. The additional annualized equipment cost was forecast as $4,824 in the NMFS model. This value was lower than all the survey based estimates.

Regarding the corrective action costs, in the NMFS model FDA estimates that the cost of responding to critical limit deviations is $1,200 per year in terms of managerial labor, analytical tests of potentially hazardous products, hazardous products destroyed or reworked, and lost production time. The occurrence of CL deviations and of corrective actions is very difficult to predict. From this study, the first year average cost for corrective actions is $24,726. This value is reduced to $8,562 per year if considering only safety related CCPs, and to $8,588 if considering only the corrective action costs attributable to the FDA regulation. The higher values of the survey results can be explained by the fact that some deviations seem to be particularly frequent only in the initial period of the HACCP implementation.

In estimating the cost of implementing HACCP, FDA assumed that no employees would be hired. This study shows instead that three companies had to hire new employees. The total cost of monitoring as forecast in the adjusted NMFS model ($15,384) was very similar to the Cost of the Minimum FDA Requirements as calculated from the survey results ($17,358). The adjusted NMFS model considers every possible cost source and this approach is proven to be effective.

***Review cost.*** The cost of reviewing records is estimated to be $900 per year in the adjusted NMFS model. The average value in this study is $7,676 per year. If considering only the safety related CCPs, the cost of record reviewing is $2,924. The incremental cost of record reviewing associated with the FDA regulation was $1,597. Though low, the adjusted NMFS model is not far from the correspondent values reported for the cost of meeting the minimum standard and cost attributable to the FDA regulation because only safety related CCPs are considered in both estimates. The record review costs were well forecast because the adjusted NMFS model is based on the experience of seafood companies. The time these sample companies spent in record review is similar to the time the interviewed companies spent on the same task. For both groups of companies, record reviewing was a new task introduced due to HACCP implementation. Hence, the two estimates are similar.

The cost of reviewing, verifying, and changing the HACCP plan after it has been put in place is estimated to be $300 per year in the adjusted NMFS model. From this study, the average cost of reviewing, verifying, and eventually rewriting a HACCP plan is $1,130 per year when the total cost of HACCP implementation is being measured, $430 per year as a part of the cost of minimum FDA requirements, and $102 per year in terms of the incremental cost of HACCP adoption attributable to the FDA regulation. The comparison between FDA estimates and survey results shows that this task was also well forecast by FDA.

*Sanitation cost.* Though the FDA regulation does not require any change with respect to GMP regulations, the adjusted NMFS model forecasted an average sanitation expenditure of $7,464. This approach is not consistent, since the FDA regulation should not be held responsible for expenses it did not require. This procedure was used in calculating the costs of the minimum FDA requirements and the incremental cost of HACCP attributable to the FDA regulation, where the sanitation costs were estimated to be zero. However, the average cost of sanitation for the interviewed companies was $31,523. Though not required, all the breaded fish companies implemented a written SSOP and had some sanitation cost.

*Validation cost.* As discussed in the previous section, the majority of the interviewed companies participated in the USDC's HACCP program and on average saved $55,495 on the validation process. The validation cost issue was not considered in the FDA cost forecast.

*Total cost.* The total first year cost predicted in the adjusted NMFS model for an average seafood company is $25,934. This value is much lower than the total cost of implementing HACCP as derived from the survey results ($113,505 or $169,000 if changes in validation costs are not considered); lower than the survey-based cost of meeting the minimum requirements ($34,323); and similar to the survey-based incremental cost attributable to the FDA regulation ($23,993).

The FDA forecast and the total cost of HACCP adoption experienced by the companies differ very widely because they are measuring different things. The Adjusted NMFS Model tries to forecast the impact of the FDA regulation by assuming that (1) every seafood company in the U.S. will implement HACCP as a result of the introduction of the new regulation and (2) they will implement only the minimum requirements. The total cost of HACCP as derived from the survey calculates the cost of implementing HACCP as the interviewed companies did it, to a large degree, independently from the FDA requirements. The real industry experience with HACCP was much more expensive than FDA predicted because the average company found that only implementing the minimum safety related requirements would not have been efficient. Instead they implemented more complex HACCP plans that included non-safety related CCPs. Thus the FDA forecast, based on implementation of plans that only met the minimum requirements, does not provide an accurate prediction and tends to significantly underestimate the total costs of HACCP as adopted by the companies.

The adjusted NMFS forecast is also low when compared to the survey-based cost of implementing the minimum FDA requirements ($34,323). Since the two estimates were made using the same assumptions, this comparison is a more direct means of evaluating the accuracy of the FDA forecast than the comparison between total and FDA forecast costs. Making this comparison, review cost and control and record-keeping cost were accurately forecast by FDA. These figures were derived from the experience of real companies and the results are very similar to the results of the survey. The cost of training was accurately forecast but did not consider the cost of internal training that companies did to make HACCP effective.

The cost of designing a HACCP plan was the most complex to predict. FDA's forecasts are based on model HACCP plans that a group of experts built for randomly selected seafood companies. This approach was successful in forecasting the time needed in record reviewing but not the time needed to develop a HACCP plan. A HACCP expert and a barely-HACCP-certified employee would spend about the same time to review daily records, but the first would develop a HACCP plan much faster than the second. Experience is essential in the first stages of HACCP development. Many companies had no experience in writing HACCP plans and had to proceed by trial and error in the development process. While an expert could write a HACCP plan in 48 hours, as predicted by FDA, it would take up to several months for an average employee to perform the same task. In its forecasts FDA assumed that companies had perfect information on how to write HACCP plans. The survey results show that the path to HACCP implementation was longer and more complicated than FDA assumed. Overall, the adjusted NMFS model effectively forecast the ongoing costs of control and record-keeping, but not the costs associated with the first stages of HACCP implementation.

An even more direct comparison is between the adjusted NMFS model and the survey-based incremental cost of HACCP attributable to the FDA regulation. This is because the two estimates have the same objective of providing an impact analysis of the HACCP regulation. The two estimates are very similar ($25,934 and $23,993). However, this does not mean that the adjusted NMFS model accurately predicted the impact of the HACCP regulation. Instead, its cost forecasts were wrong on two counts but by coincidence the two errors balance each other out. The first error is an overestimate of the costs attributable to the FDA regulation: all seafood companies are assumed to have some costs in order to comply with the requirements. This assumption is not confirmed by this study, which shows that several interviewed companies had zero costs associated with the FDA regulation because they implemented HACCP for reasons other than that it was required. The second error is an underestimation of the costs of complying with the FDA requirements for companies that did have costs, as discussed above. The estimates derived from the adjusted NMFS model and the survey-based incremental cost of HACCP attributable to the FDA regulation seek to measure the same thing but in reality do not.

By comparing the survey-based cost estimates with the FDA forecast, we can make an assessment of the forecast's accuracy. First, the Adjusted NMFS model was quite accurate in forecasting the cost of implementing the minimum FDA requirements. Though some costs were underestimated, the result was accurate because the model was very detailed and considered almost every expenditure source that the interviewed companies experienced in reality. Second, the model was not very accurate in measuring the incremental costs of the FDA regulation, which is the main target of a Regulatory Impact Analysis. Finally, the total costs of HACCP implementation, as done by companies in the industry, are many times the cost of meeting the minimum requirements or the incremental costs of the regulation.

In its Final Regulatory Impact Analysis, the FDA used a second model to forecast the costs of HACCP adoption. This model was intended to estimate the costs associated with a simple HACCP implementation. Its estimates were based on the cost of compliance for two small processors that cut and package tuna received frozen. One plant was modeled as having some sanitation deficiencies, while the other was in substantial compliance with existing GMP. As with the Adjusted NMFS Model, it was assumed that every seafood company in the U.S. would implement HACCP as a result of the new regulation and that they would implement only the minimum requirements. The forecast costs of HACCP implementation based on this "FDA Model" are shown in column five of Table 3.

In brief, for the FDA Model the total first year cost forecast for an average seafood company is $6,978. This value is much lower than any survey-based cost estimate. All expenditure sources were underestimated, particularly designing, training and monitoring costs. The underestimation occurred because the model is based on only two theoretical seafood companies with relatively simple HACCP plans. The survey results show that this model did not yield an accurate forecast of costs in the breaded fish industry. A comparison of the survey-based estimates to the two FDA models shows that the best forecast method is one based on a range of company types operating in different industry segments.

## Conclusions

The cost analysis of HACCP based on actual implementation costs is important for two reasons. First, a reliable cost analysis can be useful in analyzing actual industry response to HACCP. Data are crucial at this period when the mandatory HACCP approach is still relatively new but spreading as a regulatory tool. Second, a detailed cost analysis can give important insights into the accuracy of the forecasting methods used by the FDA in the Regulatory Impact Analysis. Forecasting HACCP costs is difficult due to the system's flexibility. Several assumptions had to be made by FDA in the forecasting process and these assumptions can now be assessed. Knowing the strengths and limitations of the forecasting methods can help in the regulatory impact analysis for future HACCP regulations.

This research presents a cost analysis of the recent HACCP regulation for seafood products. Eight breaded fish processing companies were interviewed on the cost of implementing HACCP and the cost of complying with the FDA regulation. The data were used to calculate the average total cost of HACCP adoption as the companies defined it, the cost of meeting the minimum FDA requirements, and the incremental cost of HACCP adoption attributable to the FDA regulation. The results show that cost estimates vary greatly depending on how the cost of HACCP is defined.

The average total cost of implementing HACCP for companies in the breaded fish industry was $113,505. This expenditure varied very much within the sample. The average cost for sample companies of implementing only the FDA minimum requirements was $34,323. This shows that companies generally implemented a much tougher and more expensive HACCP plan than required. The cost of implementing only the minimum FDA requirements was 30 percent of the actual costs companies incurred. Though implementing HACCP can be expensive, the results show that the incremental cost attributable to the FDA regulation, at $23,993, represents only about 20 percent of total costs. Only a minority of the interviewed companies implemented HACCP because of the FDA regulation. The majority had other reasons for doing so.

The results show that FDA generally underestimated the costs of HACCP adoption in both models used in its regulatory impact analysis.

Comparison of forecasted costs under the FDA Adjusted NMFS Model to survey-based costs show that companies' HACCP costs are higher than forecast in the model. Particularly difficult to predict were the costs of designing the HACCP plan, training, corrective actions, and sanitation. All these expenditure sources were more expensive than FDA predicted. In contrast, the on-going costs of monitoring and record-keeping, reviewing records, and reviewing the whole HACCP plan were accurately forecast. The second FDA model underestimated adoption costs to a greater degree than the Adjusted NMFS Model. An important issue in cost estimation that was not considered by either FDA model was participation in the HACCP-based NMFS certification program prior to the FDA regulation. This participation and changes in it were found to have a significant impact on companies' net HACCP implementation costs. Overall, this study shows that careful definition of the "cost of HACCP" being measured is crucial to generating reliable estimates of the costs of HACCP as adopted and the cost impact of a mandatory HACCP regulation intended to improve food safety.

**Notes**

[1]Authors are former Research Assistant and Professor, respectively, Department of Resource Economics, University of Massachusetts, Amherst, MA 01003. This research was funded by a USDA Cooperative State Research, Education, and Extension Service (CSREES) Special Grant to the Food Marketing Policy Center, University of Connecticut and by subcontract at the University of Massachusetts, Amherst.

**References**

Colatore, C. 1998. Economic Analysis of HACCP in the Seafood Industry. M.S. thesis, University of Massachusetts, Amherst, 207 pp.

Colatore, C. and J. A. Caswell. 1998. Survey Instruments for a Cost Study of HACCP in the Seafood Industry. NE-165 Working Paper #45, May (Available at http://agecon.lib.umn.edu/ ne165.html#wp).

Dun & Bradstreet. 1997. Industry Norms and Key Business Ratios: One Year. Desk-Top Edition SIC #0100-8999, Dun & Bradstreet, Inc.:31.

Kearney, A. T. 1989. Economic Impacts of HACCP Models for Breaded, Cooked, and Raw Shrimp and Raw Fish.

Kearney, A. T. 1991. Economic Impacts of HACCP Models for Blue Crab Breaded and Specialty Products, Molluscan Shellfish, Smoked and Cured Fish, and West Coast Crab.

Martin, S.A., B. J. Bowland, B. Calingaert, N. Dean, and D. Ward. 1993. *Economic Analysis of HACCP Procedures for the Seafood Industry*. Final Report, Volume 1, November. Research Triangle Institute.

Price Gittinger, J. 1973. Compounding and Discounting Tables for Project Evaluation. Johns Hopkins University Press.

U.S. Department of Health and Human Services, Food and Drug Administration. 1994. 21 CFR Parts 123 and 1240: Proposal to Establish Procedures for the Safe Processing and Importing of Fish and Fishery Products: Proposed Rule, 59 FR, No. 19, pp. 4142-4223, January 28.

U.S. Department of Health and Human Services, Food and Drug Administration. 1995. 21 CFR Parts 123 and 1240: Procedures for the Safe and Sanitary Processing and Importing of Fish and Fishery Products: Final Rule, 60 FR, No. 242, pp. 65096-65202, December 18.

U.S. Food and Drug Administration. 1994. Fish and Fishery Products Hazards and Controls Guide, Draft, February 16.

Williams, R. A. and D. J. Zorn. 1995. Final Regulatory Impact Analysis (FRIA) of the Regulations to Establish Procedures for the Safe and Sanitary Processing and Importing of Fish and Fishery Products.

Williams, R. A. and D. J. Zorn. 1994. Preliminary Regulatory Impact Analysis (PRIA) of the Proposed Regulations to Establish Procedures for the Safe Processing and Importing of Fish and Fishery Products, January 24.

Chapter 5

# Measuring the Costs and Benefits of Interventions at Different Points in the Production Process: Lessons, Questions, and Comments

Neal H. Hooker[1]

## Introduction

Recent cost benefit analyses accompanying the U.S. Department of Agriculture (USDA) Food Safety and Inspection Service's (FSIS) *Pathogen Reduction; Hazard Analysis and Critical Control Point (HACCP) Systems; Final Rule* (USDA 1996), and the U.S. Department of Health and Human Services Food and Drug Administration's (FDA) *Procedures for the Safe and Sanitary Processing and Importing of Fish and Fishery Products; Final Rule* (FDA 1995) have received some criticism for the cost assumptions and hypothesized pathogen reductions upon which they are based. The cumulative and speculative nature of the cost data is inevitable for the purpose of comparison against similarly aggregated and forecasted benefits of the regulations.

Given *actual* cost data linked to *true* plant-level *hazard* reductions associated with identifiable strategies adopted by firms in response to these rules (including interventions targeting microbial, chemical, and physical food safety concerns), some of these issues can be addressed retrospectively. Further, the relative merits of plant-level hazard interventions and the associated public health benefits accruing due to reduced food safety *risks* may be better estimated, allowing the *effectiveness* of the regulation to be determined *ex post*. This conference, and in particular this session, collects such on-going research at a key point in time, providing an opportunity to contrast projected measures to the real-world costs and benefits companies have observed following their adoption of HACCP plans. Such comparisons can guide future regulatory impact analyses and indicate potential biases in current methodologies.

In order to complete such research, it is necessary to have *representative*, *detailed* cost data which can be linked to *actual* microbiological (for example) improvements *solely* due to the particular strategy or HACCP-based system under review. In this way one can avoid (or at least minimize) the potential for confusing the *causality* issue. This arises when one incorrectly assigns costs and benefits to regulations that are more correctly due to a general trend in food safety enhancements that the plant, firm, or industry may have performed in absence of the regulation (MacDonald and Crutchfield 1996). Such marginal analysis of the two U.S. HACCP-based regulations listed above is the focus of the two of the papers discussed here.

An associated issue, which pathogen reductions are used as inputs in benefit estimations, also needs to be considered and is particularly highlighted in the work of Jensen and Unnevehr (1998). Clearly it is desirable for such food safety gains to be calculated from actual real-world changes in specified bacterial populations (among other food safety attributes of concern) observed at the individual plant-level. These reductions should not be laboratory-level performance evaluations of a strategy unless they have been validated in real-world applications. Such experiments often use inoculated samples with elevated populations of (indicator) organisms and can bias results in favor of certain interventions, suggesting unrealistic pathogen reductions. An interdisciplinary methodology is a requisite for such comparisons, as is a close working relationship with the affected stakeholders.

Related issues of the *maintenance* of pathogen reductions beyond the stage of the chain of HACCP implementation (for example a slaughter/processing facility) and the optimal stage(s), and nature of, additional food safety controls remain under studied. Indeed the title of this session suggests that such risk analysis innovations were to be discussed. This was not the case as the individual pieces of the measurement puzzle for different production processes still require enough work on their own before one can move to the more holistic implications. However, it appears somewhat misguided to trumpet the success of reducing microbial loads on freshly slaughtered/processed meat, poultry, or seafood if such gains are mitigated by downstream hazards and thereby provide no risk reduction. The research presented here needs to be expanded to other sectors and stages with the results integrated into chain-wide risk analyses. The continuing role of economics in such work is crucial. For if stage- or sector-specific risk management strategies are considered without the chain-wide determination of all economic implications, it is possible that, at best, an inefficient policy may be selected and, at worst, significant disincentives will result. Similarly, examples of *cost shifting* among segments of the chain (transfers), as opposed to "true" cost reductions, may arise with the application of HACCP-based systems (e.g., requiring more of your input supplier via a critical control point).

## Meat and Poultry HACCP

The most contentious cost issue in the USDA's regulatory impact assessment (RIA) focuses on the details of process modification required of firms to ensure compliance with the pathogen reduction standards. The rule establishes performance standards for *Salmonella* for all plants that slaughter and those that process raw ground product. Further, all slaughter establishments are to employ a generic *E. coli* testing program to validate their process. Debate has centered on the additional equipment costs required by various sized plants, the potential structural implications of the standards, and the relationship between the recurring and non-recurring elements of such process modification and other related HACCP costs.

The in-house review of costs of current pathogen reduction strategies performed by FSIS based on plant size suggests that hand hot water spraying is the lowest cost effective intervention for small slaughter facilities ($0.08 per carcass). Alternative strategies considered include a pre-evisceration acid spray system with both a pre-wash spray cabinet and a sanitizing cabinet at $0.79 per carcass for low volume use, and a trisodium phosphate (TSP)-based system at $0.85 per carcass. Also, the use of steam vacuum systems, with a non-recurring cost of $10,000 and a recurring cost around $4,500, is discussed. The poultry data are based on the use of TSP rinses, estimated to cost $40,000 per line.[2] This translates to $0.03 per broiler and $0.014 per turkey.

That the final RIA uses both high and low scenarios of the costs of process modification is indicative of the methodological problem plaguing the analysis. The low cost scenario is based on the assumption that 10 percent of the 66 large hog and beef slaughter plants would need to install a steam vacuum system to ensure compliance with the *Salmonella* standard. Further, "half of the 376 small establishments must install a hot water rinse at $0.08 per carcass" (USDA 1996, p. 38977). Conversely, the high cost scenario suggests adoption rates might be as high as half (33) for large (implying that the other plants already have such systems in place) and 100 percent for small and very small plants. No process modification costs for compliance with the *Salmonella* standard are calculated for facilities that do not slaughter, suggesting these plants "must depend on the *Salmonella* levels of their incoming product in order to meet the performance standards" (USDA 1996, p. 38977). A similar exercise for poultry suggests that the low cost scenario would have 36 large establishments installing TSP systems, with the high cost scenario increasing this number to 182 (100 large and 82 small). Finally, the process modification costs related to the generic *E. coli* sampling standard are assumed to be directly related to the *Salmonella* standard. FSIS concludes:

If the low cost scenario for compliance with *Salmonella* standards proves to be more accurate, there will likely be more separate compliance costs for generic *E. coli*. As the costs for *Salmonella* compliance go up, the likelihood of separate *E. coli* costs goes down. It is important to note that under the high cost scenario, all cattle and swine slaughter establishments are using the steam vacuum system or hot water rinse and half of all poultry slaughter establishments are using TSP systems. Under this scenario, it is difficult to imagine that any establishments would still be failing to meet the performance criteria for generic *E. coli* (USDA 1996, p. 38981-2).

So how do these numbers relate to the work of Jensen and Unnevehr (1998)? The authors estimate the minimal costs of attaining a range of pathogen standards for large pork slaughter plants. The available pathogen reduction strategies include water rinses (at three temperatures), with and without the application of a sanitizing spray (organic acid). These interventions are more traditional in nature, so their results do not directly relate to the RIA discussion of process modification. The steam vacuum strategy is not included in the model because a consistent effectiveness measure was not available. The per carcass costs of the wash and spray systems are clearly under the $0.79 and $0.85 estimates discussed above, perhaps because the Jensen and Unnevehr (1998) estimates are based on large volume plants. Even when the most restrictive pathogen standard is simulated, costs are still under $0.50 per carcass, suggesting that the RIA may have overestimated the costs, at least for large plants.

The Jensen and Unnevehr (1998) study does not attempt to test which of the two FSIS cost scenarios are more appropriate. Instead, the methodology highlights that if (and perhaps only if) the selection of an intervention strategy is made on a least cost basis then actual process modification costs may be lower than suggested in the RIA for large pork slaughter plants. When the study is expanded to include the full range of interventions including steam vacuums and multiple hurdle techniques (and extended to cover beef and poultry slaughter and processing facilities), a more complete comparison to the RIA can be made. This study will have to pay close attention to adoption rates for each strategy.

Jensen and Unnevehr (1998) present a clear framework for incorporating pathogen reduction data into their assessment of least-cost interventions. However, one needs to be very careful in applying these microbiological results. First, an honest open approach is best. As the authors admit, their data come from two separate (though small) sources. One study tested interventions in a plant environment and the other did not. Further, one inoculated samples and the other did not. The inoculation procedure effectively elevates pathogen populations to an observable level and implies that although real-world reductions (the results of interventions) will not be of the same magnitude, they will be of the same relative order. This remains an untested hypothesis for most interventions. Indeed, preliminary data from research undertaken at Texas A&M University suggests that the rankings of individual, and especially multiple hurdle pathogen reduction strategies, are very sensitive to the initial microbial load.

One cannot presume without further analysis that a certain log reduction[3] due to an intervention is an improvement over current strategies; will be achieved in all plants at all times, regardless of the "cleanliness" of animals being presented for slaughter; or will lead to a risk reduction (the maintenance of hazard reductions issue). Perhaps one should then presume the estimates generated actually report an upper bound of the resultant consumer benefits.

While not explicitly linking the suggested pathogen reductions observed for each of the candidate strategies to consumer risks, Jensen and Unnevehr (1998) suggest that interventions at slaughter plants may be preferable to, and more effective than, tighter controls earlier in the chain (i.e., producers and first handlers). However, such a conclusion again presumes the effectiveness of interventions at the slaughter plant is maintained downstream. Otherwise, perhaps, the only true risk reduction benefits would occur following HACCP implementation at the food service or retail stage.

## Seafood HACCP

Most discussions of the seafood HACCP-based rule center on a more fundamental level (as opposed to the details of process modification costs above). Colatore and Caswell (1998) provide the first detailed *ex post* review of the rule. Unlike Jensen and Unnevehr (1998), true plant-level cost data is utilized. The FDA in its HACCP RIA (FDA 1995) takes great care to stress the basis of their estimated costs by citing data from both the National Marine Fishery Service (NMFS) and an internal model and compares both data sets to "start-up" costs voluntarily reported by seafood firms. This demonstrates an admirable acceptance of the problems involved in determining compliance costs. The NMFS results (which FDA concludes "represent the best evidence available at that time" (FDA 1995, p. 65180)) suggest that first year (i.e., non-recurring and initial recurring) costs were $23,000 (this is reported in Colatore and Caswell (1998) as $25,934, presumably to account for inflation). The second model is based on the opinion of FDA experts familiar with HACCP and suggests the costs of implementation for "two small hypothetical seafood processors that the agency believes to be representative of a significant portion of the seafood industry" (FDA 1995, p. 65180). One of these plants is presumed to have certain typical hygiene deficiencies. Each of these plants has very simple processes and thus low HACCP costs. Given the process assumption, the FDA estimates may be considered a lower bound for potential implementation costs (of $6,400 or $6,978 in Colatore and Caswell (1998)). Clearly, the selection of a small sample of processors used in the FDA model beg the questions: how representative can such estimates be, how accurate is the 20 percent share of hygiene deficient plants, and does scale really impact implementation costs (FDA 1995, p. 65182)?

Given this background, consider the methodology and results reported in Colatore and Caswell (1998). Though clearly a case study, the value of the selection of breaded fish must relate to its representative nature within the industry for a fair comparison to be made. If such is true, then one must presume that two CCPs is an average number for all facilities. It appears to be overly simplistic to make this assumption without first comparing breaded fish to other seafood processes. HACCP plans must be plant specific and flexible to the unique hazards inherent to the individual plant. To presume that all breaded fish facilities have identical chemical, physical, and biological concerns seems to be another heroic assumption that suggests little or no variability within the sector -- a fact hardly supported by the data. Therefore, one should take a cautious approach in reducing compliance costs to the "mandated" two CCPs. That said, it is only in this reduction that the authors attain results close to the FDA and NMFS lower and upper bounds.

Colatore and Caswell (1998) provide an excellent detailed analysis and progression of thought surrounding the *motivation* for adoption and final coverage of the HACCP plans surveyed; however, care needs to be taken. The authors clearly found that many of the seafood processors they interviewed adopted HACCP before required to do so by the FDA. However, a protracted debate about the potential for a HACCP-based seafood regulation had been observed long before the proposed rule was announced in 1994. Knowing this, some processors pre-positioned themselves and became "first-movers." Perhaps the true cost of the final HACCP rule for these firms is not zero as suggested, but instead the costs of ensuring their "quasi-voluntary" HACCP systems conformed to the mandated system. This point highlights the difference between non-recurring (pre regulation and even *ex ante* HACCP adoption) and implementation (*ex post*) costs. The use of very detailed personal interviews surely provides a basis for future *ex post* assessments of RIAs, providing a more complete understanding of all impacts at a plant level. Further, such surveys performed *ex ante* for other sectors can provide a snapshot of current food safety costs and indicate a baseline and trend of voluntary HACCP adoption.

Finally, the FDA RIA suggests that after reviewing the start-up costs reported by some 86 firms (of which they report on a sub-sample with "more complete information" (FDA 1995, p. 65183), their theoretical range of costs roughly mirror those reported. Caution must be taken here as these are voluntarily reported numbers (questions of bias); they relate to pre-mandated HACCP costs (although the FDA suggests many HACCP plans had been verified by a third party, they cannot be linked to the mandated system); these are non-recurring costs (it is not clear if they include first year variable recurring costs as reported by Colatore and Caswell (1998)). The FDA reports that only one of the 22 firms providing detailed start-up cost information said these costs exceeded $20,000.

## Broader Economic Impacts: What Needs to be Added?

Several potential indirect impacts should be considered in the broader economic analysis of these HACCP-based regulations. First are scale effects. Clearly the FSIS RIA out-performed the FDA RIA in this regard, with the latter presuming implementation costs would not differ significantly by plant size. However, neither paper suggested how results would be affected by firm size. As HACCP-based regulations expand in their coverage (e.g., to the retail sector with many small and very small firms), it is argued that scale effects will be of paramount importance.

The food safety system put in place by a plant can also impact non-safety quality attributes, thus increasing overall efficiency (termed X-efficiency in parts of the literature -- see Unnevehr and Roberts (1997)). This result was found by Jensen and Unnevehr (1998) who suggest how multiple goal linear programming may be applied to consider these joint products. Similarly, Colatore and Caswell (1998) in effect demonstrate that HACCP implementation promotes such dual safety/non-safety firm-level benefits by identifying plants that exceed the minimal FDA requirements.

That HACCP can help limit product rejection or rework reducing the variability inherent to all production processes also deserves more attention. This benefit allows for increased customer/consumer satisfaction, reduced complaints and product return, and may increase consumer confidence. Also, international trade is clearly facilitated when "harmonized" HACCP-based regulations are adopted (Caswell and Hooker 1996). Does this promote benefits for exporters at the expense of firms uniquely producing for the domestic market? Conversely, does a nation thereby act as a "good player" by aiding equivalency evaluations for other food safety regulations for other products generating consumer benefits that outweigh possible industry costs? Should such impacts be included in future RIAs? Should nations be required to demonstrate that food safety regulations comply with *proportionality* commitments (i.e., the measures have the least international trade impact possible)?

Potential legal liability and insurance cost savings can arise from the use of innovative food safety controls. A first-mover advantage can be achieved by those plants and firms who are first to adopt a proven intervention. This can improve the overall company image, potentially providing a competitive and marketing advantage when such channels exist. Such innovation offset dynamics are discussed in Cockbill (1991), Henson (1997), and Hobbs and Kerr (1992). Again, such broader economic impacts may or may not be candidates for inclusion in future RIAs depending upon the details of the HACCP-based regulation under consideration.

## Difficulties in Forecasting Costs and Benefits for Novel Innovations

Clearly the HACCP-based regulations discussed here have an admirable degree of flexibility (minimal process standard nature). Further, the performance standard elements of the FSIS rule seem to provide some incentive to promote innovation in the pathogen reduction strategies employed. However, due in part to such success in regulatory design, this flexibility implies that *ex post* costs may differ significantly to *ex ante* estimates as more plants adopt validated pathogen reduction strategies that differ from those the USDA presumed would be used. This is further confounded when the selection of such strategies is not made on a least-cost basis.

Limited economic research exists to provide reliable estimates of costs and resultant benefits of many food safety interventions. Several pathogen reduction strategies, particularly multiple hurdle techniques, incorporate novel approaches for which only limited commercial applications exist, thus requiring a cautious approach to forecasting potential costs. Further, plant-level pathogen reduction benefits of multiple hurdle interventions are not always simply additive.

Among the approaches requiring additional analysis, we include steam and hot water pasteurization of beef and pork carcasses. Despite its increasing use as a sanitation step in the U.S. beef sector, the Frigoscandia Steam Pasteurization System has not undergone significant economic analysis. In a brief discussion, Unnevehr and Roberts (1997) suggest the recurring costs of the system range between $0.02 and $0.20 per pound (or approximately $0.20 to $2.00 per carcass) depending upon plant throughput. They also report the adoption of the process by most large Excel/Cargill plants and some 60 additional U.S. and international firms who have placed orders for cabinets.

A much more detailed estimate of the costs of the beef pasteurization process is supplied in Jensen, Unnevehr, and Gomez (1998), who suggest that a total cost of $0.266 per carcass for a large beef plant may be appropriate. The authors report that this system may be cheaper than rinses, steam vacuum, or sanitizing sprays. Similarly, Jensen and Unnevehr (1998) report a per carcass cost of $0.15 for a hog hot water pasteurizer. Both estimates are based on supplier-provided information. While this line of analysis and research is an admirable side-step of the problematic issues involved in obtaining plant-level cost data, it may be subject to certain misinterpretation. Actual costs observed by plants using these various systems will likely be higher than those reported by the equipment manufacturers. This will be due to inefficiencies in application by plant managers as they move along "learning curves," as well as inherent under reporting of implementation costs (e.g., plant redesign and training). Once again, these estimates likely provide a lower range of process modification costs and need to be verified against true plant costs. They are also only relevant (as constructed) for large establishments.

Before any meat or poultry slaughter and/or processing pathogen reduction strategy can be used in the U.S., it must be approved by FSIS. This leads to the questions of how much does this slow-up innovation, does the approval process really provide a "standard" measure of effectiveness, how does FSIS determine this effectiveness, and are the results always applicable to all types and sizes of plants?

Two candidate novel pathogen reduction strategies that require additional economic analysis are the use of ionizing radiation and high intensity light. In a recent review of the economic literature surrounding the application of irradiation to control foodborne pathogens, Morrison et al. (1997) update 1994 USDA estimates of the per-pound costs of control from $0.016 to $0.019. This measure is maintained to be the best available estimate of potential costs of irradiating meat. However, the research is sparse and does not consider possible interim strategies (e.g., the transshipment of meat to commercial irradiation facilities) and is often not linked to marketing or consumer acceptance studies. No such cost estimates are available for the use of high intensity light or other innovations currently under development.

The potential use of novel individual interventions, and innovative combinations of "traditional" interventions, clearly make the *ex ante* estimation of costs and benefits extremely difficult. It seems likely that in future RIAs, the role of pilot programs to suggest real-world impacts will be expanded. Hopefully, these studies will utilize representative firms' experiences with HACCP (or whatever food safety controls are being considered) and consider all "state-of-the-art" interventions. Where not widely adopted in the industry, special care must be taken in suggesting the impacts of any novel interventions requiring attempts to judge their effectiveness based on plant-level experiences and not just laboratory (or interested-party) assessments. At all times the effectiveness of novel interventions should be compared to current systems on a microbiological as well as a cost basis.

## Lessons for Future Regulatory Impact Analyses

We have taken pains to highlight the basis of uncertainty in certain cost benefit calculations reported in this paper. This is mostly reported as ranges, suggesting where previous estimates may more realistically provide upper or lower bounds of the true costs and benefits under analysis. As a general theme this approach may not appear too daring. If the *ex ante* estimates are off by similar magnitudes, surely the answer will be the same (i.e., benefits > costs). However, this is hardly likely mandating a closer inspection of where estimates went awry.

Framing these discussions in light of the conference title and goal to present *The Economics of HACCP: New Studies of the Costs and Benefits*, we suggest many lessons can be taken from the papers discussed here and presented in the meeting. First, one needs to be honest about what is under analysis. Is it HACCP *per se* or a regulation that is based on HACCP but also impacts other aspects of the processing environment (e.g., general plant hygiene)? Second, how can these experiences help firms (e.g., small and very small meat and poultry slaughter/processing plants) or agencies prepare for HACCP? Simply, details count to ensure all costs and benefits are fully considered.

It would be wrong to presume that just because HACCP enjoys net benefits in one industry or sector, it will do the same in all mandated applications. Finally, lets be honest and say that the economics of HACCP will not always drive all assessments of the system. RIAs will only be required (in the U.S. at least) when the regulation is "economically significant" (an annual effect of more than $100 million or some adverse effect on the economy -- see Belzer (1998)). This will not always be the case. Further, given changes in current assumptions and methodologies, it may be possible that a proposed regulation can be proven to have no net benefit yet large food safety gains. Whether such a rule is enacted depends upon the valuation placed by society on the individual components of the cost benefit analysis.

### Notes

[1] Neal H. Hooker is a Postdoctoral Research Associate, Center for Food Safety, Texas A&M University and Department of Resource Economics, University of Massachusetts. Funding is provided by the Australian Meat Research Corporation under grant MSQS.005 awarded to Texas A&M University.

[2] Large poultry establishments average two lines, small 1.5.

³A certain log reduction (e.g., 2.38-0.39) implies that the indicator pathogen population has been lowered by some 97 percent. Yet the "potency" of many pathogens is still not well understood. Hence it may take a five log reduction (a 99.999 percent fall in populations) to ensure a "safe" product.

**References**
Belzer, Richard B. 1998. HACCP Principles for Regulatory Analysis. Paper presented at the NE-165 Conference *Economics of HACCP: New Studies of the Costs and Benefits* Washington D.C., June 15-16.
Caswell, Julie A. and Neal H. Hooker. 1996. HACCP as an International Trade Standard. *American Journal of Agricultural Economics*, 78(3): pp. 775-779.
Cockbill, Charles. 1991. The Food Safety Act: An Introduction. *British Food Journal*, 93(8): pp. 4-7.
Colatore, Corina and Julie A. Caswell. 1998. The Cost of HACCP Implementation in the Seafood Industry: A Case Study of Breaded Fish. Paper presented at the NE-165 Conference *Economics of HACCP: New Studies of the Costs and Benefits* Washington D.C., June 15-16.
Henson, Spencer J. 1997. Costs and Benefits of Food Safety Regulations: Fresh Meat Hygiene Standards in the European Union. Report to the Working Party on Agricultural Policies and Markets of the Committee of Agriculture, Organization for Economic Cooperation and Development.
Hobbs, Jill E. and William A. Kerr. 1992. Costs of Monitoring Food Safety and Vertical Coordination in Agribusiness: What can be Learned for the British Food Safety Act 1990? *Agribusiness: An International Journal*. 8(6): pp. 575-584.
Jensen, Helen H. and Laurian J. Unnevehr. 1998. HACCP in Pork Processing: Costs and Benefits. Paper presented at the NE-165 Conference *Economics of HACCP: New Studies of the Costs and Benefits* Washington D.C., June 15-16.
Jensen, Helen H., Laurian J. Unnevehr, and Miguel I. Gomez. 1998. Costs of Improving Food Safety in the Meat Sector. *Journal of Agricultural and Applied Economics*, 30(1): pp. 83-94.
MacDonald, James M. and Stephen Crutchfield. 1996. Modeling the Costs of Food Safety Regulation. *American Journal of Agricultural Economics*. 78(5): pp. 1285-1290.
Morrison, Rosanna Mentzer, Jean C. Buzby, and C.-T. Jordan Lin. 1997. Irradiating Ground Beef to Enhance Food Safety. *FoodReview*, January-April: pp. 33-37.
Unnevehr, Laurian and Tanya Roberts. 1997. Improving Cost/Benefit Analysis for HACCP and Microbial Food Safety: An Economist's Overview. In *Strategy and Policy in the Food System: Emerging Issues*. Eds. Julie A. Caswell and Ronald W. Cotterill. Food Marketing Policy Center, Uni. of Connecticut and Dept. of Resource Economics, Uni. of Massachusetts: pp. 225-229.
U.S. Department of Agriculture. 1996. Pathogen Reduction; Hazard Analysis and Critical Control Point (HACCP) Systems; Final Rule. *Federal Register*, 61(144): pp. 38805-38989.
U.S. Food and Drug Administration. 1995. Procedures for the Safe and Sanitary Processing and Importing of Fish and Fishery Products; Final Rule. *Federal Register*, 60(242): pp. 65096-65202.

Chapter 6

# The Cost of Quality in the Meat Industry: Implications for HACCP Regulation

John M. Antle[1]

This paper develops a framework for measuring the plant-level cost of quality regulations, based on models of the production of quality-differentiated products. This framework emphasizes the potential importance of the impacts of regulations on both variable and fixed costs of production. Measurement of regulatory cost under performance standards and design standards is discussed. Evidence on the potential impacts of food safety regulation on variable costs of production is presented from a recent study of the meat and poultry industries.

The Food Safety and Inspection Service (FSIS) of USDA conducted a Regulatory Impact Assessment (RIA) of the HACCP and pathogen reduction regulations (FSIS 1996). This assessment indicated that the benefits of the regulations would be as high as $3.7 billion per year, whereas the costs were estimated to be about $0.1 billion per year. A subsequent study by the Economic Research Service of USDA estimated the benefits to be as high as $19.1 billion per year. Thus, these analyses imply that the benefits of the regulations would exceed the costs by a wide margin.

The FSIS cost estimates were based on an accounting of the estimated costs of quality control such as the development of a HACCP plan, costs of process modifications to implement the plan, and record keeping and testing costs. However, the FSIS assessment did not account for the impacts of the regulations on plant operating costs or productivity of slaughter and processing. The impacts of food safety regulations on productivity and cost of production was estimated in a recent study by Antle (1998b). Using these cost functions, the estimated impacts of food safety regulations on the variable cost of production in the beef, pork, and poultry industry is estimated to be in the same range as the benefits estimated by FSIS and ERS. These findings cast doubt on the conclusion of the RIA that the benefits of the recent HACCP and pathogen reduction regulations would exceed the costs by a wide margin. Indeed, these results suggest that the costs could plausibly exceed the benefits, thus casting some doubt on the social value of the regulations.

The costs of food safety regulation include the industry's cost of compliance, borne by both industry and the consumers of their products, as well as administrative costs borne by taxpayers and the deadweight loss associated with taxation. The focus here is on the plant-level costs of compliance with regulations and their implications for regulatory impact assessment.

## Product Quality and Production Structure

Analysis of the costs of food safety regulation begins at the plant level. For convenience, this discussion makes the simplifying assumption that each firm operates a single plant. For analysis of market structure and competition, the distinction between plants and firms and issues such as economies of scale and scope need to be given further consideration (Panzar 1989). This discussion focuses on the costs of statutory regulation taking the form of either performance standards or process standards. Analysis of other regulatory approaches, such as liability or product certification, involve considerations beyond the structure of the firm's production technology that are not discussed here.

Analysis of food safety requires consideration of production models that allow for quality-differentiated products (Antle 1998a). To illustrate, consider a firm operating a single plant and producing a single product, y, with quality q. For modeling purposes, quality is defined as a scalar variable, and interpreted as an index of multiple quality attributes. Wholesale markets for meat products differentiate several quality dimensions for which a buyer may be willing to pay a price premium: taste, as represented by USDA grades for red meat; safety, defined as the absence of pathogens and other hazards; wholesomeness and freshness, as related to absence of bacteria that cause spoilage and limit shelf life of fresh products; and other miscellaneous characteristics such as quality of packaging and reliability of supply.

Define the firm's production inputs as the vector x and capital stock k. The general form of the firm's production function is $f(y,q,x,k) = 0$, where $f(\bullet)$ satisfies the standard properties of multiple output technologies (Chambers 1988). In this form, quality can be interpreted as a second output of the production process, and the literature on multiple output technologies can be utilized. Two important properties of multi-product technologies are input-output separability and nonjointness in inputs. Input-output separability holds if and only if the production function can be written $f^1(y,q) = f^2(x,k)$. In effect, this property implies that the quantity of output y and output quality q can be aggregated using the function $f^1(y,q)$. Such aggregation is not useful in an analysis where the objective is to explicitly account for product quality.

Nonjointness of inputs implies that separate production functions of the form $y = f^y(\mathbf{x}^y, k^y)$ and $q = f^q(\mathbf{x}^q, k^q)$ can be defined. Some aspects of quality control, such as record keeping and product testing, are separate from the production process. Also, certain food safety processes, such as food irradiation, may operate separately from the rest of the production process. However, many aspects of quality control, such as temperature controls, cleaning of equipment, and removal of contaminated product, are integrated into the production process and may affect the overall productivity of the process (e.g., by affecting the number of hours the plant can operate and the speed at which slaughter lines can be operated). Therefore, the representation of production technologies for quality-differentiated products is likely to take the form of a multi-output process that is joint in inputs involved in the production process and in some aspects of quality control, but also may be nonjoint in other quality control inputs.

While multi-output production technologies can be utilized in the primal form, for both analysis and estimation purposes it is typically more convenient to use dual cost or profit representations of multiple-output technologies. The general nonseparable, joint representation of the variable or restricted cost function corresponding to the production function $f(y,q,\mathbf{x},k) = 0$ takes the form $vc(y,q,\mathbf{w},k)$ where $\mathbf{w}$ is a vector of prices corresponding to the input vector $\mathbf{x}$. In the case where production is nonjoint in inputs and there are distinct production functions, it then follows that dual cost functions exist of the form $vc^y(y,\mathbf{w},k)$ and $vc^q(q,\mathbf{w},k)$ (Hall 1973). As a generalization of this conventional nonjoint-in-inputs model, the preceding discussion suggests that in the case of quality-differentiated food products, the cost function may generally take the form:

(1) $c(y,q,\mathbf{w},k,\alpha,\beta,\gamma) = vc(y,q,\mathbf{w},k,\alpha) + qc(q,\mathbf{w},k,\beta) + fc(k,\gamma)$

where total cost $c(\bullet)$ is composed of a component of variable cost $vc(\bullet)$ that is joint in conventional production inputs and some quality control inputs, a component of variable cost $qc(\bullet)$ that is nonjoint in conventional inputs and certain quality control inputs (thus it is independent of y but depends on q), and a conventional fixed cost component $fc(k)$ that is independent of both output and quality. Here $\alpha, \beta$ and $\gamma$ are parameters of the respective components of the cost function.

# Measuring Costs of Performance Standards and Process Design Standards

Statutory regulation takes two basic forms, performance standards and process design standards. Following Antle (1998c), this section uses the specification of the production technology in equation (1) to show how the costs of these two types of regulation can be measured.

Performance standards impose the requirement that a firm must achieve a level of product quality, $q_g$, but do not specify the technology that the firm must use to achieve the standard. In the simplest case, a plant can efficiently achieve the higher standard with the technology that was in use before the standard was imposed, where the technology is defined in terms of the cost function parameters and the capital stock. Letting the level of product quality supplied before the regulation be $q_o$, the imposition of the performance standard $q_g > q_o$ results in an increase in the cost of production equal to:

(2) $\Delta c(y,q_o,q_g,w,k,\alpha,\beta,\gamma) = \Delta vc(y,q_o,q_g,w,k,\alpha) + \Delta qc(q_o,q_g,w,k,\beta)$.

where:

$\Delta vc(y,q_o,q_g,w,k,\alpha) \equiv vc(y,q_g,w,k,\alpha) - vc(y,q_o,w,k,\alpha)$
$\Delta qc(q_o,q_g,w,k,\beta) \equiv qc(q_g,w,k,\beta) - qc(q_o,w,k,\beta)$.

In implementing a performance standard, a firm may choose to modify its production process to achieve the mandated quality standard more efficiently, even though it is not required to do so in the regulations. Modifications of the existing plant and equipment or other operating characteristics of the plant would result in a change in the parameters of the cost function. It may also be necessary to invest in new plant and equipment, changing the firm's capital stock. Let the firm's modified production process under the performance standard be represented by the parameters $\alpha_p$, $\beta_p$, and $\gamma_p$ and the capital stock $k_p$. The change in cost of production induced by compliance with the performance standard is then:

(3) $\Delta c(y,q_o,q_g,w,k_o,k_p,\alpha_o,\beta_o,\gamma_o,\alpha_p,\beta_p,\gamma_p) =$
$\Delta vc(y,q_o,q_g,w,k_o,k_p,\alpha_o,\alpha_p) + \Delta qc(q_o,q_g,w,k_o,k_p,\beta_o,\beta_p) + \Delta fc(k_o,k_p,\gamma_o,\gamma_p)$,

where:

$\Delta vc(y,q_o,q_g,w,k_o,k_p,\alpha_o,\alpha_p) \equiv vc(y,q_g,w,k_p,\alpha_p) - vc(y,q_o,w,k_o,\alpha_o)$
$\Delta qc(q_o,q_g,w,k_o,k_p,\beta_o,\beta_p) \equiv qc(q_g,w,k_p,\beta_p) - qc(q_o,w,k_o,\beta_o)$
$\Delta fc(k_o,k_p,\gamma_o,\gamma_p) \equiv fc(k_p,\gamma_p) - fc(k_o,\gamma_o)$.

While conceptually straightforward, these alternative cases have significantly different data requirements for estimation. In the simple case (2) where the technology is not changed to comply with a performance standard, data from the time period before the regulation is imposed can be used to estimate the cost function, and this cost function can then be used to estimate the costs of the performance standard (assuming that the technology has not changed during that time interval for other reasons not related to the regulation). However, in the case (3) where the technology changes in response to the regulation, both pre- and post-regulation data are needed to estimate the cost functions and the capital stocks in order to make accurate estimates of the cost of the regulation.

Clearly, the problem posed by case (3) for conducting RIAs is that the cost estimates need to be made *ex ante* before post-regulation data are available. A more feasible strategy is to assume that the capital stock is changed but the parameters of the process are not changed. In that case, the regulatory cost is:

(4) $\Delta c(y, q_0, q_g, w, k_0, k_p, \alpha, \beta, \gamma) =$
$\Delta vc(y, q_0, q_g, w, k_0, k_p, \alpha) + \Delta qc(q_0, q_g, w, k_0, k_p, \beta) + \Delta fc(k_0, k_p, \gamma),$

In this case, regulatory costs can be estimated using the pre-regulation cost function with the addition of information about the change in the capital stock. If firms do in fact modify their production processes so as to be more efficient, then (4) will provide an upper bound approximation of (3). This observation also suggests that a useful research topic would be to obtain both *ex ante* and *ex post* data in order to assess the difference between (3) and (4) in case studies.

A strict process design standard specifies the technology that a firm must use, without specifying the outcome that must be achieved as in a performance standard. A design standard will generally require firms to modify their plant and equipment and the production process to meet the government standards. The mandated technology is represented by the capital stock $k_g$ and the cost function parameters $\alpha_g$, $\beta_g$, and $\gamma_g$. However, in the case of the process design standard, the new level of product quality that is achieved, $q_1$, is not specified in the regulation and not known *ex ante*. Following equations (2) and (3), the cost of a process design standard is given by:

(5) $\Delta c(y, q_0, q_1, w, k_0, k_g, \alpha_0, \beta_0, \gamma_0, \alpha_g, \beta_g, \gamma_g) =$
$\Delta vc(y, q_0, q_1, w, k_0, k_g, \alpha_0, \alpha_g) + \Delta qc(q_0, q_1, w, k_0, k_g, \beta_0, \beta_g) + \Delta fc(k_0, k_g, \gamma_0, \gamma_g).$

While they appear to be similar, there are important differences between the cost of a performance standard (3) and a process design standard (5). First, the performance standard establishes a level of quality or safety $q_g$ that must be achieved by every plant. In contrast, the level of safety $q_i$ achieved by a process design standard will vary across plants because each plant is actually designed and operated differently, in spite of the process standard. Second, because the performance standard allows plant managers to tailor quality control to fit their particular plant's design, there is a presumption that the cost of the performance standard will be less than the cost of the process standard in achieving a given level of safety (Antle 1995). This hypothesis could, in principle, be tested with appropriate data and methods using the relationships identified here.

Regulations may also combine elements of both performance and process design standards, as the in the case of the U.S. Department of Agriculture's recently implemented meat inspection regulations that combine a mandatory quality-control system with performance standards for certain microbial contaminants, giving a regulatory cost measured as:

(6) $\Delta c(y,q_0,q_p,w,k_0,k_g,\alpha_0,\beta_0,\gamma_0,\alpha_g,\beta_g,\gamma_g) =$
$\Delta vc(y,q_0,q_p,w,k_0,k_g,\alpha_0,\alpha_g) + \Delta qc(q_0,q_p,w,k_0,k_g,\beta_0,\beta_g) + \Delta fc(k_0,k_g,\gamma_0,\gamma_g).$

Finally, recall that for simplicity of the presentation, quality was treated here as a single dimension. In most food products there are multiple dimensions of quality, including one or more safety attributes (such as presence of multiple pathogens), taste and nutritional characteristics, degree of processing, packaging, and so forth. Some of these safety dimensions may be specified in performance standards and others may be addressed through design standards, while other aspects of quality are not regulated. The recent pathogen reduction and HACCP regulations implemented by USDA provide an example of a mixed approach to food safety regulation, wherein certain aspects of production and certain pathogens are targeted.

## The Cost of Quality in the Meat and Poultry Industries: Evidence from Estimates of Variable Cost Functions

The preceding section shows that the impacts of regulation on plant-level cost of production can be measured as changes in variable cost of production that occur in the production process, changes in variable quality control inputs that are independent of the production process, and changes in fixed costs. This section reviews recently developed methods for estimating the first of these components of cost, those associated with changes in the variable cost of production, and summarizes the findings of a recent study of the meat and poultry industry (Antle 1998b).

To empirically implement estimation of the variable cost functions $vc(y,q,w,k,\alpha)$ and $qc(q,w,k,\beta)$, conventional econometric procedures can be utilized if quality is observable. The key problem is that data on product quality (i.e., the rate of occurrence of meat products contaminated with microbes) is not usually available. Assuming that the firm is a price-setting monopolist, Gertler and Waldman (1992) developed an econometric approach to estimation of the cost function with unobserved quality. This approach uses demand variables as instruments for the quality variable in the cost function. This model requires that each firm or plant face different demand conditions so that there is variation in demand variables across observations to identify quality parameters in the cost function.

Antle (1998b) observes that a monopolistic model is not suitable for analysis of cost of production in the meat packing and food processing industry where product markets are competitive. Antle combines Rosen's (1974) model of a competitive industry producing quality-differentiated products with Gertler and Waldman's model of a quality-adjusted cost function, and shows that in addition to demand variables, the observed market price for output and other observable variables can be used as instruments to identify safety and other quality attributes.

Cost functions were estimated using data from the Census of Manufactures for slaughter and processing plants producing beef, pork, and poultry (Table 1). Most of the industry's output is produced by larger plants (a large plant is defined here as producing more than 100 million pounds of product annually). The Census of Manufactures data also show that the variable costs of these plants are dominated by the cost of animal inputs, and that variable costs represents 90 percent or more of total cost (more than 95 percent in the case of large plants). This is an important fact for analysis of the costs of regulation. It implies that if indeed regulation has an impact on both variable and fixed costs of production, the impact on variable cost is likely to be most significant simply because

**TABLE 1**

**Plant Numbers and Production by Plant Size, 1992**

(Production in billions of pounds)

|  | Small | | Large | | Total | |
| --- | --- | --- | --- | --- | --- | --- |
|  | Plants | Production | Plants | Production | Plants | Production |
| **Beef** | 52 | 2.02 | 46 | 17.83 | 98 | 19.85 |
|  | (54.7) | (10.2) | (48.4) | (89.8) |  |  |
| **Pork** | 44 | 0.87 | 33 | 11.05 | 77 | 11.92 |
|  | (57.1) | (7.3) | (42.9) | (92.7) |  |  |
| **Poultry** | 87 | 4.13 | 115 | 19.44 | 202 | 23.57 |
|  | (43.1) | (17.5) | (56.9) | (82.5) |  |  |

Note: Small plants defined as production less than 100 million lbs/year. Large plants defined as production greater than or equal to 100 million lbs/year. Percentages in parentheses.

Source: Antle (1998b).

variable costs are such a large share of total cost of production.[2]

Translog cost functions were estimated for the small and large plant groups with 1987 and 1992 data for beef, pork, and poultry plants. The hypothesis that variable cost is not a function of product safety was overwhelmingly rejected for all plant size groups and for beef, pork, and poultry plants. The estimated cost functions show that variable cost of production is an increasing function of product safety, as hypothesized. Table 2 presents the elasticities of cost with respect to safety derived from translog cost function models. The translog models show a positive elasticity of cost with respect to safety, as predicted by economic theory.

## Implications for Costs of Mandatory HACCP

In July 1996 FSIS announced new regulations for all meat and poultry plants. All slaughter and processing plants are now required to adopt the system of process controls known as Hazard Analysis and Critical Control Points (HACCP). To verify that HACCP systems are effective in reducing bacterial contamination, pathogen reduction performance standards are being established for *Salmonella*, and slaughter plants are required to conduct microbial testing for generic *E. coli* to verify that their process control systems are working as intended to prevent fecal contamination, the primary avenue of bacterial contamination. FSIS is also requiring plants to adopt and follow written standard operating procedures for sanitation to reduce the likelihood that harmful bacteria will contaminate finished products.

The RIA conducted by FSIS utilized various parts of the scientific literature on foodborne illness to estimate the potential benefits of reducing such risks (see reviews of the literature in Council for Agricultural Science and Technology 1994, and Caswell 1995). On the benefits side several key assumptions were made, including: (1) the number of unreported illnesses and deaths attributable to food pathogens; (2) the effectiveness of the regulations in reducing pathogens in meat products; and (3) a proportional relationship between pathogens in meat

### TABLE 2

Cost Elasticities for Safety Derived from Translog Quality-Adjusted Cost Functions for Small and Large Plants for Beef, Pork, and Poultry Slaughter and Processing

|  | Beef | | Pork | | Poultry | |
| --- | --- | --- | --- | --- | --- | --- |
|  | Small | Large | Small | Large | Small | Large |
| **Safety Elasticity** | 0.739 | 0.728 | 0.153 | 0.263 | 0.430 | 0.506 |

Source: Antle (1998b).

products and frequency of foodborne illness. A great deal of scientific uncertainty surrounds these assumptions. There is some scientific basis for the estimation of unreported illnesses and deaths (Council for Agricultural Science and Technology), although estimated illnesses and deaths vary widely and are controversial (Wilson). On the effectiveness of the regulations, FSIS stated, "FSIS recognizes that the actual effectiveness of the final requirements in reducing pathogens is unknown..." (FSIS 1996, p. 38968). In the regulatory impact assessment of the final rule, FSIS utilized a range of effectiveness from 10 to 100 percent. On the issue of proportionality between pathogen prevalence and foodborne illness, FSIS stated in the Regulatory Impact Assessment, "FSIS has not viewed proportional reduction as a risk model that would have important underlying assumptions that merit discussion or explanation. For a mathematical expression to be a risk model, it must have some basis or credence in the scientific community. That is not the case here. FSIS has acknowledged that very little is known about the relationship between pathogen levels at the manufacturing stage and dose, i.e., the level of pathogens consumed." (FSIS 1996, pp. 38945-38946). Thus, a high degree of scientific uncertainty is associated with the benefits estimates.

The FSIS assessment also made certain assumptions in estimating costs. The FSIS analysis of the costs of the new regulations was based on an accounting methodology that assumed that the costs of the regulations would be comprised of quality control activities, such as record keeping and product testing, and some process modifications assumed to be necessary to meet the regulatory requirements. However, no estimates of the effects of the regulations on the overall operating efficiency of the production process were included. Curiously, FSIS justified this assumption by arguing that if any additional costs were incurred in meeting the new regulations, they should not be attributed to the regulatory process because the new regulations are simply "...a more effective way of assuring that establishments meet "already established health and safety related requirements." (FSIS 1996, p. 38979). Under this assumption, the FSIS found that the recurring cost of the regulations would be on the order of $100 million per year (in 1995 dollars), or less than 0.1 cent per pound of meat product (Crutchfield *et al.*).

The limitations of the cost data and analysis did not go unnoticed by other economists. According to Belzer (1998, p. 20), "... the analysis contains several material errors in its cost assessment that severely understate the likely costs of the rule. First, the estimated cost of required SSOPs (standard sanitary operating procedures), HACCP (hazard analysis critical control points) plans, and generic *E. coli* testing includes only the cost of writing the plans themselves, training current employees, and performing the microbiological tests. *The costs associated with the operational changes necessary to comply with SSOPs and HACCP plans were not included."* (Belzer 1998, p. 20, emphasis added). Belzer also observes, "...the estimated cost of preparing HACCP plans in the analysis is unreliable. This estimate was based on a sample of nine establishments who volunteered to participate in an agency study and are not representative of the several thousand establishments regulated under this rule. Statistical inferences from even a random sample of nine are problematic, but they are obviously illegitimate from a convenience sample of nine volunteers."

The key problem in estimating the cost of food safety or other quality regulations is that safety data, such as the prevalence of pathogens in meat products, are not available at the plant level (however, the new regulations are supposed to lead to the collection of such data in the future). The principal contribution of the econometric approach described above is that it utilizes economic information, i.e., income, prices and other observable variables, to identify the relationship between safety and cost of production. This relationship provides the basis to estimate the impacts of safety regulations on variable cost of production. The potential importance of this component of the regulatory cost is highlighted by the fact that most meat and poultry products in the United States are produced by large scale, highly efficient plants whose efficiency is directly related to factors such as the speed of slaughter lines. Regulations that slow line speeds will reduce the overall operating efficiency of a plant and raise average variable cost of production. Variable cost of production represents over 90 percent of total costs in most meat and poultry slaughter and processing plants in the United States, according to the data from the Census of Manufactures.

To estimate costs, FSIS utilized an accounting methodology wherein the cost of each component of the regulation (e.g., implementation of standard operating procedures, training personnel in quality control methods, and keeping records), is estimated for representative small and large plants. These plant-level cost estimates were then used to estimate the industry-wide costs of the regulations. From an economic perspective, this accounting approach to cost estimation has a significant shortcoming. In the application of the accounting approach, the costs of producing a safer product are assumed to be variable costs independent of the production process, or fixed costs. The costs of making certain process modifications are estimated, but the effect of the regulations on the overall operating efficiency of the process is not considered, and thus is implicitly assumed to be zero. In terms of the analysis of the costs of regulation presented earlier, this is equivalent to assuming that the $\Delta vc$ terms in equations (2) through (6) are equal to zero. This assumption contradicts the findings discussed in the preceding section, where it was found that variable cost of production in beef, pork and poultry slaughter and processing plants is a function of product quality. Because variable cost is a large share of total cost, it follows that quality regulations, such as food safety regulations, can result in substantial increases in the cost of production, as will now be demonstrated in the case of meat and poultry plants.

According to the definitions given in the first section of this paper, the USDA's mandatory HACCP regulations and standard operating procedures are process design standards because they specify the process to be used, not the safety attributes of the end products. But the regulations also involve performance standards for *Salmonella* and generic *E. coli*. Following the discussion above (see equation 6), the best approach to estimate the costs of implementing combined design and performance standards is to estimate the cost function before and after the implementation of the regulation, and then utilize these pre- and post-regulation cost functions to calculate the increase in production costs, holding constant output and factor prices. Clearly, this approach is not possible before regulations have been implemented. For *ex ante* analysis of regulations, the alternative proposed here is to approximate the cost of the process design standard with the cost of an equivalent performance standard based on the pre-regulation technology. This amounts to using equation (2) in conjunction with an estimate of how effective the regulations will be in improving product safety (i.e., an estimate of the performance standard's effectiveness represented by a value of $q_g$).

Equation (2) is a special case of equation (3), wherein the technology (represented by the parameters and the capital stock) do not change. Note that the term involving changes in fixed costs is equal to zero by the assumption that the capital stock is not changed. The $\Delta qv(\bullet)$ term can be interpreted as the costs of quality control estimated by FSIS in its regulatory impact assessment. These costs of quality control are "fixed" in the sense that they do not vary with the rate of output, but as discussed above, in the interpretation of the multi-output cost function these costs vary with the level of quality produced, and so are variable costs that are non-joint with output. The term $\Delta vc(\bullet)$ is the change in variable costs of production associated with changes in input use (labor and materials) necessitated by the imposition of the higher quality standard.

As noted above, in the regulatory impact assessment of the final rule, the FSIS utilized a range of effectiveness from 10 to 100 percent because of the lack of scientific data to indicate the likely effectiveness of the regulations. The only attempt to assess the effectiveness of these regulations *ex ante* is the study by Knutson *et al.* (1995). In that study, a group of experts in food microbiology estimated that the proposed regulations would be likely to be 20 percent effective.

To estimate the cost of the new regulations, the level of product safety that was achieved before the regulations were imposed also must be estimated. As explained above, the quality-adjusted cost function can be used to provide an upper bound on the costs of safety regulations by interpreting the units of quality in the econometric model as units of safety. Because safety is unobserved, the units of safety and its base level are not defined by data contained in the model. Nevertheless, we know that prior to the new regulations, some degree of safety between zero and 100 percent was being achieved by plants in the industry. Let the level of safety prior to the new regulations be $S$, and interpret this number as a percentage so that $0 \leq S \leq 100$. It follows that if the regulations are $e$ percent effective in reducing pathogens, the observed level of safety is increased by $e(100 - S)$ percentage points or by $e(100 - S)/S$ percent. Extensive data have been collected about the prevalence of food pathogens (Council for Agricultural Science and Technology 1994). Surveys of raw meats and poultry show that prevalence of various pathogens ranges from zero to 100 percent, with many in the range of 10 to 50 percent. Therefore, in this analysis, the regulatory costs are estimated assuming the level of safety prior to the new regulations ranged from 50 to 90 percent.

Based on equation (2), the annual change in plant-level variable costs from implementing a regulation that is $e$ percent effective, starting from a safety level of $S$ percent is calculated as:

(7) $\Delta vc(e,S) = (N_{small} \cdot VC_{small} \cdot E_{small} + N_{large} \cdot VC_{large} \cdot E_{large}) \cdot e(100-S)/S$

where:

$e$ = effectiveness of the regulations (percent reduction in pathogens)
$S$ = percent degree of safety prior to imposition of the regulations
$E_i$ = elasticity of cost with respect to safety for i=small, large plants
$N_i$ = number of plants for i=small, large groups
$VC_i$ = variable cost for i=small, large plants

Using equation (7), the costs of a 5, 20, and 35 percent improvement in safety were estimated for prior safety levels of $S$ = 50, 70, and 90 percent. Table 3 shows that the estimated increase in annual total variable cost for beef, pork, and poultry plants, assuming regulations are 20 percent effective, ranges from $535 million to $4.8 billion. These data demonstrate that the costs of food safety regulations associated with increases in variable costs of production may be substantially greater than the quality-control costs estimated by FSIS to be about $100 million per year. Assuming 20 percent effectiveness, the FSIS estimate of annual benefits ranged from $198 million to $738 million; the ERS estimate were in the range of $738 million to $3.82 billion.3 Thus, the data in Table 3 indicate that the costs of higher safety standards could plausibly exceed the benefits. It is important to note that while the plants in this sample represent most of the meat and poultry produced in the United States, there are many other establishments covered by the regulations (such as small meat processing establishments) that are not included in the Census of Manufactures data utilized here. Thus, the costs in Table 3 are likely to be an underestimate of the possible total industry cost of the regulations.

The FSIS estimates imply that regulatory costs are less than 0.1 cent per pound, regardless of the effectiveness of the regulations (Crutchfield *et al.* 1997). Table 4 presents the data from Table 3 for the case of 20 percent effectiveness, translated into costs per pound. These data show that the cost per pound ranges from 0.3 cents per pound for pork with 90 percent base safety, to as high as 17 cents per pound for small beef plants and 50 percent base safety. These costs can be compared to the 1995 wholesale prices of beef, pork, and poultry in the range of $1.10 to $0.60 per pound.

A controversial issue in the debate over food safety regulations is whether they increase costs of production more for small plants than large plants. Equation (1) shows that the terms qc(•) and fc(•) do not vary with output, hence, those components of regulatory costs do decline with output. Indeed, the quality control cost estimates complied by FSIS show that these costs are about 2 percent of total cost for very small plants, and decline to about 0.3 percent of total cost for large plants (Antle 1998d). The data in Table 4 show that the impacts of regulations on average variable cost are somewhat higher for small poultry plants than large poultry plants, but there is little difference between small and large beef or pork plants. Taken together, these data show that for all but the very smallest plants, total regulatory cost per pound of product (increases in variable costs, quality control costs, and fixed costs) is somewhat higher for the very smallest plants, but is about the same for all other plants. Thus, the data suggest that the regulations are not likely to put smaller plants at a competitive disadvantage.

TABLE 3

Estimated Annual Increase in Variable Cost of Beef, Pork, and Poultry Slaughter and Processing Plants for Alternative Levels of Base Safety and Regulatory Effectiveness
(Industry Cost in Million $1995)

| Base Safety (Percent) | Effectiveness of Regulation (Percent Increase in Safety) | | |
|---|---|---|---|
| | 5 | 20 | 35 |
| 50 | 1202 | 4811 | 8420 |
| 70 | 401 | 1604 | 2806 |
| 90 | 134 | 535 | 936 |

Source: Based on data in Antle (1998b).

The data in Table 4 also show that beef plants are likely to experience a much larger increase in cost per pound than either pork or poultry plants. This higher regulatory cost for beef would accentuate the market price differential between beef and pork or poultry, and thus encourage the observed trend in consumption away from beef towards pork and poultry (Brester, Schroeder, and Mintert 1997). The significance of this differential depends on both the effectiveness of the regulations and the degree of safety attained by the industry prior to regulation.

## Conclusions

This paper begins with a discussion of the structure of cost functions for meat plants producing quality-differentiated products, focusing on the jointness properties of conventional inputs and quality control inputs. This structure is used to explore how the plant-level costs of performance standards and design standards can be measured, including difficulties in *ex ante* assessment when only parameters of the pre-regulation technologies are known. A topic identified for future research is to compare the accuracy of *ex ante* estimates of the costs of regulation with the observed *ex post* costs of regulation, to determine how reliable *ex ante* estimates are.

These results are used to investigate implications for the costs of the HACCP and related regulations being implemented by USDA, by using the model of a performance standard to estimate the implied cost of achieving a higher quality product. Estimates of variable cost functions for beef, pork, and poultry slaughter and processing plants in the United States show that variable costs of production are an increasing function of product safety and quality. Analysis based on these cost functions shows that safety regulations that significantly affect the efficiency of the

**TABLE 4**

**Estimated Increase in Variable Costs Per Pound of Production for a 20 Percent Improvement in Safety ($1995)**

|  | Beef | | Pork | | Poultry | |
| --- | --- | --- | --- | --- | --- | --- |
|  | Small | Large | Small | Large | Small | Large |
| **Base Safety = 50%** | 0.165 | 0.170 | 0.023 | 0.034 | 0.053 | 0.042 |
| **Base Safety = 70%** | 0.055 | 0.057 | 0.008 | 0.011 | 0.018 | 0.014 |
| **Base Safety = 90%** | 0.018 | 0.019 | 0.003 | 0.004 | 0.006 | 0.005 |

Note: Base safety is the level of product safety prior to the 20 percent improvement in safety.

Source: Antle (1998b).

production process can significantly raise the cost of production. These costs were not included in the FSIS regulatory impact assessment of the new food safety regulations. The analysis indicates that the costs of the regulations could exceed the estimate of the benefits contained in the FSIS regulatory impact assessment, thus casting doubt on FSIS' conclusion that the benefits of the regulations would exceed the costs by a wide margin. This conclusion is reinforced by the fact that the data used to estimate these costs represent most of the meat packing and poultry slaughter and processing industry, but do not cover other establishments beyond the packing and processing stage that are also affected by the regulations.

**Notes**

[1] J.M. Antle is a Professor, Department of Agricultural Economics and Economics, and Director, Trade Research Center, Montana State University, and a University Fellow, Resources for the Future. Part of the research in this paper was conducted while the author was a research associate at the Center for Economic Studies, Bureau of the Census. Research results and conclusions expressed are those of the author and do not necessarily indicate concurrence by the Bureau of the Census or the Center for Economic Studies. This research was supported by a grant from the National Research Initiative Competitive Grants Program, U.S. Department of Agriculture, and by the Trade Research Center at Montana State University and the Montana Agricultural Experiment Station.

[2] Some commenters have questioned this conclusion by arguing that *processing* cost (i.e., labor cost) is not a large share of total cost, and that only processing cost is affected by the regulations. This argument is correct for processing such as meat grinding that is *additively* separable from the slaughter process, but is incorrect for all aspects of production that are not additively separable. Clearly, the various steps in the slaughter process are not additively separable, even though they may be strongly or multiplicatively separable. To see why a reduction in processing efficiency of any step in the slaughter process affects overall efficiency, let variable cost be $vc(y,q,w,k) = vc_p(q,w_L,k) \bullet vc_m(w_m) \bullet y$ where $vc_p$ is the unit cost of slaughter, $w_L$ is the wage rate of labor, $vc_m$ is the unit variable cost of animal inputs, and $w_m$ is the price of meat inputs. Even though labor cost may be separable from meat input cost, the multiplicative property of the cost function means that if regulations reduce labor efficiency they will reduce overall operating efficiency.

[3] These benefits are calculated by multiplying the benefits estimates for complete elimination of all risks times 20 percent. The FSIS benefits estimates for complete elimination of risks ranged from $0.99 to $3.69 billion, the ERS estimates ranged from $6.7 to $19.1 billion.

**References**

Antle, J.M., *Choice and Efficiency in Food Safety Policy*. Washington, D.C.: American Enterprise Institute Press, 1995.

Antle, J.M., "Economic Analysis of Food Safety." B. Gardner and G. Rausser, Eds. *Handbook of Agricultural Economics*. North-Holland Pub. Co., in press, 1998a.

Antle, J.M., "No Such Thing as a Free Safe Lunch: The Cost of Food Safety Regulation in the Meat Industry." *American Journal of Agricultural Economics*, in press.

Antle, J.M. "Benefits and Costs of Food Safety Regulation." *Food Policy*, in press, December 1998c.

Antle, J.M. "Food Safety, Production Structure, and the Industrial Organization of the Food Industry." Invited Paper prepared for the 62nd EAAE Seminar and 3rd INRA-IDEI Conference on Industrial Organization and the Food Processing Industry, Toulouse, Nov. 12-13, 1998d.

Belzer, R.B. "HACCP Principles for Regulatory Analysis." Paper presented at the conference, *The Economics of HACCP: New Studies of Costs and Benefits*, June 15, 1998, Washington, D.C.

Brester, G.W., T.C. Schroeder, and J. Mintert. Challenges to the Beef Industry." *Choices*, Fourth Quarter 1997, pp. 16-19.

Caswell, J.A., *Valuing Food Safety and Nutrition*. ed. Westview Press: Boulder, 1995.

Chambers, R.G. *Applied Production Analysis: A Dual Approach*. Cambridge University Press: New York, 1988.

Council for Agricultural Science and Technology. *Foodborne Pathogens: Risks and Consequences*. Ames, Iowa, 1994.

Crutchfield, S.R., J.C. Buzby, T. Roberts, M. Ollinger, and C.-T. J. Lin. *An Economic Assessment of Food Safety and Inspection: The New Approach to Meat and Poultry Inspection*. Economic Research Service Agricultural Economic Report No. 755, Washington, D.C., 1997.

Food Safety Inspection Service, U.S. Department of Agriculture, "The Final Rule on Pathogen Reduction and HACCP." *Federal Register* July 25, 1996 (Volume 61, Number 144), pp. 38805-38855.

Gertler, P.J. and D.M. Waldman, "Quality-Adjusted Cost Functions and Policy Evaluation in the Nursing Home Industry." *Journal of Political Economy* 100(Dec. 1992):1232-1256.

Hall, R.E. "The Specification of Technology with Several Kinds of Output." *Journal of Political Economy* 81(1973):878-892.

Jones, R., T. Schroeder, J. Mintert, and F. Brazle. "The Impacts of Quality on Cash Fed Cattle Prices." *Southern Journal of Agricultural Economics* 24 (December 1992):149-162.

Knutson, R.D., H.R. Cross, G.R. Acuff, *et al., Reforming Meat and Poultry Inspection: Impacts of Policy Options*. Institute for Food Science and Engineering, Agricultural and Food Policy Center, Center for Food Safety, Texas A&M University, April 1995.

Rosen, S. (1974) "Hedonic prices and implicit markets: Product Differentiation in Pure Competition," *Journal of Political Economy* 82:34–55.

Wilson, D. "Food Poisonings' Phony Figure." *Columbia Journalism Review* May/June 1998, page 16.

Chapter 7

# HACCP Principles for Regulatory Analysis

Richard B. Belzer[1]

## Introduction

The rapid expansion of food inspection based on the Hazard Analysis and Critical Control Points (HACCP) system has created a wealth of new research opportunities for applied economists. HACCP systems are so different from prior approaches to food technology that new analytical tools and methods must be developed to enable manufacturers to gain a better grasp of the cost-effectiveness of both production innovations and new products. Similarly, as government regulators proceed to extend HACCP to additional product lines and stages in the food production and distribution system, the benefits and costs of these new regulations must be reliably assessed.

This paper draws an analogy between HACCP as applied to food technology and long-established standards for analysis of the consequences of regulatory action. To get these analyses right, a foundation for quality analysis among regulatory agencies must be established that is as obedient to fundamental analytic principles as HACCP rules require industry to behave toward the food they make. Regulatory agencies imposing HACCP principles and rules to more sectors of the food business should apply these same principles to the way they analyze the consequences of alternative regulatory approaches and design regulations. Unless regulators set such an example, their credibility among those they regulate will wither, thereby undermining the moral legitimacy of their role.

To make these points clear I use as examples the seafood HACCP rule promulgated by the Food and Drug Administration in 1995 (FDA 1995a) and the meat and poultry HACCP rule promulgated by the United States Department of Agriculture's Food Safety and Inspection Service (FSIS 1996). Both regulations were accompanied by regulatory analyses as required under Executive Order 12866 for "economically significant" rules (EOP 1993: Sec. 6). This Executive order (as well as its predecessor, Executive Order 12291 (EOP 1981)) could be thought of as a "Generic HACCP Plan for Regulatory Analysis." Both Plans were supplemented with what may be called "Generic HACCP Implementation Guides for Regulatory Analysis" (OMB 1990, OMB 1996). However, neither analysis conformed to elementary principles in these Plans and Guides, despite the fact that the Guides have been in place for almost a decade and the Plans since 1981. Had these analyses been subjected to enforcement provisions analogous to those which FDA and FSIS use to ensure compliance by food producers with HACCP regulations, both agencies would have been subject to significant sanctions. One can only speculate as to whether the regulators' analyses as "products" would have been embargoed until made compliant, recalled as defective or adulterated, or destroyed as unwholesome and unfit for consumption.

Federal food safety agencies continue to expand mandatory HACCP requirements to additional food products, such as juice (e.g., FDA 1997b, 1997c, 1998b, 1998c, 1998d), and new sectors, such as retail (e.g., FSIS 1997b). To an impartial regulatory analyst, it is troubling that these decisions are proceeding based on both extremely high expectations for their effectiveness at reducing foodborne illness and a surprisingly weak analytical foundation for claims that they actually do. Analysis suggests that successes will be limited, unsatisfying, and achieved at enormous expense and frustration, thus imperiling public confidence in both HACCP as a risk-reducing tool and the agencies as effective guardians of public health. Better compliance with established HACCP Principles for Regulatory Analysis would reduce the likelihood of these undesirable outcomes and increase the odds that regulatory action cost-effectively reduces the social costs of foodborne illness.

## Principles of HACCP

Over the last few years, HACCP seems to have become an all-encompassing food safety mantra, yet this is actually a relatively recent phenomenon. HACCP has been around for more than two decades and has been the subject of at least three supportive National Academy of Sciences reports (NAS 1985a, 1985b, 1991). However, HACCP received a cool reception from the regulatory bureaucracy throughout most of this period.

HACCP is, in fact, an adaptation of the more generalized concept of statistical process control. By identifying hazards to quality and critical steps in manufacturing, quality can be improved by reducing the variance at each step, thereby reducing the proportion of defective units. Process control is particularly attractive where a system of performance standards cannot be devised or implemented. For example, the use of performance standards obviously will not work in cases where there are no reliable instruments to measure performance objectively. Performance standards also may be inadequate if the anticipated failure rate is too small to permit inspection by sampling, if inspection itself is destructive, or if the final good is sufficiently perishable that inspection testing cannot be obtained in a timely manner.

In simplified form, HACCP involves only a handful of steps: identify and analyze those things which pose a hazard to the final product; determine the control points in the production process where these hazards may arise; establish critical limits for each control point that should not be exceeded; and set out corrective action procedures to follow when (not if) these critical limits are exceeded. Of course, this simplified exposition does not do justice to the true level of complexity involved, but complexity arises in specific applications rather than in the general theory, and each specific application is likely to be different.

## HACCP Principles for Regulatory Analysis

Something akin to HACCP could be devised for regulatory analysis, for the analogy is unmistakable and its principles self-evident. The final product is a document capturing all the relevant information necessary for rational decision making. For nearly two decades, there has been a "Generic HACCP Plan for Regulatory Analysis" (EOP 1981, 1993) and for almost half that time a "Generic HACCP Guide" to assist agency compliance (OMB 1990, 1996). "Hazards" to quality regulatory analysis are well known to those engaged in its production and consumption. They include the usual HACCP-like issues, such as the quality of raw material inputs; the ability and willingness of producers to devise and follow a comprehensive plan; and a host of places along the production process where both external and internal factors arise which, if left uncontrolled, may seriously compromise the quality of the product.

There are surely as many critical control points in the production of regulatory analysis as there are in the production of pork 'n' beans. Unlike the food business, however, there is no monitoring of critical limits. There are no corrective action plans in place that dictate how these violations should be dealt with by producers, and there is ample evidence that the regulator charged with enforcing HACCP for regulatory analysis lacks the capacity to levy significant sanctions no matter how egregious the violation. Producers cannot be fined, the product cannot be seized, embargoed, or recalled, and the "withdrawal of inspection services" is not a credible option.

### The Generic HACCP Plan for Regulatory Analysis

Executive orders governing regulatory analysis have been in place for two decades. Most recently, Executive Order 12866 formalized a set of analytical requirements for agencies to meet in support of significant federal rulemaking (EOP 1993). The degree of effort expected depends on the scope and scale of the action. Thus, these principles apply to all regulations but have particular import for "economically significant" rules.[2] These principles constitute a Generic HACCP Plan for Regulatory Analysis because they provide the framework for the analysis of each covered regulatory action ("product"), but rely on the expert knowledge of the agencies ("producers") to implement the framework in a sensible way given the particulars of the issue at hand ("plant-, process-, and product-specific concerns").

**The Generic HACCP Guide for Regulatory Analysis**

In January 1996 the Office of Management and Budget published more detailed performance standards for regulatory analysis (OMB 1996). This Generic HACCP Guide for Regulatory Analysis was prepared under the auspices of the President's Council of Economic Advisors. It reflected over two years of work by a team of regulatory economists from across the federal government, including economists from both the Food and Drug Administration and the Department of Agriculture. The contents of this document were substantially equivalent to a preceding document (OMB 1990) addressing an earlier vintage "Generic HACCP Plan" (EOP 1981). Thus, the fact that the current HACCP Guide was published in 1996 should not have posed any significant impediment to either agency's capacity to comply.

Both OMB guides provide a detailed framework for a structured, comprehensive examination of the benefits, costs, and other effects likely to arise due to the promulgation of an economically significant regulation. They also call for analysts to address several key issues critical to the development of a regulatory analysis and provide a policy-neutral interpretation of its implications. The guide states up front that regulatory analysis is intended to inform policy making rather than justify predetermined decisions.

> In particular, the [regulatory analysis] should provide information allowing decision makers to determine that:
> - There is adequate information indicating the need for and consequences of the proposed action.
> - The potential benefits to society justify the potential costs, recognizing that not all benefits and costs can be described in monetary or even in quantitative terms, unless a statute requires another regulatory approach.
> - The proposed action will maximize net benefits to society (including potential economic, environmental, public health and safety, and other advantages; distributional impacts; and equity),

unless a statute requires another regulatory approach.
- Where a statute requires a specific regulatory approach, the proposed action will be the most cost-effective, including reliance on performance objectives to the extent feasible.
- Agency decisions are based on the best reasonably obtainable scientific, technical, economic, and other information (OMB 1996).

Each of these fundamental principles is reflected in both general- and issue-specific directives and performance standards. In the next section the general principles are applied in the context of the two path-breaking efforts for the federal government to establish HACCP as the basis for controlling pathogenic food safety risks.[3]

## Applying HACCP to the Seafood and Meat and Poultry HACCP Regulatory Analyses

The Generic HACCP Plan for Regulatory Analysis sets forth broad issues which should be addressed. These include a rigorous analysis of the problem to be solved; the underlying basis for government action to solve it; a clear statement of the government's objective; a rich analysis of an array of reasonable and innovative alternative approaches for achieving this objective; and uniformly applicable principles for how benefits and costs should be estimated.

### Market or Institutional Failures as Bases for Intervention

The Generic HACCP Plan requires regulatory agencies to identify the fundamental basis for government intervention:

> Each agency shall identify the problem that it intends to address (including, where applicable, the failures of private markets or public institutions that warrant new agency action) as well as assess the significance of that problem (EOP 1993: Sec. 1(b)(1)).

Typically, some form of externality is expected to be the culprit that results in a divergence between private and social marginal costs or benefits. Indeed, market failure is a necessary (but not sufficient) condition for regulation to yield net social benefits.[4] Less frequently, intervention is premised on the existence of a natural monopoly or unusual market power. The Plan also recognizes the possibility that inadequate or (more likely) asymmetric information may be the underlying problem that government regulation is intended to overcome.

The form and type of market imperfection typically suggests the general outline of plausible solutions. For example, where an externality is present efficiency is restored when government intervention restores correct price signals. Monopoly and market power problems, in contrast, usually indicate the need for intervention to remove barriers to entry and competition. Finally, informational imperfections or asymmetries suggest the need for incentives to motivate the production and dissemination of information (where information in total is judged to be inadequate), or the alteration of rights and responsibilities with respect to disclosure (where asymmetries in information are believed to systematically disadvantage certain market actors).

*Seafood HACCP.* Based largely on work performed for the notice of proposed rulemaking (NPRM), the regulatory analysis claims that private markets have substantially failed. This market failure consists of three parts: (a) imperfections in common law which prevent injured consumers from recovering damages suffered from seafood-related foodborne illness, leading to excess risk; (b) suppressed consumer confidence in the safety of seafood resulting from inflated perceptions of the actual foodborne illness risks posed by seafood; and (c) an excess of consumer choice among products with differing degrees of safety as an identifiable product attribute.

(a) "Excess risk from imperfections in common law." Consumers who suffer seafood-related foodborne illness cannot recover damages in tort because the food responsible for transmitting it often cannot be positively identified:

> In most instances, consumers experiencing illness from food consumption are unable to link the illness to consumption of a particular food. This is because many symptoms do not occur immediately after consumption of the product. Delayed effects may vary from hours to months. To the extent that any illness is actually caused by man-made or natural contaminants in seafood, the lag may exceed ten years (FDA 1993: Sec. II).

Even where seafood can be identified as the vehicle of transmission, the specific firm responsible cannot be identified, making tort recovery impossible:

> The seafood industry differs from a large part of the food industry in that, except for certain branded fish products, almost all fresh and a large portion of frozen seafood is sold to the public unbranded or under brands that are not widely

advertised and not generally recognized. Often, when fish or shellfish is offered for sale by a supermarket or restaurant, that product has been sourced from several suppliers in order to obtain a large enough quantity to meet consumer demand. Each supplier, in turn, may source from several processors for much the same reason. Likewise, each processor may receive raw material from several harvesters and possibly import it from one or more countries. For these reasons, these products lose their source identity and are marketed generically (exceptions being canned, frozen, and branded seafood). This subsequently makes it difficult for a supermarket or restaurant to discern the source of the product involved in a consumer complaint. As a result, some firms may not be adequately motivated to provide sufficient levels of safety. Thus, it may be argued that, for the most part, the tort system does not adequately compensate consumers for illnesses derived from the consumption of seafood (FDA 1993: Sec. II).

(b) "Suppressed consumer confidence due to exaggerated risk perceptions." Consumers believe that seafood is much riskier than it actually is, thereby reducing demand for seafood below optimal levels:

Because of the negative publicity concerning water pollution and seafood safety, consumer perception of seafood safety may not be consistent with actual risk. Contamination scares cause drastic short-term drops in consumer demand for seafood products, and undoubtedly contribute to the chronic level of consumer concern about seafood safety. Thus, safety concerns about seafood are a likely factor preventing wider consumer acceptance of seafood as part of the U.S. diet.

The 1993 FDA Food Safety Survey confirms much of the previous research on consumers perception of seafood safety. Consumers in this study report that they are more careful when handling seafood than when handling meat and poultry. Given that consumption levels of fish are much lower than for meat and poultry, a disproportionately larger percentage of self-reported food illness episodes in the survey are attributed by the respondents to seafood. Although by weight, seafood consumption is only eight percent of the consumption of meat, poultry and seafood combined, consumers attributed 36 percent of their foodborne illnesses to seafood. The fact that consumers handle seafood more carefully and are more likely to attribute a food related illness to seafood than other flesh proteins suggests that consumers believe that seafood is less safe than meat and poultry (FDA 1993: Sec. VI.B, references omitted).

(c) "Excess consumer choice of alternative levels of safety." According to FDA, consumers may be better off without the freedom to choose different levels of quality. The regulatory analysis acknowledges that some processors made products safer than the minimum standard required under then-existing federal, state, and local laws and regulations, and that they charged consumers higher prices based on consumer demand for safety as a product attribute. But, the freedom to choose

> may place some consumers in a dilemma which they wish to avoid. In making their seafood purchases, some consumers may not want to be faced with the choice of a spectrum of differently priced products with different probabilities of illness. Instead, they may prefer that regulatory bodies set a minimum standard of safety that is high enough such that consumers no longer consider the risk relevant to their purchase decisions. Consumers may then take safety as given and base their purchases on other product characteristics such as price and taste (FDA 1993: Sec. II).

The first and second of these market failure arguments are inconsistent with each other, and the third is inconsistent with market failure. Even without any expectation of supporting empirical evidence, the entire market failure argument must be rejected on theoretical grounds alone.

First, imperfections in the administration of tort law do not imply the existence of a significant market failure, nor is it obvious that any material imperfections exist. Approximately 90 percent of all losses from seafood-related illness are attributed to the consumption of raw shellfish. Half of all losses consist of about 60 deaths per year from *Vibrio vulnificus* infection transmitted by raw oysters from the Gulf of Mexico. Because raw shellfish are understood to pose greater hazards than cooked seafood, it would be surprising to discover that the common law regularly assigned liability for these harms to producers. Only a handful of the remaining cases of seafood illness are severe enough to warrant bearing the cost of litigation even if the identity of the producer were known.

Risk is an inherent attribute of food, and one that, as the regulatory analysis repeatedly points out, consumers are quite capable of valuing. Its presence per se does not indicate a market failure. In any event, the regulatory analysis did not provide empirical evidence supporting the claim of *excess* risk. Rather, the analysis proceeds as if the mere existence of foodborne illness constitutes evidence of excess risk. This can only be true, of course, if the optimal level of risk is zero.

Second, suppressed consumer confidence implies that consumers *overstate* the risks posed by seafood. This contradicts the first market failure argument, which requires that consumers believe seafood to be safer than it really is. Both types of market failure cannot coexist in the same mind, however. Excess seafood-related foodborne illness implies that consumers *underestimate* the true risk, whereas insufficient consumer confidence implies that consumers *overestimate* the true risk.

FDA claimed billions of dollars per year in "consumer confidence benefits" in the preliminary regulatory analysis. These benefits, which consisted largely of reduced coronary heart disease, were presumed to result from improved nutrition attributable to the substitution of seafood for higher-fat flesh foods such as meat and poultry. FDA abandoned these quantified benefits in the final regulatory analysis in response to criticism of the asserted linkage between increased consumer confidence and improved diets. Greater consumer confidence in the *pathogenic* safety of seafood may cause consumers to substitute seafood for meat and poultry, but there is little evidence that they simultaneously choose lower-fat recipes.

Note that FDA did not abandon these "consumer confidence benefits" because they were inconsistent with the assumption that seafood is riskier than consumers expect. Rather, FDA merely left them unquantified because public commenters eviscerated the credibility of its quantitative model (FDA 1995b).[5] The final analysis continues to posit the simultaneous existence of mutually exclusive forms of market failure.

Third, the argument that consumers have too much choice with respect to the safety of seafood may be the most bizarre manifestation of market failure of all. An expansive array of choices reflects efficient markets in their most vibrant form. Any restriction on consumer choice will generally cause economic inefficiency and excess burdens. Efficiency may be enhanced through restrictions on free choice only where the exercise of choice imposes significant external costs on others. Absent any basis for or evidence of such external costs, the claim that excess consumer choice constitutes a market failure can only be regarded as specious.

***Meat and Poultry HACCP.*** Like seafood HACCP, this regulatory analysis also asserted that private markets had failed in fundamentally inconsistent ways. In fact, the regulatory analysis for the meat and poultry HACCP rule appears to have used the arguments in the seafood HACCP rule as a starting point for a much more expansive set of claims (FSIS 1996: 38949-38951):

    a) Consumers have imperfect information concerning the risks posed by pathogens in meat and poultry, resulting in a divergence between private and social marginal costs.

b) Producers rather than consumers are generally responsible for foodborne illnesses when it occurs.
c) Producers lack accountability for the foodborne illnesses they cause because they experience no reduction in either quantities demanded or profits.
d) Many firms in the food business do not use the best available pathogen reduction technologies because:
   i) entry is easy into the food business;
   ii) the industry is highly competitive; and
   iii) managers of these businesses are indifferent to the social benefits of such technology.

As the discussion below shows, these market failure arguments violate more "critical control points" for quality regulatory analysis than did FDA's analysis.

(a) "Imperfect information concerning pathogenic risks results in divergence between private and social marginal costs." FSIS' fundamental error is that all markets display imperfect information, so the absence of perfect information alone cannot justify government intervention. Otherwise, there would be no aspect of life in which regulation would not be superior per se to individual decision making founded on consumer sovereignty. Further, because regulatory agencies never possess perfect information, the same argument could be used against regulators to justify the equally valid principle that they should never act to supplant private markets. Indeed, if some standard of comparative informational richness were used as the determinative criterion, regulators could routinely come up short, for with rare exception they possess less useful information than the parties they propose to regulate.

For imperfect information about pathogenic risks to imply a significant market failure, one must show that consumers' actual willingness-to-pay (WTP) for meat and poultry products is substantially inconsistent with their knowledge or beliefs concerning such risks. If consumers expect microbial hazards to be present in raw flesh foods, WTP is diminished to account for this potential risk. Conversely, if they expect raw flesh foods to be free of pathogens, WTP is increased to account for greater safety. That is, the level of risk is simply an attribute of the commodity. The mere presence of risk is not evidence of a market failure at all. Consumers must misperceive the level of risk -- in either direction -- so egregiously that correct information dramatically alters their choices.

Even more stridently than in the seafood HACCP rule, this analysis proceeds to assume that opposite forms of market failure exist simultaneously. Consumers are said to *underestimate* the risk of foodborne illness posed by the consumption of meat and poultry products, thereby purchasing more meat and poultry products and paying higher prices for them than they would if they had proper risk information. This implies that the market-clearing price and quantity are both *above* optimal levels, and government intervention is needed to better inform consumers concerning these risks. At the same time, however, consumers are said to *overestimate* the risk of foodborne illness posed by the consumption of meat and poultry products, as evidenced by their lack of confidence in the safety of these foods. This implies that the market-clearing price and quantity are both *below* optimal levels, and government intervention is needed to restore consumer confidence in the safety of meat and poultry products.

Clearly both forms of market failure cannot coexist. The first form implies that the actual market demand curve is located upward and to the right of the efficient market demand curve. But the second form implies that the actual market demand curve is located downward and to the left. How the true demand curve can be located both above and below the observed demand curve is nowhere explained.

(b) "Producers rather than consumers are generally responsible for foodborne illnesses when it occurs." There is neither a theoretical nor an empirical basis for assuming that producers rather than consumers are always (or even mostly) "responsible" for foodborne illness. Most foodborne illness from meat and poultry results from temperature abuse, undercooking, or cross-contamination, all three of which occur under the watch of the food *preparer*. A case could be made that food *preparers* are generally responsible for foodborne illness, but producers and consumers would share responsibility in proportion to the extent of their roles as food preparers.

Under these circumstances, assigning liability to producers is efficient only if one of two conditions applies. The first condition is that producers are able to control the behavior of food preparers. The second condition has two parts: (i) producers are able to mitigate pathogenic risks more cheaply than can distributors, retailers, consumers and others in the roles as food preparers; and (ii) holding producers (strictly) liable does not create a significant moral hazard among these other actors. The first condition clearly does not hold, for producers cannot monitor -- much less control -- the actions of consumers and other food preparers. For the second condition to hold, the cost of preventative actions must be greater for consumers and other food preparers than for food producers, net of the costs of moral hazard. There is no evidence that this condition holds, either.

(c) "Producers lack accountability for the foodborne illnesses they cause." The notion that producers generally "cause" foodborne illness already has been debunked, so "causation" is assumed here to be limited to foods prepared by regulated producers and ready-to-eat at the point of sale. Even in this limited context, however, producers as a class do experience lost sales when foodborne illness occurs. All other things equal, consumers purchase smaller quantities and pay lower prices for ready-to-eat products that they believe pose health risks. All producers of such products lose sales, and no producer can recoup these losses unless it can convince consumers that its products are safer than average. This potential price premium creates an incentive for some firms to offer safer products, provided that they can successfully market this safety attribute. But competition in pathogenic risk reduction is inhibited by FSIS regulations prohibiting producers from making valid product claims involving risk. Thus, there may be an informational defect in the market, but it is largely one of the government's own making rather than the result of deficient market processes.

Given government restrictions against truthful promotion of reduced-risk products, one logical market response would be to cultivate brand names as implicit proxies for the prohibited claim. Interestingly, FSIS' analysis acknowledges that this may have occurred. The analysis then asserts, however, that brand names are in fact ineffective proxies for reduced risk because *all branded* products are not produced utilizing the best available pathogen reduction technology. This rebuttal is merely a *non sequitur*. Use of the best available pathogen reduction technology is not a necessary condition for a brand name to truthfully transmit a message of reduced risk. A name brand performs this function if it poses lower risks on average and consumers recognize it. It need not transmit a message of zero risk, nor must it transmit a message of the lowest technically feasible level of risk.

(d) "Many firms in the food business do not use the best available pathogen reduction technologies because (i) entry is easy, (ii) the industry is highly competitive, and (iii) managers of these businesses are indifferent to the social benefits of such technology." As noted above, there is no reason to expect all firms to use the best pathogen reduction technology in perfectly competitive markets. The fact that many do not offers neither evidence of market failure nor any efficiency basis for government intervention. Technology can be expected to vary simply because pathogenic risks vary by species, product, type and size of establishment, extent of market, and a host of other factors. Further, the extensive (though not universal) use of best available pathogen reduction technology absent a government mandate indicates that safety is a quality attribute upon which producers would like to compete.[6] Similarly, easy entry and competitive behavior in the food business both argue against a diagnosis of market failure, particularly in the form of market power. Competition creates incentives to improve safety as long as firms are permitted to market validated safety claims and consumers value such safety improvements more than the cost of providing them.[7]

The analysis offers no evidence supporting the rather startling assertion that food company managers are indifferent to food safety. If indifference actually ruled, then few, if any, producers would use best available pathogen reduction technology without both a government mandate and an oppressive enforcement regime. A fair reading is that the analysis is bereft of logic, and on this point it stoops to careless slander of even the best actors in the food business.

In sum, the regulatory analysis provides a most unusual examination of market failure. It is replete with internally conflicting representations of the direction by which efficient markets are said to be distorted, and other specious arguments. If the demonstration of a market failure matters as a critical control point for a regulatory analysis for a regulation purporting to yield billions of dollars in net social benefits, then this one has substantially violated the critical level. Deafening alarms should be clanging already.

**Analysis of Reasonable Regulatory and Non-Regulatory Alternatives**

The Generic HACCP Plan for Regulatory Analysis requires an assessment of the benefits, costs and other effects for a reasonable range of regulatory and non-regulatory alternatives:

> Each agency shall identify and assess available alternatives to direct regulation, including providing economic incentives to encourage the desired behavior, such as user fees or marketable permits, or providing information upon which choices can be made by the public (EOP 1993: Sec. 1(b)(3)).

The Generic HACCP Guide for Regulatory Analysis offers an extensive elaboration on this simple principle. It calls for regulatory analyses to examine an array of both regulatory and non-regulatory alternatives, including performance standards, differential requirements and/or effective dates, alternative levels of stringency and methods of ensuring compliance, informational measures, and market-based incentives (OMB 1996). As indicated earlier, these requirements are not new; they reiterate substantially identical principles that have been in place since at least 1990 (OMB 1990).

In addition, the Generic HACCP Plan also directs agencies to analyze the incremental effects of exceeding minimum statutory requirements. Where agencies have the discretion to exceed minimum statutory requirements,

> agencies should select those approaches that maximize net benefits (including potential economic, environmental, public health and safety, and other advantages; distributive impacts; and equity), unless a statute requires another regulatory approach (EOP 1993: Sec. 1(a)).

*Seafood HACCP*. The regulatory analysis considers only two options besides the one selected: (a) an approach focusing on high risk seafood products only, and (b) the establishment of differential standards for small businesses. The first option was rejected without analysis because it was imperfect, a standard to which the agency's preferred approach was not also subjected:

> The first option is inconsistent with the objective of this regulation, to control all physical, chemical or microbiological hazards reasonably likely to be found in seafood products (FDA 1995a: 65179).

The second option was indirectly rejected, for all seafood processors were given two years to come into compliance. The preamble to the final rule implies that the absence of any serious consideration of alternatives reflected a conscious FDA policy:

> FDA confirms its tentative view, reflected in the proposal, that HACCP should be the norm, rather than the exception, for controlling safety related hazards in the seafood industry. Existing standards for such contaminants as drug residues, pesticides, and industrial contaminants, are established to ensure that their presence in foods does not render the food unsafe. Processors of fish and fishery products are obliged to produce foods that meet these standards (FDA 1995a: 65118).

A number of reasonable alternatives could have been analyzed, such as a program targeted on raw shellfish where 90 percent of the risks reside. Alternatively, FDA could have considered a voluntary HACCP regime in which participation in the program allowed preferential labeling options. But the absence of any credible analysis of reasonable regulatory alternatives constitutes an obvious and severe violation of this critical control point in the production of quality regulatory analysis.

*Meat and Poultry HACCP*. Unlike the seafood HACCP analysis, the regulatory analysis for meat and poultry HACCP includes no substantive examination of alternatives at all. Several "strawman" alternatives are described and evaluated based on subjective criteria against which only the agency's preferred alternative could possibly succeed. These alternatives also appeared in the NPRM. At least one commenter on the NPRM called this a "sham" analysis, a charge that the agency duly reported but did not rebut (FSIS 1996: 38988).

FSIS identified four approaches in the regulatory analysis, but analyzed only one of them. These alternatives were: (a) market incentives, (b) education programs, (c) voluntary industry standards, and (d) uniform mandatory government standards. FSIS summarily rejected each of the first three alternatives. Only the option of mandatory government standards was left standing after FSIS dispatched the others.

Market incentives. The regulatory analysis rejected market incentives on the ground that existing markets were imperfect. Of course, how well existing markets function has nothing whatsoever to do with the merits of market incentives. A prominent example of a potential market-based incentive is the establishment of a special labeling program open only to those establishments that adopted HACCP (or some other pathogen-reducing production system). If HACCP is indeed a lower-risk food production process and consumers are willing to pay price premiums for reduced risk, then differential labeling offers a reasonable alternative to a one-size-fits-all federal mandate.

Education programs. The analysis rejected education programs because

> experience has shown that education alone has limited effectiveness in reducing foodborne illness. The effectiveness of education for food safety, and, indeed, for improving diets and other food related behavior, has not been demonstrated (FSIS 1996: 38950).

The implications of the rejection of education programs is itself revealing. First, it betrays a desire to reform consumers' behavior rather than merely overcome an alleged failure of private markets. In effect, the agency disapproves the preferences of the people it has been hired to serve. Second, consumer education remains a significant element of the agency's activities, one that the agency aggressively defends in budget debates. If education is in fact as ineffective as the agency now claims, then funding for these programs could be cut or eliminated without significant adverse effect.

Voluntary industry standards. The agency rejected voluntary industry standards on the ground that such standards would be more expensive than government standards and they would not be readily enforceable by the government. However, the analysis provided no empirical evidence of this greater expense, nor did it offer a logical argument to support the claim that FSIS enforcement is necessary. Limitations on the agency's capacity to enforce such standards is only a problem if non-regulatory enforcement mechanisms fail. Yet, the widespread use of voluntary industry standards elsewhere suggests that non-regulatory enforcement mechanisms exist and work reasonably well.

Uniform mandatory government standards. According to the analysis, a preference for uniform mandatory government standards is the inevitable result of having rejected each of the other three approaches. But the analysis did not subject the uniform mandatory government standards alternative to the same criteria that were used to reject the others. For example, if the same standard used to reject market-based incentives were applied to uniform mandatory government standards, the latter would have to be rejected as well. FSIS currently operates a comprehensive inspection system that, by the agency's own acknowledgment, has completely failed to address pathogenic risks. Unlike the unsupported claim that private markets have failed and thus cannot be relied upon to solve the problem of pathogenic risks, the agency's acknowledgment that its existing inspection regime has failed amounts to a signed confession that government regulation does not work.

Similarly, the criticism that education programs are less than fully effective applies with at least as much force to uniform mandatory government standards. Thus, holding the agency's preferred solution to the same standard used to reject the consumer education alternative would cause uniform mandatory government standards to be rejected as well.

Process control. Separate from the HACCP-related provisions of the regulation are additional requirements for microbiological testing of outputs, ostensibly to verify process control. These requirements are fundamentally inconsistent with the theory behind HACCP, which is that performance standards verified through end-product testing are either infeasible or undesirable. If performance standards (such as *Salmonella* tests) capture the essence of pathogenic risk reduction, then the entire HACCP program is largely superfluous. Establishments would choose to adopt HACCP if it offered the least-cost method of compliance with these performance standards. Similarly, if other performance standards (such as the generic *E. coli* criteria) properly measure the effectiveness of an establishment's sanitation program, then the requirement for sanitary standard operating procedures (SSOPs) is redundant as well. However, if performance standards for either sanitation or pathogen loads in final products cannot be devised or effectively implemented, thus arguing for HACCP instead, then microbiological testing will be incorporated into HACCP plans only when it makes sense as a method for verifying compliance with such plans. By imposing these testing requirements, FSIS conveys a preference for redundancy or an implicit distrust of the HACCP model.

**Benefit and Cost Assessment**

For each alternative, the Generic HACCP Plan for Regulatory Analysis requires a full assessment of its likely benefits and costs:

> Each agency shall assess the benefits and costs of the intended regulation (EOP 1993: Sec. 1(b)(6)).

Again, much more detail can be found in the companion Generic HACCP Guide for Regulatory Analysis (OMB 1996: Sec. III(A)). In fact, the Guide can be thought of as providing a long list of "generic critical control points for regulatory analysis," including:
a) Counting benefits and costs correctly, including non-monetized benefits and costs.
b) Identifying and using an appropriate baseline to estimate incremental benefits and costs.
c) Evaluating each alternative fairly according to identical, appropriate criteria.
d) Discounting future benefits and costs.
e) Analyzing and presenting uncertainty and variability in estimates of benefits, costs, and other effects.
f) Assessing and valuing variable and uncertain risks.
g) Revealing all assumptions.
h) Accounting for international trade effects.
i) Explicitly and separately describing or quantifying equity considerations.

Clearly, each of these critical points is not equally critical across all regulatory analyses. The same can be said for critical points in food production, of course. Just as food safety regulators desire evidence that producers have carefully thought about each possible control point to determine, based on both theory and empirical data, which of them should be critical and which should not, an analogous process ought to be followed by regulatory analysts.

*Seafood HACCP*. FDA's analysis of benefits and costs violates a wide array of critical control points for regulatory analysis. Many of these violations are so severe that the document should be considered wholly unreliable as a summary of the rule's likely effects. The following discussion highlights but a handful of these problems.

Benefit assessment. FDA builds its estimate of the expected safety benefits in three parts: (a) estimates of the baseline incidence of foodborne illness from seafood; (b) estimates of the costs of morbidity and mortality associated with various pathogenic infections; and (c) estimates of the proportion of cases of foodborne illness that would be averted because of the regulation. The analysis reports safety benefits to the nearest dollar (FDA 1995a: 65187, Table 10) despite uncertainties in the estimation methodology that, at best, suggest confidence to the nearest $10 million.

The analysis also claims substantial benefits other than enhanced safety, including $20 million per year in cost savings to U.S. exporters needing HACCP certification, another $20 million per year in reduced enforcement costs, plus the unquantified benefits of reduced rent-seeking and increased consumer confidence (FDA 1995a: 65187-8, 1995b).[8] As indicated earlier, benefits from increased consumer confidence (resulting from consumers' risk perceptions exceeding actual risk) are incompatible with benefits from increased safety (resulting from actual risk exceeding consumers' risk perceptions). Rent-seeking might decline under the rule, but only if prospective rent-seekers believed this rule was the "last word" in federal regulation of seafood. The estimate of cost savings to U.S. exporters appears to be based on the assumption that the Department of Commerce's existing certification program was inadequate (despite the fact that it also addressed non-safety concerns excluded from FDA's seafood HACCP program) combined with a "strawman" program of entry-by-entry inspection in EU ports (FDA 1995a: 65188).[9] Finally, enforcement costs seem likely to increase rather than decline, for without aggressive enforcement, the seafood HACCP rule stands to experience widespread noncompliance.

The use of an expert panel of four agency scientists to estimate (c) -- the proportion of cases of foodborne illness likely to be averted due to the seafood HACCP regulation -- is an innovative analytical approach to dealing with what is clearly significant scientific uncertainty.[10] Unfortunately, the procedures used by this panel were not documented. Without such documentation, the validity and reliability of the resulting predictions cannot be evaluated (although they can be tested *ex post*).

A necessary condition for expert judgment to be valid and reliable as a critical input into regulatory analysis is that the experts must make fully transparent their procedures, assumptions, models, and data. Especially important is a clear delineation of the precise mechanisms by which specific regulatory provisions result in the effects predicted. Transparency enhances accountability because experts have professional reputations that they value and seek to protect. Conversely, accountability is lost when experts make compromise or "consensus" predictions, for errors can be disavowed as the product of others' input.

Special problems arise where experts are subject to conflicts of interest, and conflicts were clearly present here because the experts were employed by the regulatory agency. Where internal experts are used, special procedures must be devised to insulate them from a host of political, bureaucratic, and professional pressures leading them to shade their judgments in ways beneficial to the agency. But the act of insulating them from these pressures sacrifices transparency, for the ability to persuasively deny responsibility for the resulting prediction is an essential feature of insulation.

Having followed the internal-expert judgment approach, the analysis claims that the regulation will avert between 18 and 52 percent of approximately 114,000 seafood-related cases of foodborne illness per year, and between 18 and 47 percent of the $245 million in associated costs. Approximately 90 percent of all cases of seafood-related foodborne illness are attributable to the consumption of raw shellfish, and a similar fraction of safety benefits derive from reduced risks from raw shellfish consumption. More than half of all safety benefits are attributed to the prevention of between 20 and 50 percent of the 60 annual deaths from *V. vulnificus* infection, deaths which occur only in persons with severe liver disease who consume infected raw molluscan shellfish harvested from the Gulf of Mexico (FDA 1995a: 65185-6). Perhaps unwittingly, the analysis hints that these benefits may in fact be illusory because no proven technology existed to control this pathogen:

> *Vibrio vulnificus* is a naturally occurring, ubiquitous, marine organism. The lower and upper bound numbers reflect the fact that controls are newly emerging for this organism and still have uncertainties associated with them.

The analysis implies that its estimates are merely suggestive: "FDA has made *a preliminary attempt* in this analysis to *explore costs and benefits of future actions* which may occur to control this hazard" (FDA 1995b, emphasis added). Thus, half of the claimed benefits of the seafood HACCP rule are speculative. FDA's benefits assessment thus violates numerous critical control points for regulatory analysis, for there is simply no documented linkage between the provisions of the seafood HACCP rule and the benefits claimed.

Cost assessment. Given the magnitude of safety benefits attributed to this regulation, one might expect dramatic changes in seafood processing technology. Such changes might indeed be necessary to comply with the seafood HACCP regulation, but the costs of implementing them are largely ignored in the regulatory analysis. In effect, the regulatory analysis uses fundamentally inconsistent baselines for estimating benefits and costs. Benefits are estimated from a baseline that roughly corresponds to current practices and incidence of seafood-related foodborne disease. However, costs are estimated from a very different baseline in which the vast majority of seafood processors need only write HACCP plans, make minor expenditures to actually comply with these plans, and keep better records.[11]

The analysis used two alternative methodologies for estimating costs, one based on a survey conducted for the National Marine Fisheries Service (NMFS) and another model derived from unspecified FDA inputs and "agency expertise." This latter approach yielded estimates one-fourth as large and reflected the likely costs of operating a HACCP plan for a pair of small seafood processors. FDA assumed that one of these "model plants" substantially complied with current good manufacturing practices (CGMPs) and assumed that the other had "some CGMP deficiencies ... typical of those displayed by seafood processors." Cost estimates for the latter firm included some costs for improved CGMP compliance but no costs for any operational changes necessary to comply with a HACCP plan. Estimates for the first firm included none of these costs.

The analysis states that the most important difference between the two cost models is that in the NMFS-based model, 80 percent of plants must make operational changes to comply with the CGMPs, whereas only 20 percent must do so in the FDA-based model. This four-fold difference highlights the importance of identifying the most appropriate baseline for analysis, for the larger the baseline compliance rate, the lower the expected costs of the regulation.[12] Indeed, the analysis asserts that any costs attributable to compliance with the sanitation requirements in the seafood HACCP rule are properly attributed to existing regulatory requirements:

> Because FDA holds that [the existence of sanitation deficiencies] must be corrected under existing requirements, the costs associated with these corrections will be borne by processors regardless of whether sanitation provisions are included in the seafood HACCP regulations or somewhere else (FDA 1995a: 65183).

If this were true, of course, then the sanitation provisions in the seafood HACCP rule would be superfluous, imposing no costs and yielding no benefits.

Another noticeable aspect of these models is that they assume a negligible number of critical control points. The FDA-based model, for example, assumes a single critical control point for a production line involving frozen tuna steaks and zero critical control points for one processing imported frozen orange roughy filets.[13] Thus, this cost model assumes away the very problem that the seafood HACCP rule must address in order to generate public health benefits -- the risks posed by raw shellfish. According to FDA's most recent HACCP Guide, producers engaged in marketing raw shellfish must deal with a long list of hazards and critical control points (FDA 1998a). Thus, any cost estimate based on low-risk seafood products rather than raw shellfish is simply disingenuous.

*Meat and Poultry HACCP.* Like the seafood HACCP analysis, FSIS' analysis of benefits and costs violates a wide array of critical control points for regulatory analysis.

Benefit assessment. FSIS' analysis includes several material errors that severely overstate the likely benefits of the rule. The most obvious of these errors is that the analysis simply assumed that the rule would reduce both the incidence of foodborne illness and losses which result from these illnesses by as much as 90 percent. The agency provided no risk assessment or other scientific basis for this assumption and quite readily acknowledged the absence of any credible basis for it:

> The link between regulatory effectiveness and health benefits is the assumption that a reduction in pathogens leads to a proportional reduction in foodborne illness. [The Food Safety and Inspection Service] has presented the proportional reduction calculation as a mathematical expression that facilitates the calculation of a quantified benefit estimate for the purposes of this final [regulatory impact analysis]. FSIS has not viewed proportional reduction as a risk model that would have important underlying assumptions that merit discussion or explanation, *For a mathematical expression to be a risk model, it must have some basis or credence in the scientific community. That is not the case here.* FSIS has acknowledged that very little is known about the relationship between pathogen levels at the manufacturing stage and dose, i.e., the level of pathogens consumed (FSIS 1996: 38945-38946, emphasis added).

Obviously, this is inappropriate for any regulatory analysis. It makes a laughingstock of the entire analytic enterprise. If it is acceptable to simply assume the existence of billions of dollars worth of benefits, no one should expect analysis to be useful for decision makers. A defensible scientific basis for estimates of risk reduction legitimately attributable to the rule constitutes an obvious critical control point for any credible regulatory analysis.

Cost assessment. FSIS' analysis contains several material errors in its cost assessment that severely understate the likely costs of the rule. First, the estimated cost of required SSOPs, HACCP plans, and generic *E. coli* testing includes only the cost of writing the plans themselves, training current employees, and performing the microbiological tests. The costs associated with the operational changes necessary to comply with SSOPs and HACCP plans are not included. However, if no operational changes are in fact required, then the requirement to develop these plans becomes superfluous. All benefits attributable to both SSOPs and HACCP must be related to specific operational changes that reduce the variance (and possibly the mean) in the level of pathogens in meat and poultry products. If no such changes are actually necessary, then the maximum value of benefits legitimately attributable to both SSOPs and HACCP plans is zero.

Second, the analysis excludes the cost of training new employees in SSOPs, HACCP, and other regulatory provisions. Many parts of the food industry experience high turnover, which will require recurring expenditures on new employee training. Third, productivity losses associated with diverting employees from production to training were not estimated. Again, both SSOPs and HACCP plans require extensive training if they actually change the way food is made. But every hour spent in training represents an hour not engaged in the production of food, which is regrettably a real cost. Fourth, the estimated cost of preparing HACCP plans in the analysis is unreliable on its face. This estimate was based on a sample of nine establishments that volunteered to participate in an agency study and are not representative of the several thousand establishments regulated under this rule. Statistical inferences from even a random sample of nine are problematic, but they are obviously illegitimate from a convenience sample of nine volunteers.[14]

## Identifying the Hazards to Quality Regulatory Analysis

So far this discussion of the regulatory analyses of the seafood HACCP and meat and poultry HACCP rules has focused on what could be called critical control points for quality regulatory analysis and the myriad ways in which FDA and FSIS violated these critical control points. But it would be incomplete if it did not address the underlying hazards to quality analysis for which critical control points for regulatory analysis stand as sentinels. Analysts, perhaps especially government analysts, experience numerous pressures to shade their work in service of what their bosses present as higher purposes. These blandishments fool only a few competent analysts, but even they lack the political and institutional support to resist. Quality analysis is the neglected stepsister of government rulemaking; agencies tolerate it only so long as it does not interfere with their agendas and welcome it only when it can be used to advance these agendas. Frequently they manipulate analysis instead.

**Analytical Capacity**

Agencies vary in their capacity to perform quality analysis because not all analysts are equally competent. This is particularly true in government agencies where getting rules issued is more highly valued than getting them right. Some agencies simply lack personnel with the ability to perform quality regulatory analysis, and the staff they do have may not be capable of adequately supervising contractors. These agencies also have serious problems retaining good analysts, who will tend to migrate to organizations where competence is rewarded.

Thus, an obvious hazard to quality regulatory analysis occurs when those entrusted with the job are not capable of performing it well. Where competence is a necessary (but not sufficient) condition for the production of a credible work product, a lack of competence is undoubtedly sufficient to ensure failure. Incompetence is as likely to succeed in producing high quality regulatory analysis as it is in producing wholesome and unadulterated food.

**Non-Economic Inputs**

Both HACCP analyses illustrate the limitations that arise when scientific information is either limited or completely absent. Both analyses required competently performed risk assessments as inputs but did not have any available. Frequently, risk assessments are available but have been produced under conditions that make them inappropriate for economic analysis. Common examples include chemical risk assessments, which focus only on upper-bound expressions of hazard or exposure (or both). Where scientific information is not available, particularly information defining a credible link between specific regulatory provisions and risk reductions, quality regulatory analysis may be simply impossible. This does not argue for forsaking regulatory analysis, however, for without it agencies can be assured that the worthy objectives they seek to accomplish will materialize only by chance. Like inputs to a food production process, bad inputs lead to bad outputs -- there is no "reworking process" that will convert contaminated inputs into a wholesome product.

The approach taken in the seafood HACCP analysis to utilize expert judgment to estimate the likely effectiveness of the regulation represents a welcome innovation. There were problems with the implementation of this approach, however, insofar as the process relied on internal FDA personnel, lacked documentation, and could not be independently audited. Improvements in the design of expert panel processes must be made so that they are transparent and imbued with incentives deterring strategic behavior.[15]

**General Counsel Office Objectives**

The role of agency lawyers cannot be readily discerned from either analysis, but experience teaches that the effects of legal inputs can be easily underestimated. In general, agency lawyers take policy objectives as given, strive to deter aggrieved parties from litigating, and maximize agency enforcement discretion. Because the nature of legal risks varies across regulations and cannot be readily generalized, specific hazards to quality regulatory analysis associated with agency counsels will depend on the case at hand.

Obviously, agency lawyers pose a hazard to quality regulatory analysis whenever it could conflict with the lawyers' mission. Regulations are easier to defend when they have more benefits than costs, so analysts face pressure to manufacture such results in the service of litigation strategy. Regulations also may be difficult to defend where there are well-documented uncertainties or broad areas of scientific ignorance. Where a convincing case can be made that an agency lacks a solid scientific, analytical, or empirical basis for its actions, an aggrieved party will be better able to prove to a court's satisfaction that the agency's actions are arbitrary and capricious. In some cases, competent regulatory analysis may pose a clear and present danger to the achievement of agency objectives. In such cases, agency lawyers will strongly prefer that no analysis be performed at all.

**Political Level Objectives**

The political managers of FDA and FSIS were determined to press forward with these rulemakings despite lacking competent risk assessments or any analytic bases for expecting net social benefits to accrue from these regulations. One can only speculate about their motives. One possibility is that they found themselves "ahead of the curve," perhaps having publicly committed to take these actions long before the analyses had been conceived, much less completed, and could not bring themselves to abandon such ill-considered promises. Whatever the explanation in the case of these two rules, it seems clear that where political-level objectives conflict with those set forth in the Generic HACCP Plan for Regulatory Analysis, this conflict poses a lethal hazard to the analysis and not to political-level objectives. Notice and comment and a variety of review procedures have been devised over the years, but none has provided an effective antidote.

# Conclusion

HACCP remains a promising model for improving the safety of the nation's food supply. It is based on the belief that process control represents a superior approach than performance standards because defects are rare, unobservable through reasonable sampling protocols, and impossible to detect in a timely manner. The HACCP model also can be applied to regulatory analyses, such as those prepared in support of major HACCP regulations. When HACCP principles are applied, however, it becomes readily apparent that these regulatory analyses would not pass muster. They would be rejected as adulterated products unfit for human consumption, and it is an open question whether the agencies that produce these products would be allowed to remain in business.

The analogy to HACCP is legitimate because both rules were advertised as solutions to significant real-world problems. If government risk estimates are true, thousands of people lose their lives each year due to foodborne illness and countless others suffer preventable illness. Yet, based on the regulatory analyses prepared in support of these actions, one should expect that neither regulation will achieve but a fraction of the promised public health benefits and at substantially greater than promised costs. There is no moral argument in support of wasting scarce resources, nor can there be an ethical justification for promising effective regulatory solutions that cannot be delivered or resorting to misleading or incompetent analyses to support them.

Regulators have an ethical obligation to follow the rules. Because of their awesome power, this obligation exceeds that of any citizen or firm they regulate. For almost 20 years, the rules have required regulatory agencies to perform credible analyses of the likely consequences of the exercise of regulatory power. When regulators flagrantly break these rules, as FDA and FSIS did in the two HACCP rules described above, they undermine their own moral legitimacy and render suspect everything else they do.

**Notes**

[1] Visiting Professor of Public Policy and Regulatory Program Manager, Center for the Study of American Business, Washington University in St. Louis. The author reviewed both of the regulations discussed in this chapter in his previous position as staff economist in the Office of Information and Regulatory Affairs, Office of Management and Budget.

[2] See OMB 1996. An *economically significant rule* is defined as one that has an annual effect on the economy of $100 million or more or adversely affects in a material way the economy, a sector of the economy, productivity, competition, jobs, the environment, public health or safety, or state, local, or tribal governments or communities. See EOP 1993: Sec. 6(a)(3)(C). Both FSIS' meat and poultry HACCP rule and FDA's seafood HACCP rule were determined to be economically significant.

[3] A comprehensive analysis based on the issue-specific principles set forth in the guide would add considerable richness to the discussion, but it would only reinforce the conclusions reached here based on an examination of only the general principles.

[4] Conversely, the presence of net social benefits implies the existence of a significant market failure. The greater the estimate of net social benefits, the larger must be the market failure. Market failure is not a sufficient condition for government intervention, however, because poorly crafted government action may not remedy the problem, and in some cases it may exacerbate it instead.

[5] This health benefit claim skirted extremely close to (and possibly beyond) one that would be prohibited if made by a seafood producer. 21 CFR 101.71(e) prohibits any health claim with respect to omega-3 fatty acids and coronary heart disease. 21 CFR 101.75 permits similar health claims with respect to dietary saturated fat and cholesterol, but only after successfully leaping a long succession of procedural and substantive hurdles. The preliminary regulatory analysis only alluded to this evidentiary burden and did not attempt to actually meet it.

[6] Some firms using best available pathogen reduction technology may prefer not to compete based on reduced risks, but instead use advanced technology as a weapon against competitors in a regulatory environment. Such firms are better off

competitively if regulations mandating such technology increase competitors' costs by a larger amount.

⁷Ironically, a much stronger case for market failure would be possible if in fact all firms used identical food production technology. Uniformity could be explained as a natural monopoly phenomenon or market power.

⁸The regulatory analysis supporting the notice of proposed rulemaking claimed benefits of $3 billion to $14 billion per year in improved health resulting from increased consumer confidence in seafood safety, leading to greater seafood consumption. As indicated earlier, these benefits were deleted from the final regulatory analysis in the face of critical public comment. The final regulatory analysis continues to assert that increased consumer confidence will result from the seafood HACCP rule, but does not quantify these benefits.

⁹The alternative of voluntary HACCP for firms exporting seafood to the EU was not examined.

¹⁰The experts were Dr. George P. Hoskin, Dr. Karl C. Klontz, Dr. Kaye 1. Wachsmuth and Dr. Thomas C. Wilcox. See FDA 1995a: 65185.

¹¹In the preliminary regulatory analysis, FDA estimated the cost of corrective actions taken in response to violations of critical control limits at $1,000 per year per plant. This estimate was severely criticized as low by many public commenters. Only one commenter is noted as believing that this estimate was reasonable, a fact which was used to justify raising the estimate to just $2,000 per plant per year in the final analysis (FDA 1995b). In a similar vein, the cost of HACCP plan verification was assumed to be $1,000 per year per plant (FDA 1995b).

¹²Benefits should decline as well, of course. However, as the previous subsection shows, no similar adjustment was made in the estimate of benefits because of uncertainty about the appropriate baseline.

¹³The number of critical control points assumed in the NMFS-based model is not reported in the regulatory analysis, but appears to be similarly low based on the magnitude of the cost estimates.

¹⁴A sample of nine is the largest sample exempt from public notice, comment, and oversight by the Office of Management and Budget under the Paperwork Reduction Act. A fair conclusion is that this sample was chosen for the express purpose of avoiding OMB and its attendant public oversight.

¹⁵In contrast, FDA's "alternative" cost analysis was developed using precisely the wrong form of expert judgment. The critical assumptions, data, and inferences which went into the model were not disclosed, and the identities of the seafood experts on whose judgment the model rested were kept secret.

## References

Executive Office of the President (EOP) 1981. Executive Order No.12291 ("Federal Regulation"). *Federal Register* 46:13193 (February 17, 1981).

Executive Office of the President (EOP) 1993. Executive Order No.12866 ("Regulatory Planning and Review"). *Federal Register* 58:51735-51744 (October 4, 1993).

Food and Drug Administration (FDA) 1993a. Proposal to Establish Procedures for the Safe Processing and Importing of Fish and Fishery Products; Proposed Rule. *Federal Register* 58:4142.

National Academy of Sciences (NAS) 1985a. Meat and Poultry Inspection: The Scientific Basis of the Nation's Program. Washington, D.C.: National Academy Press.

National Academy of Sciences (NAS) 1985b. An Evaluation of the Role of Microbiological Criteria for Foods and Food Ingredients. Washington, D.C.: National Academy Press.

National Academy of Sciences (NAS) 1991. Seafood Safety: Committee on Evaluation of the Safety of Fishery Products. Washington, D.C.: National Academy Press.

Office of Management and Budget (OMB) 1990. "RIA Guidance." In: Regulatory Program of the United States Government, April 1, 1990 -- March 31, 1991. Washington, D.C.: Office of Management and Budget.

Office of Management and Budget (OMB) 1996. *Economic Analysis of Federal Regulations Under Executive Order 12886.* Washington, D.C.: Office of Management and Budget.

U.S. Food Safety and Inspection Service (FSIS) 1996. Pathogen Reduction; Hazard Analysis and Critical Control Point (HACCP) Systems; Final Rule. *Federal Register* 61:38805-38989 (July 25, 1997).

U.S. Food Safety and Inspection Service (FSIS) 1997a. Generic HACCP Models and Guidance Materials Available for Review and Comment. *Federal Register* 52:32053-32054 (June 12, 1997).

U.S. Food Safety and Inspection Service (FSIS) 1997b. HACCP-Based Meat and Poultry Inspection Concepts. *Federal Register* 62:31553-31562 (June 10, 1997).

U.S. Food and Drug Administration (FDA) 1993. Preliminary Regulatory Impact Analysis of the Proposed Regulations to Establish Procedures for the Safe Processing and Importing of Fish and Fishery Products. Washington, D.C.: Food and Drug Administration (Docket Nos. 90N-0199 and 93N-0195, http://vm.cfsan.fda.gov/~djz/lcfudpria.txt).

U.S. Food and Drug Administration (FDA) 1994. Fish and Fishery Products Hazards and Controls Guide; Availability. *Federal Register* 59:12949.

U.S. Food and Drug Administration (FDA) 1995a. Procedures for the Safe and Sanitary Processing and Importing of Fish and Fishery Products. *Federal Register* 60:65096-65202.

U.S. Food and Drug Administration (FDA) 1995b. Final Regulatory Impact Analysis: Procedures for the Safe and Sanitary Processing and Importing of Fish and Fishery Products. Washington D.C.: Food and Drug Administration (Docket No. 93N-0195, http://vm.cfsan.fda.gov/~lrd/haccpria.txt).

U.S. Food and Drug Administration (FDA) 1997a. Fish and Fishery Products Hazards and Controls Guide; Availability. *Federal Register* 62:7465-7467.

U.S. Food and Drug Administration (FDA) 1997b. Retail Food Program Standards; Notice of Grassroots Meetings. *Federal Register* 62:31611-31612 (June 10, 1997).

U.S. Food and Drug Administration (FDA) 1997c. Fruit and Vegetable Juice Beverages: Notice of Intent to Develop a HACCP Program, Interim Warning Statement, and Educational Program. *Federal Register* 62:45593-45596 (August 28, 1997).

U.S. Food and Drug Administration (FDA) 1998a. Fish and Fishery Products Hazards and Controls Guide. http://vm.cfsan.fda.gov/~dms/HACCP-2.html.

U.S. Food and Drug Administration (FDA) 1998b. Hazard Analysis and Critical Control Point (HACCP); Procedures for the Safe and Sanitary Processing and Importing of Juice, Food Labeling; Warning Notice Statements; Labeling of Juice Products; Proposed Rules. *Federal Register* 63:20449-20486 (April 24, 1998).

U.S. Food and Drug Administration (FDA) 1998c. Food Labeling: Warning and Notice Statements; Labeling of Juice Products. *Federal Register* 63:20486-20493 (April 24, 1998).

U.S. Food and Drug Administration (FDA) 1998d. Preliminary Regulatory Impact Analysis and Initial Regulatory Flexibility Analysis of the Proposed Rules to Ensure the Safety of Juice and Juice Products; Proposed Rule. *Federal Register* 63:24253-24302 (May 1, 1998).

Chapter 8

# Benefit-Cost Analysis of Reducing Salmonella Enteritidis: Regulating Shell Eggs Refrigeration

Hyder Lakhani[1]

## Introduction

The existing literature on estimates of incremental benefits of health costs of foodborne diseases such as Salmonellosis report damages associated with all food products combined instead of a *specific* disease associated with a *specific* food item. For example, the pioneering research of Buzby and Roberts (1995, 1996, 1997) report the economic benefits of avoidance of disabilities and deaths associated with Salmonellosis caused by consumption of food. This paper analyzes a *specific* food item (eggs) and a *specific* foodborne disease Salmonellosis associated with *Salmonella* Enteritidis (SE). Results of such an analysis would enable policy makers to regulate safety of shell eggs so as to mitigate the number of disabilities and deaths associated with their consumption. As noted recently by Arrow et al. (1996), it is necessary to relate the incremental benefits of a proposed regulation to its incremental costs of compliance to determine if the regulation is economically efficient.

An opportunity to develop such a study was provided by the U.S. Congress. In 1991, as part of the Food, Agriculture, Conservation and Trade Act Amendments of 1991 (Public Law 102-237), Congress amended the Egg Products Inspection Act (EPIA) of 1970. The amendments required that egg handlers store and transport shell eggs packed for consumer use under refrigeration at an ambient temperature of no greater than 45°F (7.2°C). The Food Safety and Inspection Service (FSIS) is preparing a final rule that would require that shell eggs packed for consumer use be stored and transported at 45°F (7.2°C). This paper is based on the final rule estimates of economic benefits and costs.

The next section reports selected microbiological data. The estimates of benefits associated with refrigeration of shell eggs are reported in section 3. The costs of compliance with the refrigeration requirements are reported in section 4, along with incremental benefit-cost ratios for four alternative scenarios of benefits. The last section outlines conclusions.

## Microbiological Data and Method

While Salmonellosis is associated with both food and nonfood sources, Crutchfield et al. (1997) report that between 87 and 96 percent of all cases of Salmonellosis are foodborne (non-typhoid). They estimated the number of cases of illnesses associated with *Salmonella* from 696,000 to 3,840,000 and the number of deaths from 870 to 1,920 per year. The economic value of health costs associated with Salmonellosis ranged from $900 million to $12.3 billion.

Some microbiologists suggest that an egg's contents can become contaminated with SE before the egg is laid. Although the mechanism is still not well understood, SE will infect the ovaries and oviducts of some egg laying hens, permitting "transovarian" contamination of the interior of the egg while the egg is still inside the hen (Humphrey and Whitehead 1993). After an infected egg is laid, SE contamination tends to grow inside the egg (Humphrey and Whitehead 1993). Humphrey (1990) suggests that refrigerating during storage can prevent such a growth. Other measures of preventing the growth include refrigeration during transportation and retail sales, reducing shelf life of egg cartons at retail, thorough cooking, and pasteurization, and processing shell eggs into frozen, liquid, or dry egg products (Hammack et al. 1993; Whiting 1998).

The Centers for Disease Control and Prevention (CDC) reported that recent outbreaks of SE infection were associated with consumption of raw shell eggs (Centers for Disease Control and Prevention 1996). Hennessy (1996) reported a national outbreak of SE infections from ice cream containing raw eggs. The CDC data on National Salmonella Surveillance System reveal an increasing trend in SE serotype over the last twenty years. For example, the proportion of S serotype that was SE increased from 5 percent in 1976 to 24.5 percent in 1996. The number of reported outbreaks, associated cases, hospitalizations, and deaths associated with SE also increased during this period. Bean et al. (1997) report that during the 1988-92 period, an increasing number of Salmonella outbreaks were caused by SE (e.g., 47 percent in 1988 versus 75 percent in 1992). They noted that SE was also the most frequently reported cause of foodborne outbreaks during this period, accounting for 14 percent of all outbreaks and 33 percent of outbreaks for which an etiology (i.e., food vehicle) was determined. They add that SE also resulted in more deaths (27) than any other pathogen: 23 (85 percent) of these deaths occurred among residents of nursing homes, which reflects the seriousness of SE infections in elderly persons, many of whom may be immunocomprised.

As noted above, the estimates of Buzby and Roberts (1997) pertain to damages associated with S from all food sources. The CDC (1994) data reveal that from 1985 to 1993, "of the 233 outbreaks for which epidemiological evidence was sufficient to implicate a food vehicle, 193

(83 percent) were associated with eggs" (CDC, *Mortality and Morbidity Weekly Report*, Vol. 43, No. 36, September 16, 1994). The data from CDC do not directly report the number or the percentage of outbreaks of SE associated with eggs. Therefore, the percentage of human outbreaks of SE was estimated from equation 1. Data for the three components of the right-hand side of the equation were available and were multiplied to obtain the left-hand side of the equation as follows:

(1) % Human Outbreaks of SE = [% H & NH outbreaks of S for which a food vehicle is identified]
X [% S isolates for Humans]
X [% SE outbreaks of H&NH that are due to eggs]

where H = human; NH = non-human

The left-hand side of this equation was determined by multiplying three values available for the right hand side of the equation (e.g., 45% x 83% x 25% = 9.33%). Mishu et al. (1994) reported that a food vehicle was identified in 45 percent of outbreaks of SE infection. The CDC data referred to above revealed that in 83 percent of those outbreaks the implicated food contained lightly cooked or raw eggs. Also, 25 percent of all *S* isolates are from humans. In short, almost 10 percent of all outbreaks of H&NH *S* constitute the SE strain traced to eggs as a vehicle. This estimate is comparable with an assumption of egg-related Salmonellosis cases at 10 percent of total Salmonellosis cases (U.S. Department of Agriculture/Economic Research Service 1993). Therefore, this estimate is employed in the following tables to evaluate economic benefits of avoidance of shell-egg related SE by requiring refrigeration of shell eggs at 45°F during storage and transportation.

## Benefits of Mitigating SE by Refrigerating Shell Eggs

Table 1 shows the number of deaths and disabilities associated with *Salmonella* from consumption of all food items. The number of deaths ranges from a low of 870 per year to a high of 1,920 per year. The number of annual disabilities ranges from a low of 696,000 to 3,840,000 (Buzby and Roberts 1996). These estimates are comparable to the number of cases ranging from 1.92 million to 2.96 million and the number of deaths varying from 31.9 to 1,920 for Salmonellosis, nontyphoid, in a report from the Council for Agricultural Science and Technology (1994).

Based on the avoidance of medical costs, Buzby and Roberts (1997) estimated the economic values of prevention of these cases. Their estimates of health costs of human Salmonellosis-linked diseases and deaths were at $900 million and $4.9 billion, respectively (1996 dollars);

their high estimates were $3.5 billion and $12.2 billion (1996 dollars). The wide variation in this range of estimates is attributed both to the wide range in estimates of the number of cases and the economic methods used for the analysis.

Since the preceding ranges of estimates for Salmonellosis-related deaths and diseases are based on Salmonellosis from all food sources, it is necessary to adjust the estimates downwards to obtain only the egg-related disabilities and deaths associated with SE. The estimate from equation 1 above as well as Lin and Roberts (1993) revealed that egg-related SE cases represented 10 percent of total foodborne human Salmonellosis cases. SE was isolated from samples collected in the majority of SE cases involving eggs that were investigated by health officials. Table 1 shows the total number of egg-related SE cases and corresponding economic values. These range from a low estimate of $90 million to $490 million, to a high estimate of $350 million to $1.22 billion.

In order to determine the benefits of refrigerating eggs at 45°F during storage and transportation, it is necessary to determine the resulting percentage of reduction in the total number of egg-related deaths and

### TABLE 1
### Health and Economic Benefits of Refrigerating Shell Eggs at 45°F

| | | Annual Deaths (No. of Cases) | Annual Disabilities (No. of Cases) | Lower Bound Estimate of Health Costs (1996$) | Upper Bound Estimate of Health Costs (1996$) |
|---|---|---|---|---|---|
| **Egg-Related** *Salmonella* **Enteritidis Cases**[a] | *Low* | 87 | 69,600 | $90 million | $490 million |
| | *High* | 192 | 384,000 | $350 million | $1.22 billion |
| | | No. of Deaths Avoided | No. of Disabilities Avoided | Lower Bound Estimate of Economic Benefits (1996$) | Upper Bound Estimate of Economic Benefits (1996$) |
| **Reduction in Egg-Related SE Cases due to Refrigeration at 45°F**[b] | *Low* | 7 | 5,568 | $7.2 million | $39.2 million |
| | *High* | 15 | 30,720 | $28 million | $97.6 million |

[a]Estimated at 10 percent of all human *S* cases. Jean C. Buzby and Tanya Roberts, "ERS Updates U.S. Foodborne Disease Costs for Seven Pathogens," *Food Review*. (September-December 1996): 20-25.

[b]Estimated 8 percent. Personal communication with Dr. Richard C. Whiting, Microbial Food Safety Research Unit, USDA/Agricultural Research Service, Eastern Regional Research Center, Wyndmoor (Philadelphia), PA. The estimated 8 percent reduction in *SE* associated with refrigeration at 45°F is based on his risk assessment model updated in April 1998.

disabilities from SE cases referred to above. Whiting (1998) revealed that the refrigeration at 45°F would reduce the risk of SE by 8 percent. Table 1 shows the estimated benefits of reduction in SE cases associated specifically with refrigeration of shell eggs. These are the incremental social benefits of the rule. These estimates range from a low of $7.2 million to $39.2 million in Table 1, to a range of $28 million to $97.6 million in Table 2 (in 1996 dollars). Arrow et al. (1996), as well as Executive Order 12866, stipulate that for the regulatory requirements to be cost-effective, the value of incremental social benefits should exceed the value of incremental social costs. Therefore, these social benefits estimates would be juxtaposed against estimated social costs. The social costs are identified and discussed below.

## Compliance Costs

The incremental social costs associated with the refrigeration requirements include the first year fixed capital costs and the annual recurring costs of compliance to be incurred by the egg storage and transportation industry. The first year costs include the costs of replacing or retrofitting refrigeration units, compressors, and coils. These capital costs are required for storing shell eggs at 45°F or below after washing and packing. The capital costs to the industry would also include the costs of replacing or retrofitting transportation vehicles that have refrigeration units capable of producing air at 45°F or below. The annual recurring costs would encompass the energy costs of maintaining ambient temperatures in storage facilities and transportation vehicles at 45°F or below. These capital and recurring costs would be incurred either by shell egg producers or by their contractors for storage and transportation.

To estimate potential costs of compliance, Agricultural Marketing Service (AMS) conducted a survey of egg industry components engaged in producing, storage, and transportation of eggs. Based on this survey, AMS estimated the first-year capital costs of compliance at $40.67 million. The

**TABLE 2**

**Annual Incremental Benefit-Cost Estimates of Refrigerating Shell Eggs at 45°F**

| Benefits Estimate Scenarios | Incremental Benefits (1996$) | Benefit-Cost Ratio |
|---|---|---|
| *Low Benefits Estimates* | | |
| Lower bound | $7.2 million | 0.65 |
| Upper bound | $39.2 million | 3.56 |
| *High Benefits Estimates* | | |
| Lower bound | $28 million | 2.56 |
| Upper bound | $97.60 million | 8.87 |

Note: Incremental recurring costs = $11 million.

annual recurring costs were estimated at $10 million per year. As noted earlier, the capital costs would be incurred for replacing or retrofitting existing refrigeration units with larger size compressors or coils for the existing units. The recurring costs would be energy costs of maintaining ambient temperatures in storage facilities and transport vehicles at 45°F or below.

The costs estimated in 1992 have to be adjusted upwards because of inflation over the last six years. To adjust for this increase, the $40 million capital costs were inflated by 8 percent (based on U.S. Department of Commerce, Bureau of Economic Analysis, Price Index of Transportation and Related Equipment Index, 1992 = 100, 1997 = 108.5). This adjustment increased the capital cost estimate from $40 million to $43.2 million.

The fixed capital costs are sunk costs that do not enter in setting price of eggs by a firm in this competitive industry. Therefore, only the recurring annual costs are compared with the annual benefits. The recurring costs of compliance, estimated at $10 million per year in the 1992 survey, were assumed to comprise mostly the energy costs of refrigeration. These estimates were updated for inflation over the last six years to $10.9 or $11 million approximately (based on U.S. Department of Commerce, Bureau of Economic Analysis, Price Index of Electricity and Gas, 1992 = 100, 1997 = 108.98, or by 9 percent).

Table 2 reports a comparison of the incremental recurring costs and alternative scenarios of the benefits of the rule. The last column of this table reveals that the benefit-cost ratios range from a low of 0.65 to a high of 8.87. Three out of four of these benefit-cost ratios exceed one, i.e., the incremental social benefits exceed the incremental costs of the rule. The three positive values of the ratios suggest that the net social benefits of the proposed standards are likely to range from two to nine times the costs of compliance.

The costs of compliance to the industry are reduced by four factors. First, the proposed performance standards exempt small producers with flocks of 3,000 layers or less. There are approximately 80,000 such small egg producers that would not be required to comply with the proposed refrigeration provision. These producers account for only 1 percent of the nation's eggs. Second, of the 757 producers currently registered with USDA for purposes of egg surveillance, 329 are major producers with flocks of 75,000 or more who produced 94 percent of U.S. table eggs in 1996. These producers are members of United Egg Producers (UEP), an organization that provides a variety of services to member egg producers. The UEP already has a quality assurance program that recommends refrigerating eggs at 45°F or below as quickly as possible after washing and grading and that the same temperature be maintained during transportation. A letter from UEP to USDA FSIS indicated that a number of these producers have already started refrigerating at 45°F or below. Therefore, these producers are unlikely to incur significant incremental costs of compliance.

Many states have already enacted laws requiring specified ambient air temperatures for shell egg storage and transportation (states may have their own laws governing eggs, as long as they are not inconsistent with federal laws). Approximately one-half of the states require 45°F or less for storage and transportation. Several states, including Oregon, Washington, Louisiana, Maryland, Ohio, Arkansas, Florida, and Georgia, have enacted laws requiring shell egg refrigeration for storage and transportation at 45° F or below since 1992. Some of these states are among the largest producers of eggs. For example, Ohio had a flock of about 25 million layers out of about 300 million layers in the country in 1996. Therefore, the costs of compliance in the states that have enacted similar laws are likely to be considerably less than the estimated costs of approximately $40 million.

Finally, the higher capital costs of compliance are likely to be offset by the lower energy costs associated with the newer energy efficient vintages of refrigeration units used in storage facilities and transportation vehicles.

## Conclusions and Future Research

This paper estimated four scenarios of incremental benefits of refrigerating shell eggs at 45°F or less and compared them with the estimated costs of compliance. The annual benefits are $7.2 million, $39.2 million, $28 million, and $97.6 million, depending on the method used. Fixed capital costs could be as high as $43 million. The variable costs are estimated at $11 million per year. The incremental benefits-cost ratios for the four scenarios are 0.65, 3.56, 2.56, and 8.87. As three out of four of these ratios exceed one, it is concluded that the final rule is likely to bring about net social benefits in the long run.

### Notes

[1]Hyder Lakhani is an Economist, U.S. Department of Agriculture, Food Safety and Inspection Service, Washington, DC. The author wishes to thank Ms. Rachel Edelstein, Ms. Margaret Glavin, and Ms. Patricia Stolfa for helpful comments on an earlier draft. This paper is written in the author's personal capacity and neither the USDA nor the FSIS is responsible for the views expressed in this paper.

### References

Arrow, K.J., M. L. Cropper, G.C. Eads, R.W. Hahn, L.B. Lave, R.G. Noll, P.R.Portney, M.Russel, R. Schmalensee, V.K.Smith and R.N.Stavins (1996). *Benefit-Cost Analysis in Environmental, Health and Safety Regulation: A Statement of Principles.* The AEI Press, Washington, DC.

Bean, N.H., J.S. Goulding, M.T. Daniels, and F.J. Angulo (1997). "Surveillance for Foodborne Disease Outbreaks – United States, 1988-1992." *Journal of Food Protection*, Vol.60, No.10, 1265-1286.

Buzby, J.C., and T. Roberts (1995). "ERS Estimates Foodborne Disease Costs." *Food Review*. USDA, Economic Research Service, Vol.18, Issue 2, May-Aug.: 37-42.

Buzby, J.C., and T. Roberts (1996). "ERS Updates U.S. Foodborne Disease Costs for Seven Pathogens." *Food Review*, September-December: 20-25.

Buzby, J.C. and T. Roberts (1997). "Guillain-Barre Syndrome Increases Foodborne Disease Costs." *Food Review*. September-December: 36-42.

Centers for Disease Control and Prevention (1996). "Outbreaks of *Salmonella* Serotype Enteriditis Infection Associated with Consumption of Raw Shell Eggs: United States, 1994-1995. *Mortality and Morbidity Weekly Report.* Vol.45: 737-742.

Centers for Disease Control and Prevention (1994). *Mortality and Morbidity Weekly Report*, Vol.43, No.36, September 16, 1994.

Chalker, R. and M. Blaser, "A Review of Human Salmonellosis: III. Magnitude of *Salmonella* Infection in the United States." *Review of Infectious Diseases.* 10 (1) (1988): 111-124.

Council for Agricultural Science and Technology (1994). *Foodborne Pathogens: Risks and Consequences.* Task Force Report. September, p.46, Ames, Iowa.

Crutchfield, S.R., J.C. Buzby, T. Roberts, M. Olinger, and C.T. Jordan Lin (1997). *An Economic Assessment of Food Safety Regulations: The New Approach to Meat and Poultry Inspection.* U.S. Department of Agriculture/Economic Research, Agricultural Economic Report Number 755, Washington, DC, July, p. 3.

Hammack, T. et al. (1993). "Research Note: Growth of *Salmonella* Enteritidis in Grade A Eggs During Prolonged Storage." *Poultry Science.* Vol.72: 373-377.

Hennessy, T., et al. (1996). "A National Outbreak of *Salmonella* Enteritidis Infections from Ice Cream." *New England Journal of Medicine.* Vol.334: 1281-1286.

Humphrey, T.J. (1990). "Growth of *Salmonella* in Intact Shell Eggs: Influence of Storage Temperature." *Veterinarian Record: 1236-1292.*

Humphrey, T. and A.Whitehead (1993). "Egg age and the growth of *Salmonella* Enteritidis PT4 in Egg Contents." *Epidemiological Infections.* Vol.111: 209-219.

Jordan Lin, C.T., and Tanya Roberts (1993). "Producing Safer Poultry: Modernizing the Methods," *Agricultural Outlook.* July: 33-38.

Mishu, B. et al. (1994). "Outbreaks of *Salmonella* Enteritidis Infections in the United States, 1985-1991." *Journal of Infectious Diseases.* Vol.169: 547-552.

Whiting, R.C. (1998), Microbial Food Safety Research Unit, USDA/Agricultural Research Service, Eastern Regional Research Center's microbiological risk analysis model estimated in April 1998. Personal communication with author.

Chapter 9

# The Distributional Effects of Food Safety Regulation in the Egg Industry

Christiana E. Hilmer, Walter N. Thurman and Roberta A. Morales[1]

## Introduction

In light of recent outbreaks of foodborne diseases, Americans have become increasingly concerned with food safety. High profile cases of *Salmonella enteritidis* (*S.e.*) poisoning have focused attention on the safety of the egg industry. In 1995, there were 29 outbreaks of salmonella poisoning in the United States, affecting 857 individuals, leading to 28 hospitalizations and five deaths. Given such statistics, the USDA has recognized the importance of regulating the egg and poultry production industries in which there is a potential for salmonella transmission.

In response to growing concerns over the level of *S.e.* in table eggs, the USDA has been forced to consider new procedures for regulating the egg production process. Hazard Analysis and Critical Control Point (HACCP) is a relatively new concept in the food safety industry. Originally developed to ensure food quality in the space program, HACCP is a seven step procedure that establishes a means by which individual establishments identify and evaluate hazards that affect the safety of their products, institute controls to prevent those hazards from occurring, monitor the performance of those controls, and maintain records of that monitoring (MacDonald and Crutchfield 1996). In 1996, the USDA's Food Safety and Inspection Service (FSIS) announced that HACCP would be mandatory for all meat and poultry plants under its jurisdiction and that each plant would be required to develop a HACCP plan subject to FSIS inspection (Roberts et al. 1996). Consequently, by late 1996, HACCP was being used by about 10 percent of the meat and poultry plants in the northeast (Unnevehr and Jensen 1996).

There is some debate over the value of implementing a mandatory HACCP program. Previous inspection programs have regulated food safety by attempting to detect safety and health hazards. It has been argued that HACCP may be preferable to such programs for a number of reasons. Antle (1996) argues that the previous methods have been incapable of detecting disease-causing microorganisms or chemical contamination because "they primarily rely on organoleptic surveillance methods (i.e., sight, smell, and touch)." Similarly, Unnevehr and Jensen (1996) contend that HACCP is superior because it relies on science-based risk assessment and prevention by building safety controls into the production process itself instead of relying on detection after contamination has occurred. In their view, HACCP implementation will enhance the productivity of existing inputs in producing product safety. On the opposite side, the Safe Food Coalition is concerned that HACCP will result in a relaxation of safety inspection standards by replacing on-site inspections with record keeping throughout food production industries (Becker 1992). Further, Antle argues that HACCP supporters may not realize the implications of making HACCP a mandatory program. Prior to the 1996 announcement, many firms had been employing HACCP programs, but they had been doing so on a voluntary basis. As Antle points out, there are important differences between a voluntary program that is designed and implemented by a firm and a regulation that is mandated and enforced by government bureaucrats and inspectors.

To empirically determine the value of implementing HACCP, it is necessary to compare the expected benefits of implementing such a system with the expected costs. The main economic benefit of HACCP is that society will save money by reducing foodborne illnesses and increase utility from disease avoidance. Estimates of such benefits vary widely depending on the degree to which HACCP is assumed to decrease foodborne illnesses and deaths. Using a conservative assumption of a 20 percent decrease, Crutchfield et al. (1997) estimate the benefit to be at least $1.9 billion over the next twenty years while using a liberal assumption of 90 percent the estimate balloons to $170 billion.[2] The degree to which HACCP succeeds in reducing foodborne illnesses and deaths depends in large part on industry's response to the regulation.

In this paper we focus on the egg industry's response to the implementation of mandatory HACCP regulation. We start by developing a model of the egg industry that accounts for heterogeneity in sanitary propensity. We then derive comparative statics that demonstrate how firms respond to changes in regulations and discuss the welfare implications of different views of HACCP regulation. Finally, we examine the effect that firm heterogeneity has on the manner in which egg producers market their product.

## The Welfare Effects of HAACP Regulation

The key economic issue in deciding whether to implement HACCP regulation is whether such regulation can be efficient. The efficiency of food safety regulation depends vitally on the type of regulation to be used. Broadly, food safety regulations can either take the form of performance standards or process standards. Performance standards specify a quality level that a firm's output must meet, but allow the firm autonomy over its production process. Process standards attempt to determine conditions for each step in the production process required to produce output of the desired quality then require all firms in the industry to follow those conditions. As an example, suppose that the goal of regulation is to decrease the fraction of eggs testing positive for *S.e.* to one tenth of 1 percent. A performance standard would simply require that upon end testing, no more than one tenth of 1 percent of a firm's eggs test positive for *S.e.* Such a standard would allow the firm to make its own decisions regarding initial egg temperature, wash water temperature, wash water pH, bacterial load in wash water, sanitizer concentration, egg temperature, and egg storage temperature. Instead of relying on end testing, a process standard would determine the appropriate minimum levels for each intermediate measure and require all firms to meet or exceed those standards.

There has been much debate in literature over the relative efficiency of performance standards and process standards. Economic theory argues, neglecting enforcement costs, that performance standards are more efficient because they allow firms to tailor the standards to their specific production method, whereas process standards dictate industry-wide regulations that may be inappropriate for individual firms. The regulatory community does not share this view, however. According to Antle (1995, 1996), food safety regulators generally prefer to implement process standards because they believe that process standards are the only way to prevent risk and they deem performance standards too costly due to end product testing. The question becomes which of these views is correct. Antle (1995) argues that a mandatory HACCP program will be inefficient because its mandatory nature makes it a form of process regulation. MacDonald and Crutchfield (1996) counter with the argument that mandating HACCP may overcome the high information cost that would be required to enforce a pure performance standard. Antle agrees that if implemented as a set of performance standards for different stages in the production process, the mandatory HACCP program would be similar to a voluntary HACCP program because it would allow firms to design quality control systems that suit their particular production systems. Similarly, Caswell and Hooker (1996) note that HACCP will primarily be a performance standard if regulators require companies to develop and implement a HACCP plan but don't specify individual elements of the plan in detail.

As these observations indicate, process standards can be performance standards for specific process outputs. While the pure performance standard may be costly to implement due to the high costs of measuring, in this case, *S.e.* risk, some of the benefits of performance standards may be achieved by imposing standards on production sub-processes with measurable output. There remains the question of to what extent acceptable HACCP plans resemble such an efficient tradeoff between measurement costs and process standard inefficiency.

The discussion above concerns economic efficiency, equally weighting dollar losses and gains across individuals. A separate concern over HACCP is the disproportionate effect that its implementation may have on small firms. Unnevehr and Roberts (1996) point out that HACCP involves a large fixed investment to develop the plan and train the employees, and may even include investment in new capital equipment, but that the variable costs of HACCP are not usually consequential. They argue that as a result, HACCP will have a much greater impact on small rather than large firms because fixed costs are a much larger portion of total costs for small firms than for large firms. Antle (1995) supports this view by noting that the extensive monitoring and record keeping required under HACCP are primarily fixed cost in nature resulting in higher average cost per unit of production for smaller firms. In support of this view, Ollinger et al. (1997) report that 60 to 90 percent of small firms (less than 25 employees) in the beef and pork slaughter and processing industries fail to survive five years after entry.[3] Such findings suggest that in modeling the effect of HACCP implementation in the egg industry, it is important to account for heterogeneity in firm size.

Yet another source of firm heterogeneity is important to acknowledge in order to understand the effects of food safety regulation, both efficient and distributional. We call it variation in sanitary propensity. Some firms, by nature of locale and the input costs they face, are more inclined to a higher sanitary standard than others. We explore the effects of this variation here. Our development expands the homogenous firm model of industry equilibrium described in Morales and Thurman (1996).

Consider egg-producing firms with fixed capacity; each produces one dozen eggs. The eggs produced can be sold on either or both of the shell and breaker egg markets. Shell egg production requires more careful handling and a higher sanitary standard. Shell eggs receive a premium relative to breaker eggs. Breaker eggs are sold at price P and shell eggs are sold at price $P + \phi$. The proportion of a firm's production sold on the shell market is denoted $\rho = q_s/(q_s+q_b)$, where $q_s$ and $q_b$ are the quantities produced of shell and breaker eggs.

Firm profit is written as:

(1) $$\pi = P + \rho\phi - f(\alpha, \beta, \rho) w,$$

where f(.) is the unit input requirement and w is the wage of the single input.

The specification of f(.) is crucial. Here it is written as a function of firm parameters $\alpha$ and $\beta$, and decision variable $\rho$. The parameters $\alpha$ and $\beta$ vary across firms in ways to be specified below. For production purposes, an $(\alpha, \beta)$ combination uniquely identifies a firm. The parameter $\alpha$ will represent sanitary propensity or, observably, a firm's tendency to produce shell eggs. The parameter $\beta$ will represent a more general efficiency in the production of eggs. Unit input requirement is assumed to depend on $\rho$ such that, for any firm, $f_\rho > 0$ and $f_{\rho\rho} > 0$.

Because the scale of a producing firm is fixed, the only decision for a firm is that of $\rho$, shell egg proportion. The optimal choice is characterized by:

(2) $$\frac{\partial \pi}{\partial \rho} = 0 \quad \text{or} \quad \frac{\partial f}{\partial \rho}\left(\alpha, \beta, \rho^*\right) = \phi / w .$$

where $\rho^*$ is the optimal proportion.

At a given set of market prices (P, $\phi$, and w) some firms' maximum profits will be positive, others will not. Marginal firms, characterized by zero profit, are those $(\alpha, \beta)$ combinations satisfying:

(3) $$\pi = P + \rho^* \phi - f(\alpha, \beta, \rho^*) w = 0.$$

The zero-profit condition determines an implicit relation between $\alpha$ and $\beta$. If higher levels of both $\alpha$ and $\beta$ are cost reducing, then the zero-profit relation between $\alpha$ and $\beta$ will be a negative one. Differentiating the zero-profit condition with respect to the parameters yields:

(4) $$\left.\frac{d\beta}{d\alpha}\right|_{d\pi=0} = -\frac{f_\alpha}{f_\beta}.$$

Consider now the following specific functional form for unit input requirement:

(5) $$f(\rho) = (A - \alpha)\rho + \frac{k}{2}\rho^2 + (B - \beta).$$

The parameters $\alpha$ and $\beta$ index firms; the parameters A, B, and k are constant across firms. Firms with higher $\alpha$ have lower marginal costs of increasing $\rho$ (and lower costs overall). Firms with higher $\beta$ have lower unit costs, but have no lower marginal costs of increasing $\rho$.

We will require the first and second partial derivative with respect to $\rho$ to be positive:

(6) $\quad f_\rho = (A - \alpha) + k\rho > 0$ and $f_{\rho\rho} = k > 0.$

The second derivative condition implies that the parameter k is positive. In order for the first derivative condition to hold for all values of $\rho$, we will require that $\alpha < A$: the largest-$\alpha$ firm has a value of $\alpha$ less than A. This gives A the interpretation of the maximum possible value for $\alpha$.

To restrict values for $\beta$, we will require that the unit input requirement be positive, for all values of $\rho$ and for all firms. Because we will require that $\beta < B$ and so B is interpreted as the maximum possible value for $\beta$ across firms. These restrictions imply the shape of the unit input requirement function in Figure 1.

The firm's optimum choice of $\rho$ with this specification of $f(\alpha, \beta, \rho)$ is given by:

(7) $\quad (A - \alpha) + k\rho = \phi/w,$ or $\rho^* = \dfrac{\phi/w - (A - \alpha)}{k}.$

The solution is illustrated in Figure 2 for two firms: one with $\alpha = \alpha_0$, the other with $\alpha = \alpha_1$, $\alpha_0 < \alpha_1$. The higher-$\alpha$ firm chooses a higher level of $\rho$. Figure 2 also illustrates how corner solutions can arise. For firms with sufficiently low $\alpha$, the $f\rho$ curve intersects the vertical axis above $\phi/w$ and the optimal choice of $\rho$ is zero. Similarly, for firms with sufficiently high $\alpha$, the $f\rho$ curve intersects the $\rho=1$ line below $\phi/w$ and the firm's optimal choice of $\rho$ is one.

Having determined the firms' optimal choices and their dependence on

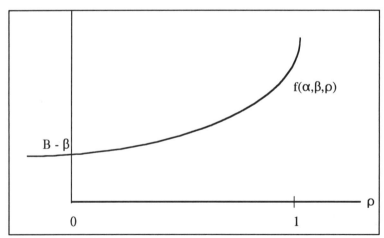

FIGURE 1. Unit input requirement.

firm characteristics ($\alpha$ and $\beta$) and market prices (P, $\phi$, and w), we turn to the solution for market supply of shell and breaker eggs. For a given set of prices, the firms that will produce are those with positive profits. Firms with negative maximum profits will not produce. The marginal firms are characterized by ($\alpha$, $\beta$) combinations for which profit is zero:

(8) $$\pi = P + \rho^*\phi - f(\alpha, \beta, \rho^*) w = 0.$$

For simplicity of exposition, assume that the optimal choice of $\rho$ for each producing firm is an interior solution, strictly greater than zero and less than one. In this case, the $\pi=0$ contour in ($\alpha$, $\beta$) space can be written explicitly as:

(9) $$\tilde{\beta}(\alpha, P, \phi, w) = B - P - [\phi - (A - \alpha)]\rho^* + \frac{k}{2}\rho^{*2}$$
$$= B - \frac{P}{w} - \frac{(\phi/w)^2}{2k} + \frac{\phi/w}{k}(A-\alpha) - \frac{1}{2k}(A-\alpha)^2.$$

Figure 3 depicts the $\pi=0$ contour in ($\alpha$, $\beta$) space. The slope of any iso-$\pi$ contour can be shown to equal $-\rho^*$.

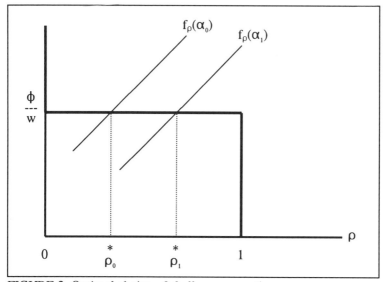

FIGURE 2. Optimal choice of shell egg proportion.

To translate the information in Figure 3 into a supply function requires integrating the density of firms over the region of producing firms. Letting $d(\alpha, \beta)$ denote the density function of firm characteristics, we have:

$$Q_S = \int_0^A \int_{\tilde{\beta}(\alpha,P,\phi,w)}^B d(\alpha,\beta)\rho^*(\alpha,P,\phi,w)\,d\beta\,d\alpha, \text{ and}$$

(10)

$$Q_B = \int_0^A \int_{\tilde{\beta}(\alpha,P,\phi,w)}^B d(\alpha,\beta)[1-\rho^*(\alpha,P,\phi,w)]\,d\beta\,d\alpha.$$

Because the optimal choice of proportion, $\rho^*$, does not depend upon the general cost characteristic, $\beta$, the previous expressions simplify to:

$$Q_S = \int_0^A \rho^*(\alpha,P,\phi,w) \int_{\tilde{\beta}(\alpha,P,\phi,w)}^B d(\alpha,\beta)\,d\beta\,d\alpha$$

$$= \int_0^A \rho^*(\alpha,P,\phi,w)N(\alpha,P,\phi,w)\,d\alpha$$

(11)

$$Q_B = \int_0^A \left[1-\rho^*(\alpha,P,\phi,w)\right] \int_{\tilde{\beta}(\alpha,P,\phi,w)}^B d(\alpha,\beta)\,d\beta\,d\alpha$$

$$= \int_0^A \left[1-\rho^*(\alpha,P,\phi,w)\right]N(\alpha,P,\phi,w)\,d\alpha,$$

where $N(\alpha,P,\phi,w)$ is the density of producing firms with sanitary characteristic $\alpha$.

Closed form supply functions can be calculated from these expressions given a distribution of firm characteristics. If, for example, the distribution of firm characteristics were uniform over the rectangle $\alpha \in (0, A)$ and $\beta \in (0, B)$, then:

(12) $$N(\alpha, P, \phi, w) = B - \tilde{\beta}(\alpha),$$

where $\tilde{\beta}(\alpha)$ is given above in equation (9). The role of $\tilde{\beta}(\alpha)$ is key here because it determines the equilibrium tradeoff between cost efficiency ($\beta$) and sanitary propensity ($\alpha$). Firms with higher $\alpha$ choose higher proportions of shell egg production and are more profitable as a result. Thus, low-$\beta$ firms will produce only if they also are high-$\alpha$ firms.

Comparative statics from the model can be derived graphically and analytically. The exogenous variables are P, $\phi$, and w. Note, however, that the three prices enter expression (7) for $\rho^*$ and expression (9) for $\tilde{\beta}$ only through the two relative prices P/w and $\phi$/w. Therefore, in what follows, normalize w to equal one and understand that variations in P and $\phi$ represent independent variations in P/w and $\phi$/w.

Figure 4 displays $\rho^*$ and N as functions of $\alpha$. The optimal shell proportion, $\rho^*$, increases linearly with firms' $\alpha$ levels. The density of firms with a given a level of $\alpha$, N, increases quadratically with $\alpha$ reflecting the fact that higher $\alpha$ firms are more profitable. At each $\alpha$, the slope of the N($\alpha$) curve equals $\rho^*$. Equilibrium production of shell eggs is given from equation (11) as the area in Figure 4 under $\rho^*$N. Production of breaker eggs is the area between N and $\rho^*$N. Total egg production is the sum of areas $Q_S$ and $Q_B$.

If both the shell and breaker egg prices increase, but the shell premium

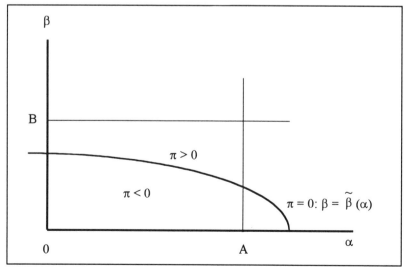

FIGURE 3. Marginal profit contour.

is constant, both shell and breaker egg production will increase. In Figure 4, an increase P would imply a shift upward in $N(\alpha)$ ($\partial N / \partial P = 1$) but no shift in $\rho^*(\alpha)$. Both the $Q_S$ and $Q_B$ areas would increase.

The supply effects of an increase in $\phi$, the shell egg premium, are more complicated. In sign they are straightforward enough in the case of shell egg production. From equation (11) or Figure 4 it can be seen that $Q_S$ is the integral under $\rho^* N$. Further, both $\rho^*$ and N increase with an increase in the unit input requirement, f. Therefore, $Q_S$ increases with $\phi$. The comparative static effect of a change in $\phi$ on $Q_B$ is indeterminate in sign, however, because of counteracting substitution and scale effects. From equation (11), $Q_S$ is the integral under $(1-\rho^*)N$. Further, it can be shown that:

$$(13) \quad \frac{\partial \left(1-\rho^*\right) N}{\partial \phi} = \rho^* - \frac{3}{2}\rho^{*2} - \frac{P}{k} .$$

Expression (13) can be either positive or negative and takes a maximum at the value $\rho^* = 1/3$. Therefore, the integral of (13) over the range of $\alpha$ can either be positive or negative and the effect of an increase in $\phi$ on $Q_B$ is ambiguous.

The comparative statics of the model are directly relevant to assessing the welfare effects of HACCP regulation. To the extent that the

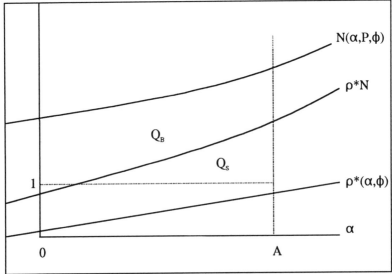

FIGURE 4. Optimal shell egg proportion and firm density as functions of $\alpha$.

regulations apply only to shell egg production, because breaker egg products are pasteurized and not a food safety threat, the effect on firms is equivalent to a decrease in $\phi$. All firms will decrease their proportion of shell eggs produced. Total shell egg production will decrease; breaker egg production could increase or decrease.

The welfare costs of the food safety tax are spread throughout the industry. For producing firms, the envelope theorem gives:

$$\text{(14)} \quad \frac{\partial \pi}{\partial \phi} = \rho^*$$

The direct incidence of the tax, ignoring the effects of equilibrium price changes, is higher for higher-$\alpha$ (higher-$\rho$) firms.

The other view of food safety regulation is as a minimum $\alpha$ standard: firms with $\alpha$ below $\alpha_{min}$ would not be allowed to produce shell eggs. For firms above this level, the supply decision rules would be unchanged. Firms below this level could choose either not to produce or to produce only breaker eggs, with $\rho$ set equal to zero. In this case, breaker egg production unambiguously would increase and shell egg production would decrease. Demand conditions would cause breaker egg prices to fall and shell egg prices to rise. Thus, with the minimum standard view of the regulation, there are created two classes of firms: those who clearly are harmed, the below-minimum-$\alpha$ firms, and those who benefit, the above-minimum-$\alpha$ firms. There is no such clear distinction in the case of a regulation that acts like a tax on the shell egg premium. One's view of the distributional welfare effects of HACCP regulation, and the political economy of its support, depends upon which view of the implemented regulation is more nearly accurate.

## The Food Safety Effects of Non-Regulatory Government Action: Commodity Promotion Boards and Name Branding

The second section focused on the economic cost of regulation designed to promote food safety in the egg industry. Here we consider a symmetric situation: the food safety effects of regulation designed to promote industry profit.

Implicit in the argument for HACCP, and other food safety regulations, is a claim of market failure: consumers are willing to pay more for food safety than it costs firms to provide it, but these potential gains to trade are left unexploited. If the claim is true, there are impediments to trade between consumers and firms. While assessing government action (regulation) that circumvents private transactions, we should try to understand the sources of the impediments.

In the second section, we argued that an important characteristic of the egg market is that egg producers differ by sanitary propensity. This heterogeneity not only influences production decisions, but also has implications for the manner in which egg producers choose to market their output. Specifically, the diversity in size and sanitary propensity implies diversity in the incentive to name brand. However, American egg producers have developed a system under which commodity promotion boards force uniformity of marketing identity and diminish the incentive to develop product quality reputations.

Egg producers currently invest in national advertising and promotion through a mandatory contribution to the American Egg Board (AEB) of five cents for every case (360 eggs) of table eggs sold. The AEB is charged with promoting the industry, conducting industry-specific research, and communication with the egg production/processing community. A notable fact is the size of the AEB's advertising budget relative to the advertising budgets of egg substitutes. In 1991, the AEB spent $7.5 million promoting their product in contrast to the $13 million spent promoting Dunkin Donuts and the combined $44 million spent promoting Eggo Waffles and Pop Tarts (Forker and Ward 1993). This disparity may be explained in large part by the fact that such substitutes are name branded whereas eggs, for the most part, are not. If true, it becomes important to examine the relationship between brand naming and the incentive to advertise.

Klein and Leffler (1981) first discussed the economics of name brand advertising by analyzing the non-governmental repeat-purchase contract enforcement mechanism. Under such a mechanism, two-party contracts can only be terminated by discontinuation of the business relationship and not by order of a third party such as a court of law. Further, it is assumed that the product's true quality level is initially unknown to the consumer and information about the product's quality cannot be determined costlessly before purchase. Pre-purchase inspection reveals only whether quality is below some minimum level. In these respects, the industry hypothesized by Klein and Leffler resembles the egg industry. One key difference is that they also assume that the consumers know the exact identity of the firm. Clearly, this is not the case for the egg industry where output, typically, is sold generically.

Turning to the consumer, Klein and Leffler make the critical assumption that consumers are able to costlessly communicate. This assumption implies that if a firm cheats by supplying a lower than contracted quality product the information is spread throughout the market and all future sales are lost. Consequently, selling low quality merchandise that is promoted as high quality merchandise results in only a one-time wealth increase as consumers will permanently boycott the firm for all future periods.

Even with the perfect inter-consumer communication assumption, a supply of high quality product is not guaranteed, however. A firm will decide whether to cheat on quality by comparing the present value of future non-cheating income to the one-time gain from cheating. If the one-time gain from cheating is greater, the firm will decide to cheat. Hence, in order to prevent cheating, firms must receive quasi-rents in the form of higher than perfectly competitive prices for above minimum quality goods. Klein and Leffler demonstrate that under very general cost conditions, a price premium will exist that motivates competitive firms to honor high quality promises because the value of satisfied customers exceeds the cost savings of cheating them. This price premium must not only compensate the firm for the increased average production costs incurred when producing a quality product but must also yield a normal rate of return on the forgone gains from exploiting consumer ignorance.

Within this framework, there is a minimum price required to assure that the good being produced is of the minimum quality. Clearly, consumers will not purchase from firms attempting to sell the good at prices below the quality-assuring price. Accordingly, competition to dissipate the economic profits earned by existing firms must occur in non-price dimensions. A prime example of such a non-price dimension is brand name advertising. Expenditures on brand name advertising only benefit a firm if the firm consistently produces a high quality product that consumers are willing to pay a premium to purchase. If the quality of the firm's output falls below the desirable level, consumers will no longer be willing to pay the price premium and the firm's advertising expenditures will have been for naught. Therefore, Klein and Leffler note that advertising expenditures are in essence a type of collateral that the firm loses if it supplies output of less than anticipated quality.

The magnitude of a firm's investment in brand name advertising indicates the magnitude of the price premium that the firm stands to receive. A sufficient investment in advertising implies that a firm will not engage in short-run quality deception, because the added expense increases production costs thereby necessitating that the firm earn a price premium to remain profitable. Thus, the mere fact that a particular brand is advertised could be taken as a signal that the brand is of high quality (Nelson 1970). This implies that when consumers pay a price premium for a branded product, they are paying for an implicit guarantee that the good is of superior quality (Png and Reitman 1995).

The impact that brand name advertising has on consumers differs depending on whether the good can be classified as a search good or an experience good. A search good is a good for which the consumer can discern quality prior to purchase (McClure and Spector 1991). An experience good is a good for which consumers can only discern quality after purchase. Examples of search goods are picture frames or trash cans.

Examples of experience goods are brand name canned tuna fish or car repair at a gas station. Given these characteristics, it can be seen that advertising will differ markedly in the amount and type of information provided to the consumer. For search goods, information is direct as advertising provides specific information about the quality of the good being produced. For experience goods, information is indirect, as specific quality information is not provided, and product quality is implied through the firm's willingness to pay to promote its product. Nelson (1974) demonstrates that the difference in the character of information leads to greater advertising expenditures for experience goods than for search goods but argues that consumers acquire greater marginal benefits from search goods than from experience goods due to the nature of information conveyed.

The above discussion provides insight into the relative lack of name branding in the egg industry. From a consumer's perspective, eggs are an experience good in that the quality of a particular egg can only be ascertained once it is consumed. Further, the main difference between the market analyzed by Klein and Leffler and the egg market appears to us to be that egg consumers lack information about the producers of generically marketed egg. As a result, price premia cannot function to assure quality. A plausible explanation for the dearth of name brand identities in eggs, compared to say poultry meat, is generic promotion, which may crowd out firm-level name branding. We now are exploring the implications of this story.

## Conclusion

In response to growing concerns over foodborne illnesses, such as Salmonella enteritidis, the USDA announced in 1996 that it was implementing a mandatory HACCP program for all meat and poultry plants under its jurisdiction. We discuss the effects that HACCP implementation will have on the U.S. egg market. We argue that an important aspect of the egg industry is sanitary propensity; the inclination of firms, absent regulation, to produce to high sanitary standards. This propensity, we argue, varies across firms. We explore the implications of this view of the egg industry and conclude that the welfare effects of HACCP regulation can be skewed in the cross section and can bestow benefits, in equilibrium, on some firms while imposing costs on others. We also explore the food safety consequences of a non-regulatory government sponsored program: industry promotion through mandatory check offs. We discuss the possibly inhibiting effect of generic promotion on the incentives for firms to develop name brand identities, which could serve as private guarantees of food safety. It is important to understand that this effect is not a criticism of check off programs for reasons that they fail to do what they are set up to do. Rather, it is an unintended consequence of a program that may well enhance industry profit in the way that its proponents claim.

**Notes**

[1] Christiana E. Hilmer is a graduate student in Economics at North Carolina State University, Walter N. Thurman is Professor of Economics and Agricultural and Resource Economics at North Carolina State University, and Roberta A. Morales is Assistant Professor of Veterinary Medicine at the University of Maryland. This research is funded through a grant from the U.S. Department of Agriculture.

[2] As Crutchfield et al. point out, these estimates may tend to understate the true benefit because they only include the six primary pathogens for which disease data available and additional economic benefits would accrue if other pathogens were also abated. Furthermore, Roberts et al. (1996) contend that a true estimate of the benefits of HACCP would include the value of risk reduction to all consumers in addition to the costs imposed on those who happen to become ill (Unnevehr and Jensen). However, this cost-of-control approach to measuring the benefits of regulation has been extensively criticized in the literature (e.g., Harrington and Portney, Atkinson and Crocker, and Mullahy and Portney). Tolley et al. contend that the fundamental problem with this approach is that it is hard to identify and measure all of the inputs that affect health thus suffer from bias due to omitted variables.

[3] This is not to imply that small firm failures are due to HACCP regulations. Certainly, other factors affect the whether a small firm survives more than five years after entry. However as Ollinger et al. (1997) argue, it is hard to dissociate the low survival rate of small firms with the effect of implementation of HACCP regulations.

**References**

Antle, John. 1996. Efficient Food Safety Regulation in the Food Manufacturing Sector. *American Journal of Agricultural Economics* 78 (December): 1242-1247.

Antle, John. 1995. *Choice and Efficiency in Food Safety Policy*. Washington, DC: The AEI Press.

Caswell, Julie and Neal Hooker. 1996. HACCP as an International Trade Standard. *American Journal of Agricultural Economics* 78 (August): 775-779.

Crutchfield, Steve, Jean Buzby, Tanya Roberts, Michael Ollinger, and C.-T. Jordan Lin. 1997. An Economic Assessment of Food Safety Regulations: The New Approach to Meat and Poultry Inspection. Economic Research Service/USDA. 1997 (July).

Forker, Olan and Roland Ward. 1993. *Commodity Advertising: the Economic and Measurement of Generic Programs*. New York, NY: Lexington Books.

Klein, Benjamin and Keith Leffler. 1981. The role of markets forces in assuring contractual performance.*Journal of Political Economy*. 89(41): 615-641.

MacDonald, James and Stephen Crutchfield. 1996. Modeling the Costs of Food Safety Regulation. *American Journal of Agricultural Economics* 78 (December): 1285-1290.

McClure, James and Lee Spector. 1991. Joint Product Signals of Quality. *Atlantic Economic Journal*, 19: 38-41.

Morales, Roberta and Walter Thurman. 1997. Equilibrium Analysis of Food Safety Regulation: *Salmonella enteritidis* and Eggs. Working Paper. Department of Agricultural and Resource Economics, North Carolina State University, Raleigh, NC.

Nelson, Phillip. 1970. Information and Consumer Behavior. *Journal of Political Economy*. 78: 311-29.

Nelson, Phillip. 1974. Advertising as Information. *Journal of Political Economy*. 81: 729-54.

Png, I.P.L. and David Reitman. 1995. Why are Some Products Branded and Others Not? *Journal of Law and Economics* XXXXVIII: 207-224.

Roberts, Tanya, Jean Buzby and Michael Ollinger. 1996. Using Benefit and Cost Information to Evaluate a Food Safety Regulation: HACCP for Meat and Poultry. *American Journal of Agricultural Economics* 78 (December): 1297-1301.

Unnevehr, Laurian and Helen Jensen. 1996. HACCP as a Regulatory Innovation to Improve Food Safety in the Meat Industry. *American Journal of Agricultural Economics* 78 (August): 764-769.

Unnevehr, Laurian and Tanya Roberts. 1996. Improving Cost/Benefit Analysis for HACCP and Microbial Food Safety: An Economist's Overview. *Strategy and Policy in the Food System: Emerging Issues*. Washington, DC: June 20-21.

Chapter 10

# The Costs, Benefits and Distributional Consequences of Improvements in Food Safety: The Case of HACCP

Elise H. Golan, Katherine L. Ralston, Paul D. Frenzen and Stephen J. Vogel[1]

## Introduction

The costs and benefits of implementing a Hazard Analysis and Critical Control Point (HACCP) regulatory program for meat and poultry slaughterhouses and processors are distributed throughout the economy. The costs of implementing HACCP are paid initially by the meat and poultry industry, while the benefits of controlling foodborne illness are distributed initially among consumers. However, the ultimate impact of these costs and benefits extend well beyond the initial payers and beneficiaries, with economic ramifications for many different segments of the economy.

In order to examine the full economic ramifications of HACCP implementation, we use a Social Accounting Matrix (SAM) framework to investigate the impact that HACCP and reductions in foodborne illness have on the level and distribution of consumption, production, and income in the U.S. economy. We find that the ultimate distribution of the costs and benefits of HACCP differs substantially from the initial distribution. General equilibrium effects have not been examined in previous cost-benefit studies of HACCP.

The SAM accounting of the final impact of costs and benefits of HACCP provides useful information for policy makers by indicating the direction and magnitude of the economic flows resulting from regulation costs and reductions in foodborne illness. Use of the SAM framework also focuses attention on the difficulty of assessing the economic value of health. The SAM analysis demonstrates the usefulness of the cost-of-illness approach in deciphering the economic distortions caused by health shocks to the economy. It also highlights the danger in equating income changes with changes in well being.

## Which Costs and Benefits?

In any cost-benefit study, analysts must decide which costs and benefits to include in the analysis and which methodology to use to evaluate non-market goods and services. For this study, we use the benefit estimates of the reduction of foodborne illness calculated by USDA's Economic Research Service (Buzby et al. 1996) and the cost estimates of the implementation of HACCP calculated by USDA's Food Safety and Inspection Service (USDA, FSIS 1996). These estimates were the basis for the official regulatory impact analysis of the HACCP regulatory program (USDA, FSIS 1995 and USDA, FSIS 1996). These cost and benefit estimates are examined in detail in Crutchfield et al. (1997). We use the summary estimates prepared by Crutchfield et al. in our analysis. All dollar amounts have been converted to 1993 dollars.

The cost estimates prepared for the Food Safety and Inspection Service (FSIS) cost-benefit analysis of HACCP[2] measure the additional costs to FSIS and to the meat and poultry industry of implementing HACCP. These cost estimates depend on numerous assumptions, including assumptions about the structure of the processing industry, wages, modification costs, costs of training, supply and demand conditions, and timing of implementation. FSIS estimates the costs of HACCP at $1.04 billion to $1.23 billion over 25 years (Crutchfield et al. 1997). Other studies have used different assumptions and have produced different estimates (for example, Knutson et al. 1995; Jensen et al. 1998). For this study we chose to use the official FSIS estimates, though the methodology we develop could be applied to any of the other cost estimates. Depending on the types of costs included in the estimates, the level and distribution of economic effects could differ substantially.

The benefit estimates of the HACCP regulatory program reported in Crutchfield et al. range from $1.80 billion to $162.39 billion over 25 years. These benefit estimates include calculation for four foodborne pathogens transmitted by meat and poultry: *Salmonella, E. coli* O157:H7, *Campylobacter jejuni* or *coli* and *Listeria monocytogenes*. There are four primary reasons for the large variability in the estimates. First, there is uncertainty about the incidence of foodborne illness transmitted by meat and poultry, and as a result, benefit estimates reflect a wide range of incidence estimates. Second, there is uncertainty about the efficacy of the HACCP rule in reducing foodborne pathogens due to meat and poultry. The highest benefit estimate reported in Crutchfield et al. incorporates an efficacy rate of 90 percent while the lowest estimate uses a rate of 20 percent. The third reason for the variation in the benefit numbers is the use of two different discount rates. Crutchfield et al.'s lower estimates use a relatively high discount rate of 7 percent to reflect private valuations, while the higher estimates use a discount rate of 3 percent to incorporate a more societal viewpoint. The fourth, and most critical source of variation in the benefit estimates, is the use of two different methods for assigning economic value to the improvements in health and longevity resulting from reductions in foodborne illness. The higher benefit estimates reported in Crutchfield et al. use hedonic wage estimates (Viscusi 1993), while the lower use the cost-of-illness methodology (Landefeld and Seskin 1982).

For this study we use mid-range estimates of $4.46 billion to $22.19 billion for 25 years of benefits. These estimates are calculated with a HACCP efficacy rate of 50 percent, a discount rate of 7 percent, and the Landefeld and Seskin cost-of-illness valuation approach.[3] We chose the moderate efficacy rate and the steeper interest rate because we wanted to use conservative estimates. We chose the cost-of-illness approach because it provides a measure of the economic distortions arising from adverse health and premature death (or, in this case, a reduction in both).

Cost-of-illness estimates measure two types of costs: direct medical expenses and human capital costs. The direct medical costs of illness are expenditures for medical goods and services such as doctor visits, hospitalization, residential care, and medications. Human capital costs of illness are the present value of wages (and non-wage benefits) forgone as a result of an adverse health outcome. The cost-of-illness approach produces an accounting of the dollars that are spent differently as a result of illness or premature death.

The cost-of-illness method for evaluating the costs of illness and premature death has been criticized in the health economics literature, primarily because it does not incorporate valuations for pain and suffering and other non-market commodities. For more accurate appraisal of the changes in welfare resulting in changes in health and longevity, economists prefer the willingness-to-pay approach (for a review and critic of valuation methodologies for health cost-benefit analysis, see Tolley et al. 1994). However, willingness-to-pay amounts do not measure economic distortions. Though they may be useful for indicating how much a society should pay to avoid adverse health outcomes and premature death, they do not measure the economic impact of such outcomes. The willingness-to-pay estimates that are used in the upper range of HACCP benefit estimates were derived by observing the wage premium paid to workers for risky jobs. These wage premiums, and the attitudes towards risk and health that they reveal, do not shed light on the effect of illness or premature death on the level or distribution of economic activity.

The cost-of-illness approach traces the economic flows associated with an adverse health outcome. It accounts for the drop in productivity resulting from illness, accident or premature death, and it accounts for the shift in consumer expenditure from more general consumption goods and savings and investment, to medical goods and services. The cost-of-illness approach provides an accounting of the dollars spent on medical expenses and the labor dollars that are forgone as a result of illness, accident or premature death. When combined with a general equilibrium analysis, such as a Social Accounting framework, the cost-of-illness approach provides the first step in deciphering the full impact of health shocks to the economy. This is important information for policymakers interested in gauging the extent and distribution of the costs of foodborne illness due to meat and poultry and the benefits of the HACCP program.

## The Social Accounting Framework

A SAM is a form of double-entry accounting in which national income and product accounts and Input-Output production accounts are represented as debits (expenditures) and credits (receipts) in balance sheets of activities and institutions. Activities are industries and services, and institutions are households, firms, government, and the rest of the world. Entries in the SAM include intermediate input demand between production sectors; income (value added) paid by production sectors to different types of labor or capital; the distribution of wages across different household groups; and the distribution of household expenditures across savings, domestic consumption, and imports. Unlike the Input-Output framework, the SAM framework endogenizes income and consumption, thereby permitting an appraisal of the full effects of specific changes to the economy. In addition to providing a snapshot view of the circular flow of accounts of an economy, a SAM also provides the basis for a SAM multiplier model. The SAM multiplier model is a linear, general equilibrium model of the economy that traces the impact of exogenous change on every endogenous account in the economy.

The first task in constructing a SAM is to identify the important activities and institutions in the economy with respect to the policy issues under consideration. Aggregation of the industries, services, households, government agents, and rest-of-the world accounts of the economy into major accounts makes the model more manageable and serves to focus the investigation. The aggregation scheme determines the flows that the model will be able to trace explicitly. If the aggregation is done correctly, the major flows in the economy, both positive and negative, will be evident. Otherwise, the impact of policy will be blurred, with negative and positive flows occurring within a single account. In the HACCP SAM, the accounts were constructed to focus the model on the primary activities and institutions affected by foodborne illness and by HACCP regulation. Particular attention was paid to the construction of the industrial and household accounts.

The industrial aggregation in HACCP SAM highlights the three major areas of the economy most directly impacted by HACCP and foodborne illness: the meat and poultry production and distribution system, the health care system, and the health insurance system. With the HACCP SAM industrial aggregation, meat products (and regulatory costs) can be traced from the Livestock sector, through the Food Processing sector, the Wholesale Trade sector, the Food Retail Trade sector, and eventually to final demand. Medical costs can also be traced from the account of payment (household and insurer) to the account of receipt (pharmaceuticals, medical services, etc.).

The grouping of households for the HACCP SAM was designed with respect to two primary considerations. First, the distribution of foodborne illness varies according to age and is more common among the very young and very old (Council for Agricultural Science and Technology 1994). Households with young children and older adults are therefore expected to incur a disproportionately large share of the total expenses due to foodborne illness. Second, the expenditure and savings patterns of households depend on the income level and age composition of the household. This is particularly true with regard to medical expenditure and the Medicaid and Medicare programs. The three household categories for the HACCP SAM are: (1) households headed by persons aged 65 or older, (2) households with heads under age 65 and one or more children under age 18, and (3) childless households with heads under age 65. In the population tabulations incorporated in the 1993 HACCP SAM, households with children account for the largest share of the population (54 percent). Each household category is further divided into households above and below the official poverty level because income affects both the propensity to spend on health care and eligibility for Medicaid. In the population tabulations incorporated in the 1993 HACCP SAM, the poverty rate is slightly higher for members of households with children than for other persons.[4]

For this analysis, we construct a SAM and SAM multiplier model based on a 1993 SAM derived from a Computable General Equilibrium (CGE) model of the U.S. economy developed at the Economic Research Service (ERS), USDA (Hanson et al. forthcoming). The underlying data for the CGE model and our HACCP SAM's are the 1987 benchmark input-output accounts prepared by the Bureau of Economic Analysis (U.S. Department of Commerce 1994).[5] Using the SAM multiplier model, we first examine the economic ramifications of reducing foodborne illness and then examine the economic impact of the costs of implementing HACCP for meat and poultry.

## The Benefits of HACCP -- The Costs of Foodborne Illness

In order to trace the economic impact that reductions in foodborne illness have on the economy, the different types of benefits embedded in the HACCP benefit estimates must be differentiated. Extrapolating from information in Buzby et al. (1996) we estimate that the mean mid-range benefit estimate of $13.32 billion is composed of $5.25 billion due to the reduction in premature deaths; $3.15 billion due to the reduction in work-loss days (productivity costs due to time lost from work because of nonfatal illness); and $4.92 billion due to reductions in the direct medical costs of illness such as expenditures on physician visits, hospital and nursing home care, drugs, and medical tests and procedures.

**Initial Distribution of the Benefits of a Reductions in Foodborne Illness**

The initial distribution of the costs of foodborne illness is established by tracing the incidence and severity of illness in each household category. In order to measure the distribution of illness, we relied on respondent reports of foodborne illness and acute health conditions resembling foodborne illness derived from the National Health Interview Survey (NHIS). Other sources of data on foodborne illness based on medical records underestimate the incidence of illness because most cases are never seen by physicians. In addition, other data sources provide little or no information about the socioeconomic characteristics of the persons who became ill.[6]

The NHIS is a nationally-representative annual survey of the U.S. civilian noninstitutional population that inquires about health conditions in a sample of approximately 49,000 households (Benson and Marano 1994). Respondents are asked to report about the health of other household members as well as their own health during the two weeks preceding the interview, yielding information on approximately 120,000 persons. The NHIS also collects information about family size and composition, income, employment, health insurance coverage, and the impact of illness on daily activities. We pooled the 1992, 1993, and 1994 NHIS annual samples for this analysis in order to obtain more stable estimates of the incidence of acute conditions for our household groupings. The pooled sample includes information on 354,000 persons, representing nearly 14,000 person-years of exposure to the risk of foodborne illness.

The NHIS estimate indicates that there were approximately 13.5 million cases of foodborne illness and other acute conditions potentially due to foodborne pathogens each year in the U.S. civilian noninstitutional population during 1992-94. This estimate is similar to the Buzby et al. (1996) estimate of 12 to 15 million annual cases due to six pathogens from all sources. However, the two estimates are not comparable for three reasons. First, the NHIS counts only those cases severe enough to require at least half a day of restricted activity or a physician visit, whereas the Buzby et al. estimate includes all cases regardless of severity. This difference suggests that the cases identified by the NHIS are likely to be more severe on average than the cases included in the Buzby et al.

estimates. In fact, 35 percent of all cases of foodborne illness and other acute conditions potentially due to foodborne pathogens identified by the NHIS were severe enough to require a visit to a physician, in contrast to only 15 to 27 percent of the cases examined by the Buzby et al. estimates.

The second reason why the NHIS estimate differs from the one calculated by Buzby et al. is that the NHIS respondents' reports of acute health conditions tend to represent symptoms rather than medically diagnosed diseases unless respondents visited a physician who diagnosed the condition. NHIS medical coders classify these reports using a "Short Index" relating symptoms to specific diseases (National Center for Health Statistics 1990). Preliminary analysis of the NHIS data indicated that the coders placed most symptoms potentially due to foodborne pathogens in four general disease categories: "intestinal infections due to other organisms, not elsewhere classified," "food poisoning -- unspecified," "infectious colitis, enteritis, and gastroenteritis," and "infectious diarrhea." Therefore, in our analysis we examine all acute conditions classified in these general categories, as well as those classified in the specific categories corresponding to the six pathogens included in the baseline estimates.[7] As a result, our definition of illness due to the seven pathogens is broader than the definition employed in the Buzby et al. estimates, and undoubtedly includes some illnesses due to other pathogens.

The final reason why the NHIS estimate differs from the one calculated by Buzby et al. is that the NHIS does not cover the institutionalized population. The NHIS estimate consequently omits all cases of foodborne illness occurring among persons in institutions, notably nursing homes, whereas the Buzby et al. estimates include such cases.

Despite these differences, the NHIS is the still the best data source for this study because it provides information about the socioeconomic distribution of foodborne illness unavailable from any other data source. For this study, we assume that the distribution of cases among households revealed by the NHIS is similar to the distribution of cases of foodborne illness due to the four pathogens included in the HACCP benefit estimates. In the absence of more comprehensive data on socioeconomic variations in foodborne illness rates, this seems a reasonable assumption.

The NHIS indicates that the distribution of foodborne illness and other acute conditions potentially due to foodborne pathogens varies by household type (Table 1). The average annual number of cases per 1,000 persons during 1992-94 was highest in households with children (70.0), a result probably due to the higher incidence of foodborne illnesses among young children. In contrast, the annual incidence rate was lowest in households with elderly heads (15.3). The reason for the low incidence rate in this household category is not entirely clear, although one factor may be the exclusion of institutionalized persons from the NHIS sample. Elderly persons in nursing homes are likely to be in poorer health and therefore at greater risk of foodborne illness than the noninstitutionalized elderly, so the exclusion of the institutionalized elderly from the NHIS probably leads to an underestimation of the incidence rate of foodborne illness among the elderly.

The NHIS also indicates that the average annual incidence of foodborne illness and other acute conditions potentially due to foodborne pathogens was slightly higher among the poor (60.1) than the nonpoor (53.0). However, this difference was not significant.

In contrast to the incidence of illness, there was little difference in the proportion of cases seen by physicians by either household type or poverty level. One explanation for this pattern is that there may have been little difference in the degree of severity of illness. Alternatively, the propensity to visit a physician after becoming ill may have varied within the population in a way that masked differences in the severity of illness.

The NHIS estimates provide a detailed picture of the distribution of foodborne illness and other acute conditions potentially due to foodborne pathogens that are severe enough to require physician care. However, the NHIS does not reveal which cases resulted in hospitalization or death. Since hospitalizations and deaths account for a substantial proportion of

### TABLE 1
#### Incidence of Foodborne Illness and Other Acute Conditions Potentially Due to Foodborne Pathogens, 1992-94[a]

| Household Characteristic | Average Annual Number of Conditions Per 1,000 Persons | Percent of Conditions Medically-Attended |
|---|---|---|
| **Household Type** | | |
| With children | 70.0 (3.7)[b] | 35.5 (3.6) |
| Without children | 40.2 (3.7) | 33.6 (6.2) |
| Elderly head | 15.3 (3.3) | 41.5 (16.5) |
| **Income** | | |
| Above poverty | 53.0 (2.7) | 35.1 (3.5) |
| Below poverty | 60.1 (7.2) | 36.4 (8.4) |
| **Health Insurance Coverage** | | |
| Public coverage | 38.1 (4.2) | 44.4 (8.7) |
| Private coverage | 60.6 (3.7) | 33.7 (4.1) |
| Uninsured | 44.0 (7.7) | 27.2 (10.4) |
| **Total** | **52.9 (2.4)** | **35.3 (3.1)** |

[a]Source: 1992-1994 NHIS.
[b]Standard errors shown in parentheses.

the total costs of foodborne illness, assumptions about the distribution of hospitalizations and illness within the population will have a major impact on conclusions about the share of costs borne by different groups.

In order to determine the distribution of hospitalizations and deaths within the population, we assume that the actual risks of hospitalization and death for persons who became sick enough to visit a physician are the same throughout the population. We also assume that these risks are equal to the national-level risks implied by the estimates of physician-attended cases, hospitalizations, and deaths reported by the ERS baseline studies (Buzby et al. 1996). Using these assumptions we allocate the total hospitalizations and deaths reported by Buzby et al. by household category. We distribute the initial benefits arising from reductions in the costs of illness according to this distribution. The first two columns of Table 2 present the distribution of human capital costs of foodborne illness. In keeping with the theoretical basis of the human capital approach, the costs of both work-loss days and premature death are distributed only among households headed by a working-age adult.[8] The costs of direct medical expenses (column 3) are distributed across all household categories.

### TABLE 2
Initial Distribution of the Benefits of a Reduction in Foodborne Illness by Household Type, 25 Year Estimates ($Billions 1993)

| Household | Benefits of Reduction in Premature Death | Benefits of Reduction in Work-Loss Days | Benefits of Reduction in Medical Expenditures | Total Benefits |
|---|---|---|---|---|
| **With Children** | 3.99 (76%) | 2.39 (76%) | 3.54 (72%) | 9.92 (74%) |
| Above poverty | 3.26 | 1.95 | 2.87 | 8.08 |
| Below poverty | 0.73 | 0.44 | 0.67 | 1.84 |
| **Without Children** | 1.26 (24%) | 0.76 (24%) | 1.13 (32%) | 3.15 (24%) |
| Above poverty | 1.12 | 0.67 | 1.01 | 2.80 |
| Below poverty | 0.14 | 0.09 | 0.12 | 0.35 |
| **Elderly** | 0 | 0 | 0.25 (5%) | 0.25 (2%) |
| Above poverty | 0 | 0 | 0.22 | 0.22 |
| Below poverty | 0 | 0 | 0.03 | 0.03 |
| **Total** | 5.25 (100%) | 3.15 (100%) | 4.92 (100%) | 13.32 (100%) |
| Above poverty | 4.38 | 2.62 | 4.10 | 11.10 |
| Below poverty | 0.87 | 0.53 | 0.82 | 2.22 |

Note: Percentages may not total to 100 due to rounding.

The final distribution of benefits depends on households' economic reactions to the initial benefits and households' linkages with the rest of the economy. Next we examine the economic reactions to this initial distribution of the reductions in the costs of foodborne illness.

## The Final Distribution of the Benefits of a Reduction in Foodborne Illness

Direct medical costs and human capital costs have different kinds of impacts on the economy. Medical expenditures have direct and immediate impacts. These expenditures circulate throughout the economy triggering economic activity and growth in some industries and reductions in others. Unlike direct medical costs, human capital costs do not entail economic flows that can be traced from one industry to another. Instead, these costs mark a pure drop in economic activity. In this section we use the multiplier model to trace the impact of medical costs and human capital costs. For both types of costs we attempt to identify those industries and households of the economy that ultimately benefit from reductions in the costs of foodborne illness.

*Economic Impact of Premature Death.* In the first experiment, we use the SAM model to trace the economic ramifications of the benefits of reductions in productivity losses due to premature deaths. In this case, household income is increased by $5.25 billion according to the cost distribution reported in the first column of Table 2. In other words, in keeping with the theoretical underpinnings of the human capital approach, the reduction in premature death resulting from HACCP translates into an increase in national income. However, the initial increase in national income does not incorporate the full impact of the reduction in productivity losses due to premature death because households respond to the initial increase in income by expanding consumption and savings. This expansion triggers increases in economic activity extending far beyond the originally affected households.

The SAM multiplier model traces the impact of the initial increase in household income to its positive effects on consumer demand, industrial output, and factor payments. After the SAM model accounts for the general equilibrium impacts, the original growth in household income due to the reduction in premature death results in a $14.31 billion increase in industrial output, a $6.80 billion increase in factor payments, and a total increase of $10.08 billion in household income. Thus, every dollar of income gained due to reductions in premature death results in an economy-wide income gain of $1.92. These results demonstrate that premature death imposes substantial costs on society as a whole: reductions in premature death lead to an increase in household income nearly double the size of the initial increase.

There are also important differences between the initial and final distribution of the benefits of reductions in premature death by household category. In the final benefit distribution, households with children gain a smaller percentage of benefits than in the initial distribution, while childless households and elderly-headed households gain a higher percentage (Table 3). In fact, although elderly-headed households are not allocated any of the initial benefits of reductions in premature death, they receive 6 percent of the final benefits. These differences arise because, unlike the initial distribution of benefits, the final distribution does not mirror disease incidence, but depends instead on the linkages between households and the economy. The same results appear when households above and below poverty are compared. Poor households realize 17 percent of the initial increase in income due to reductions in premature death, but only 9 percent of the final increase. This result is understandable since lower income households have weaker factor-payment linkages to industrial production than other households. Conversely, upper income households with strong factor-payment linkages

TABLE 3

Final Distribution of the Impact of the Reduction in Foodborne Illness on Household Income by Household Type, 25 Year Estimates ($Billions 1993)

| Household | Benefits of Reduction in Premature Death | Benefits of Reduction in Medical Expenses Paid by Households | Benefits of Reduction in Medical Expenses Paid by Ins/Govt |
|---|---|---|---|
| **With Children** | **5.75 (57%)** | **-0.61 (46%)** | **-0.74 (47%)** |
| Above poverty | 4.95 | -0.60 | -0.68 |
| Below poverty | 0.80 | -0.01 | -0.06 |
| **Without Children** | **3.73 (37%)** | **-.70 (53%)** | **-0.80 (51%)** |
| Above poverty | 3.58 | -.69 | -0.72 |
| Below poverty | 0.15 | -0.01 | -0.08 |
| **Elderly** | **0.60 (6%)** | **-0.02 (1%)** | **-0.03 (2%)** |
| Above poverty | 0.60 | -0.02 | -0.03 |
| Below poverty | -- | 0 | -- |
| **Total** | **10.08 (100%)** | **-1.33 (100%)** | **-1.57 (100%)** |
| Above poverty | 9.16 | -1.31 | -1.40 |
| Below poverty | 0.92 | -0.02 | -0.17 |

Note: Percentages may not total to 100 due to rounding. A "—" denotes a quantity more than zero but less than 0.01 billion.

are directly affected by changes in the returns to labor and capital.

*Economic Impact of Work-Loss Days.* Just as in the case of benefits of reductions in premature death, it is likely that the impact of the initial distribution of the benefits of reduced work-loss days will be diffused and amplified once the general equilibrium effects of these productivity gains are calculated. However, the economic impact of time lost from work due to illness is more complex and difficult to interpret than the impact of premature deaths. Column two of Table 2 shows the initial distribution of benefits of reduced work-loss days based on incidence rates, but clearly, some, if not all, of the gain in productivity due to fewer work-loss days is absorbed by industries.

The economy-wide impact of productivity gains from reductions in time lost from work depends on the ultimate allocation of these benefits between industry and households. This allocation in turn depends on a myriad of industry specific characteristics. The task of modeling the relationship between industry and labor is beyond the scope of this study. However, whether these productivity gains are passed on to households

TABLE 3 (continued)

**Final Distribution of the Impact of the Reduction in Foodborne Illness on Household Income by Household Type, 25 Year Estimates ($Billions 1993)**

| Household | Costs of HACCP | Total Impact on Household Income[a] |
|---|---|---|
| **With Children** | 0.35 (47%) | 5.43 (58%) |
| Above poverty | 0.32 | 4.63 |
| Below poverty | 0.03 | 0.80 |
| **Without Children** | 0.37 (49%) | 3.35 (36%) |
| Above poverty | 0.36 | 3.23 |
| Below poverty | 0.01 | 0.12 |
| **Elderly** | 0.03 (4%) | 0.60 (6%) |
| Above poverty | 0.03 | 0.60 |
| Below poverty | -- | -- |
| **Total** | 0.75 (100%) | 9.38 (100%) |
| Above poverty | 0.71 | 8.51 |
| Below poverty | 0.04 | .87 |

Note: Percentages may not total to 100 due to rounding. A "--" denotes a quantity more than zero but less than 0.01 billion.

[a]For the total calculations, the benefits of a reduction in medical expenses is calculated as the mid-point between columns 3 and 4.

through labor income, capital income or lower prices, the result will likely be an increase in economic activity similar to the one modeled with reductions in premature death.

*Economic Impacts of Direct Medical Expenses.* We next use the SAM multiplier model to trace the economy-wide impact of reductions in medical expenditures due to foodborne illness. In this experiment, $4.92 billion, the cost of medical goods and services, is deducted from the "medical supply" industries of the economy and redistributed to general consumption and savings activities at the household level according to the original distribution of medical expense reductions reported in Column three of Table 2. In this pure expenditure switching experiment, we allocate the additional consumption of the other goods according to household consumption coefficients for each good. The reduction in payments across medical supply sectors is extrapolated from information reported in Buzby et al. (1996). We reduce payments of $4.0 billion from the Medical Services sector for medical care, $0.89 billion from the Chemicals sector for pharmaceuticals, $1.0 million from the General Manufacturing sector for medical equipment, and $30.0 million from the Educational Services sector for rehabilitation and special education.

After the SAM model accounts for the general equilibrium effects of the decrease in medical expenditures, there are net *decreases* of $1.6 billion in industry output, $1.42 billion in factor payments, and $1.34 billion in household income. Thus, every dollar of medical expenses saved as a result of HACCP leads to an economy-wide income *loss* of $0.27. In other words, the consumption of medical goods and services due to illness triggers growth in the economy that outweighs the economic decrease due to reduced household spending on non-medical goods and services. The medical expenditures precipitated by foodborne illness lead to an increase in economic activity. Redirecting these expenditures to other goods and services results in a decrease in economic activity. The explanation for this result is that in general, medical goods and services use a very high proportion of domestically produced inputs.

The seemingly perverse effect of defensive expenditures on national accounts has been well documented by environmental economists (for example, see Lutz 1992). The decrease in income resulting from the reduction in medical expenditures does not necessarily make households worse off. This result points out the fundamental difference between the human capital costs and the medical costs of foodborne illness, and highlights the need for refinements in methodology to account for changes in well being that are not captured by income measures alone.

The decrease in household income triggered by the decrease in medical expenditures is distributed differently than the initial distribution of the reduction in medical expenses. Higher income households with stronger factor-payment links to the economy bear a larger share of the decrease in economic activity than lower income households with weaker links. In fact, households with incomes below the poverty level bear only 1 percent of the decrease in household income triggered by increased medical expenditures, although their members comprise 16 percent of the population.

For many households, direct medical expenses are paid through medical insurance, thus softening the effects outlined above. In order to examine the economic impact of the reduction in medical expenses when they are initially paid through private or public medical insurance, we use additional information from the NHIS to classify households into one of three health insurance categories based on the coverage of individual household members.[9] This classification distinguishes households whose health care costs are wholly or partially subsidized by public programs from households protected by private insurers and households lacking any kind of coverage. Public coverage takes precedence in the classification in order to identify all households receiving public funds. The three household categories are:

1. Households with public coverage: one or more household members has Medicaid, Medicare, or other public health coverage, regardless of whether any members has private coverage.
2. Households with private coverage: at least one household member is covered by a private health plan, and all other members are uninsured.
3. Households without coverage: no household member has either public or private coverage.

Medicare is considered public coverage because most Medicare beneficiaries elect optional Part B coverage, which is subsidized by the federal government. This approach differs from the classification developed by Paulin and Weber (1995), which treats Medicare as private coverage. Military health coverage is treated as private coverage because military dependents and retirees included in the NHIS sample receive coverage as an employment benefit. Single-purpose hospitalization plans covering only hospital charges are also counted as private coverage following Bloom et al. (1997).

The majority of non-elderly households falls into the private insurance category. Sixty-five percent of households with children have private coverage, 23 percent public coverage, and 12 percent no coverage. Seventy-four percent of households without children have private coverage, 11 percent public coverage, and 16 percent no coverage. In contrast, elderly households depend almost exclusively on public health insurance coverage, reflecting the role of Medicare in providing health care for the elderly. Three percent of elderly households have private coverage, 96 percent public coverage, and 1 percent no coverage.[10]

We use the information from the NHIS on the distribution of illness by household insurance category (Table 1) to distribute the $4.92 billion dollars in medical expenditure savings. Households with private coverage account for a much larger share of the total savings ($3.19 billion) than households with public coverage ($1.39 billion) or households without coverage ($0.34 billion). The availability of health insurance changes the linkages examined in the earlier SAM experiment. Most importantly, the fact that nearly one-third of medical expenses are incurred by households with public coverage or no coverage links these savings to tax payers.

We use the SAM multiplier model to trace the impact of reductions in direct medical costs when "third-party payers" (private insurance or the government) pay the bills. The initial drop in medical expenses for publicly-insured and uninsured households is deducted from medical sectors and distributed back to households as "tax cuts." Specifically, the $1.73 billion reduction in the medical expenses of publicly-insured and uninsured households is distributed back to households above poverty. These households increase their consumption and saving accordingly. The initial impact of the reduction in medical costs for privately-insured households is represented by a $3.19 billion decrease in operating costs for the insurance sector. The decrease in costs for the insurance sector is modeled by diverting sector expenditures from the purchase of medical goods and services to general expenditures as indicated by the expenditure coefficients in the SAM.

The final impact of the decrease in medical expenses paid through third parties is again a decrease in economic activity. In fact, the decrease in output is larger when medical expenses are paid by third-party payers ($5.04 billion) than when they are paid out of household income ($1.6 billion). However, this larger decrease in production did not translate into larger decreases in factor payments or household income. When medical expenses are paid by third-party payers, the decrease in factor payments and total household income are $1.68 and $1.57 billion, respectively. In contrast, these decreases are $1.42 and $1.99 billion, respectively, when expenses are paid out of household income. Every dollar of medical expenses paid by third-party payers results in an economy-wide income loss of $0.32, as opposed to a loss of $0.40 when households pay expenses out of pocket.

The final distribution of the decrease in household income resulting from third-party payments of medical expenses differs from the initial distribution of foodborne illness for two reasons. First, medical expenses are paid by insurance companies and taxpayers rather than households, thus diffusing *initial* cost reductions throughout the economy. Second, the decrease in economic activity resulting from lower medical expenditures is distributed back to households through factor payments, thus diffusing the *final* decrease in income throughout the economy. When medical expenses are paid by third-party payers, the link between the initial distribution of illness and the distribution of the economic impacts is broken because both the initial and final impact of foodborne illness are diffused throughout the economy. As a result of the greater diffusion, the final distribution of economic impacts differs from that when expenses are paid out of household income (Table 3).

The final impact of medical expenses on the economy probably falls between the two cases analyzed here: neither households nor third-party payers pay all medical expenses. However, regardless of the exact mix between household payments and insurance and government payments, the SAM multiplier experiments indicate that the ultimate impact of a reduction in medical expenses is a decrease in economic activity.

## The Costs of HACCP

The initial costs of HACCP accrue to meat and poultry slaughterers and processors for increased production costs and to FSIS for supervision costs. Crutchfield et al. (1997) calculate the distribution of costs according to the first two columns of Table 4 (only the mid-point estimates are examined here).

The expenditures entailed with the regulatory activities listed in Table 4 include a wide range of goods and services. On the industry side, the major expenditures are for increased labor. Additional expenditures include document storage, travel to classes, and specimen collection supplies. For FSIS, most of the increased expenditures are also primarily for labor. Table 4 outlines our estimates of the specific expenditures arising with HACCP implementation. These estimates are extrapolated from FSIS's regulatory impact analysis for HACCP (USDA, FSIS 1995, and USDA, FSIS 1996).

Like medical expenditures, the costs of implementing HACCP have direct and immediate impacts on the economy. These expenditures circulate throughout the economy triggering economic activity and growth in some industries and reductions in others.

We model the initial impact of these costs on the economy in two steps. First, we trace the $1.1 billion increase in costs due to HACCP to the industries or factors supplying the goods and services to the meat and poultry slaughterers and processors and to FSIS. We estimate that $66 million of the increased costs associated with HACCP go to paying Medical Services (laboratory labor), $8 million to Chemicals, $54 million to General Manufacturing (laboratory supplies), $4 million to Other Services, $9 million to Transportation and $997 million to Labor. Second, we assume that all of the funding for these cost increases is paid by consumers of beef and poultry. In most cases, it is a reasonable long-run assumption to trace costs back to consumers. In the case of HACCP, this assumption is even more reasonable in light of the low per-pound costs of the regulation. In its analysis of HACCP costs, FSIS estimates that HACCP would cost industry $0.00002 per pound of meat and poultry. This translates into an average expenditure increase for consumers of less than $0.50 a year (Food and Chemical News 1996). We model this increase by raising household consumption expenditure on beef and poultry by $1.1 billion. In order to absorb this increase, we reduce household expenditure on other goods, services, and savings according to the expenditure shares in the SAM.

After the SAM model accounts for general equilibrium effects, the ultimate impact of these costs is a *decrease* in output of $.36 billion, an *increase* in factor payments of $.83 billion, and an *increase* in household income of $.75 billion. Though diverting expenditures from general goods and services and savings results in lower output, the increased use of labor for HACCP implementation counterbalances this decline in output to increase both labor factor payments and household income. Somewhat surprisingly, every dollar spent on HACCP results in an economy-wide income *gain* of $0.66.

TABLE 4
HACCP Costs -- Present Value Estimate of 25 Year Costs
(1993 Dollars)

| Regulatory Component | Cost Estimates[a] (Millions) | Expenditures[b] % of Component |
|---|---|---|
| Sanitation SOPs | $175.12 | Storage (1%) |
|  |  | Labor (99%) |
| Microbial Testing -- | $174.68 | Lab supplies (18%) |
| Generic *E. coli* |  | Lab labor (37%) |
| Testing |  | Labor (45%) |
| Compliance with | $153.29 | Chemicals (5%) |
| *Salmonella* Standards |  | Lab supplies (15%) |
|  |  | Labor (80%) |
| **HACCP Plan** |  |  |
| Plan Development | $55.79 | Labor (97%) |
| Annual Plan Review | $9.03 | Travel (2%) |
| Record-Keeping | $449.03 | Storage (1%) |
| Initial Training | $23.11 |  |
| Recurring Training | $22.47 |  |
| Additional Overtime | $17.84 | Labor (100%) |
| FSIS Costs | $57.62 | Labor (99%) |
|  |  | Lab supplies (1%) |
| Total | $1,1138.00 |  |

[a]Crutchfield et al. (1997) 25 year average cost estimates converted to 1993 dollars.
[b]Extrapolations from (USDA, FSIS 1995, and USDA, FSIS 1996).

The distribution of this increase in household income reflects the labor market ties of the household groupings (Table 3). Households below poverty only enjoy 6 percent of the increase in economy-wide income though their members comprise 16 percent of the population, and elderly households receive only 4 percent of the increase though their members comprise 20 percent of the population.

## Conclusion

The SAM multiplier analysis reveals the ultimate economic impact of the benefits and costs of HACCP and highlights the qualitative differences in the mechanisms by which these amounts impact on the economy. It is shown that the economic impact of human capital costs differs fundamentally from the impact of defensive expenditures. On the benefit side, the SAM experiments indicate that every dollar of income saved by preventing a premature death from foodborne illness results in an economy-wide income *gain* of $1.92, every dollar of household income saved by reduced medical expenses results in an economy-wide income *loss* of $0.27, and every dollar of private and public insurance expenses saved by reduced medical expenses results in an economy-wide income *loss* of $0.32. On the cost side, these experiments indicate that every dollar spent on HACCP results in an economy-wide income *gain* of $0.66. The net economic impact of the costs and benefits of HACCP is an increase in production output of $10.63 billion, an increase in factor payments of $6.08 billion, and an increase in household income of $9.38 billion (1993 dollars).[11] These net benefits would be larger if the benefits of reduced work-loss days were included.

The SAM framework extends the initial cost-benefit analysis to account for the full economic impact on producers and consumers. Such an accounting indicates who ultimately benefits from improved health outcomes and who ultimately pays the costs of food safety regulation. The SAM experiments indicate that the ultimate distribution of the costs and benefits of HACCP differs substantially from the initial distribution. HACCP triggers economic activity in industries supplying HACCP inputs and an increase in the demand for labor at slaughterhouses and process plants. Conversely, the reduction in foodborne illness results in a decrease in economic activity for medical services and supply industries. The ultimate increase in economic activity and economy-wide income is distributed back to households with strong factor linkages with the economy. In our analysis, economic feedback effects and private and public insurance diffuse the benefits of reductions in foodborne illness throughout the economy. Households with children receive 58 percent of the increase in income, households without children receive 35 percent, and elderly households receive 6 percent (Table 3). Poor households receive only 9 percent of the increase although their members compose 16 percent of the population.

The SAM accounting of the final impact of costs and benefits of HACCP provides useful information for policy makers by indicating the direction and magnitude of the economic flows resulting from regulation costs and reductions in foodborne illness. Use of the SAM framework also focuses attention on the difficulty of assessing the economic value of health. The SAM analysis demonstrates the usefulness of the cost-of-illness approach in deciphering the economic distortions caused by health shocks to the economy and the danger of equating changes in income with changes in well being.

**Notes**

[1] Elise H. Golan, Katherine L. Ralston and Stephen J. Vogel are economists and Paul D. Frenzen is a demographer with the Food and Rural Economics Division of the Economic Research Service, USDA. This work was initially supported through a cooperative agreement between the Economic Research Service and the Department of Agricultural and Resource Economics, University of California at Berkeley.

[2] Here, we use the term "HACCP" or "HACCP rule" to denote the whole package of rules promulgated in 1996, including the Standard Operating Procedure requirements and the *E. coli* and *Salmonella* testing requirements.

[3] The Landefeld and Seskin (1982) approach is a slightly modified cost-of-illness approach. They added an individualized element to their human capital calculations by computing earnings net of taxes, including non-labor income, using an individual, rather than a social discount rate, and including a risk aversion factor. Buzby et al. (1996) adjusted the Landefeld and Seskin measures of lifetime after-tax income by averaging across gender, interpolating between age groups, and updating to 1993 dollars.

[4] The income calculations for the poverty classification exclude all in-kind assistance, Earned Income Tax Credits (EITC), Supplemental Security Income (SSI), Aid for Families with Dependent Children (AFDC), and general assistance payments in order to focus on the household's ability to achieve an adequate income without government assistance.

[5] The equation for the multiplier model is $y=Mx$. In addition to intermediate demands, the SAM multiplier matrix $M$ also incorporates factor income flows and household expenditures.

[6] The detailed study of the incidence of foodborne illness by Steahr (1994) does not examine variations by income or employment sector. Other studies of foodborne illness based on mortality data reveal little about socioeconomic variations because only a small proportion of cases result in death, and because death certificates provide little information about socioeconomic characteristics.

[7] The NHIS classifies diseases using the ICD-9 system (Benson and Marano 1994). The specific ICD-9 categories included in our analysis were 003.0, 003.1,

003.2, 003.8, 003.9, 005.0, 005.2, 005.9, 008.0, 008.41, 008.43, 008.8, 009.0, 009.2, 27.0, and 130.0-130.9. The Council for Agricultural Science and Technology (1994) lists 40 known foodborne pathogens. Garthright et al. (1988) discuss some of the issues involved in using NHIS respondent reports of illness to measure the incidence of intestinal infectious diseases.

[8]This distribution rests on the assumption that all labor force participants are aged 18 to 64 and that all members of elderly-headed households are out of the labor force. This is a simplifying assumption. Some persons over age 64 remain in the labor force, and a small proportion of labor force participants aged 18 to 64 were members of elderly-headed households (4 percent in 1992-94). We distributed reductions in productivity losses to the two other household groups (households with children and households without children) on the basis of incidence of foodborne illness. It would have been more faithful to the spirit of the human capital approach if we had conditioned this distribution by labor force participation rates.

[9]The focus on health insurance reduces the size of the NHIS sample available for analysis by approximately one-sixth because survey questions about health insurance coverage were not administered during the first half of 1993.

[10]It is important to note that the three health insurance categories we use capture only some of the differences in sources of payment for health care. Further research will be useful to refine this part of the analysis. Many households with public coverage also have private coverage, notably "medigap" policies for costs not covered by Medicare. Some households with private coverage pay less out-of-pocket for health care than others because they have more comprehensive policies, or because their employers pay a larger share of the premium. Finally, some uninsured households may have better access than others to health care providers who reduce their fees for low income patients, and then shift the unreimbursed cost to public payers (through government subsidies or charitable deductions) or private payers (through higher charges).

[11]We averaged the results of experiments two and three for these calculations.

**References**
Benson, V. and M.A. Marano. 1994. Current Estimates from the National Health Interview Survey, 1993. *Vital and Health Statistics*. National Center for Health Statistics, Series 10, No. 190.
Bloom, B., G. Simpson, R.A. Cohen,, and P.E Parsons. 1997. Access to Health Care, Part 2: Working Age Adults. *Vital and Health Statistics*, National Center for Health Statistics, Series 10, No. 197.
Buzby, J., T. Roberts, C.-T. J. Lin, and J. MacDonald. 1996. Bacterial Foodborne Disease: Medical Costs and Productivity Losses. *Agricultural Economic Report* No. 741, Food and Consumer Economics Division, Economic Research Service, U.S. Department of Agriculture.
Council for Agricultural Science and Technology. 1994. Foodborne Pathogens: Risks and Consequences. *Task Force Report* No. 122, September.

Crutchfield, S., J.C. Buzby, T. Roberts, M. Ollinger, C.-T. J. Lin. 1997. An Economic Assessment of Food Safety Regulations: The New Approach to Meat and Poultry Inspection. *Agricultural Economic Report* Number 755, U.S. Department of Agriculture, Economic Research Service.

Food and Chemical News. 1996. HACCP Costs, Benefits 'Reasonably' Estimated by FSIS, GAO Concludes, But Texas A&M's Knutson Disagrees. March 4.

Garthright, W.E., D.L. Archer, and J.E. Kvemberg. 1988. Estimates of the Incidence and Costs of Intestinal Infectious Diseases in the United States. *Public Health Reports* 103:107-115.

Hanson, K. Forthcoming. A Computable General Equilibrium Framework for Analyzing Welfare Reform. Food and Rural Economics Division, Economic Research Service, U.S. Department of Agriculture.

Jensen, Helen H., Laurian J. Unnevehr and Miguel I. Gomez. 1998. Costs of Improving Food Safety in the Meat Sector. Presented at the Southern Agricultural Economics Association meetings, Little Rock, Arkansas. January 31 - February 3.

Knutson, Ronald D., H. Russell Cross, Gary R. Acuff, Leon H. Russell, John P. Nichols, Larry J. Ringer, Jeff W. Savell, Asa B. Childers, Jr., Suojin Wang. 1995. Reforming Meat and Poultry Inspection: Impacts of Policy Options. Institute for Food Science and Engineering, Agricultural and Food Policy Center, Center for Food Safety, Texas A&M University System.

Landefeld, J.S., and E.P. Seskin. 1982. The Economic Value of Life: Linking Theory to Practice. *American Journal of Public Health* 6:555-566.

Lutz, E., ed. 1992. *Toward Improved Accounting for the Environment.* Washington DC: World Bank.

National Center for Health Statistics. 1990. *Public Use Tape Documentation, Part III - Medical Coding Manual and Short Index.* NCHS: Hyattsville, MD.

Paulin, G.D., and W.D. Weber. 1995. The Effects of Health Insurance on Consumer Spending. *Monthly Labor Review* 118:34-54.

Steahr, T.E. 1994. Foodborne Illness in the United States: Geographic and Demographic Patterns. *International Journal of Environmental Health Research* 4:183-195.

Tolley, George, Donald Kenkel, and Robert Fabian, editors. 1994. *Valuing Health for Policy.* Chicago, IL: University of Chicago Press.

U.S. Department of Agriculture, Food Safety and Inspection Service. 1995. Pathogen Reduction; Hazard Analysis, and Critical Control Points (HACCP) Systems; Proposed Rule. *Federal Register*, Part II, 60 (23): 6774-6889 Feb, 3.

U.S. Department of Agriculture, Food Safety and Inspection Service. 1996. Pathogen Reduction; Hazard Analysis, and Critical Control Points (HACCP) Systems; Final Rule. Supplement-Final regulatory Impact Assessment for Docket No. 93-016F, May, 17.

U.S. Department of Commerce, Bureau of Economic Analysis. 1994. Benchmark Input-Output Accounts for the U.S. Economy, 1987. *Survey of Current Business*, 74(4).

Viscusi, W.K. 1993. The Value of Risks to Life and Health. *Journal of Economic Literature* 31(4): 1912-1946.

Chapter 11

# Market Influences on Sanitation and Process Control Deficiencies in Selected U.S. Slaughter Industries

Michael Ollinger[1]

The production of wholesome red meat and poultry that is free of pathogens, harmful bacteria, rodent hairs, and other contaminants has long been a major concern of the Food Safety Inspection Service (FSIS) of the United States Department of Agriculture (USDA). Traditionally, FSIS has assured red meat and poultry wholesomeness by visually inspecting animals for obvious defects, such as body sores, and monitoring plant sanitation and process controls. Some experts suggest that, in the absence of FSIS monitoring, the chance of including contaminants in output would rise because plants would reduce production and investment costs by relaxing sanitation and process control standards and decreasing investments in new food safety technologies.

Caswell and Mojduszka (1996) and Akerlof (1970) argue that sellers of food products are better informed about quality attributes than consumers, suggesting that firms can sell off-quality products without consumer knowledge. However, other economists, such as Weiss (1964), argue that consumers of low-cost, repeat purchase products, such as meat, can punish suppliers of off-quality products by changing suppliers. For example, economic theory suggests that, if there were one beef products producer, then consumers would punish that firm by switching to other food products if they consumed beef containing contaminants. However, moral hazard (Holmstrom 1979 and 1982) arises when producers of off-quality products cannot be identified. If the beef industry consisted of thousands of equally sized firms producing a homogenous product that became mixed before reaching the consumer, then the identity of any single producer would be obscured. As a consequence, consumers would punish all firms equally when switching to different food products, and the cost to any single beef supplier would be much smaller.

U.S. meat and poultry production consists of neither a large number of equally-sized small plants nor a single large producer. Rather, with a concentration ratio exceeding 70 percent for cattle slaughter and 50 percent in hog slaughter, a share of production by plants with at least 400 employees exceeding 70 percent for the cattle slaughter industry, and over 100 plants of various sizes in both the hog and cattle slaughter industries, there are both very large and very small plants (Ollinger, MacDonald, Nelson, and Handy 1996). Given these differences in plant size, there should be different incentives to avoid sanitation and process control deficiencies.

The purpose of this paper is to examine the effect of market influences on sanitation and process control deficiencies (hereafter, deficiencies). It is hypothesized that moral hazard and the costs of a lost reputation affect deficiencies. Below, we provide an economic framework, discuss the role of FSIS inspectors, provide an empirical model, describe the data and estimation procedure, and present results.

## Economic Framework

Economic theory suggests that meat and poultry firms will supply higher quality meat if it is profitable for them to do so. Quality includes fat content, taste, and product contaminants, including rodent feces, bacteria and pathogens. One objective of firms is, therefore, to maintain a reputation for selling products with fewer contaminants. Loss of this reputation can be costly. As reported by many news services, the owner of several large hamburger production plants was forced to sell its hamburger operations after one of its plants was found to have produced hamburgers with *E. Coli 01757: H7*.

To maintain a positive reputation, a plant could continuously clean its facilities and test each animal for excessive bacteria and pathogens. However, the costs of maintaining such rigorous standards are extremely high and may not be necessary. Holmstrom (1979 and 1982) reminds us that moral hazard is an asymmetry of information among individuals that results from an inability to observe individual actions; and, Barzel (1982) argues that, since the measurement of product attributes is costly, buyers will not be able to learn all quality attributes. It follows that, since plants want to minimize costs, they will invest sufficient resources in product quality to avoid consumer identification as a producer of products containing contaminants.

Meat and poultry contaminants are detected either at the plant by production personnel or FSIS inspectors, or after purchase by a consumer.[2] If contaminants are detected at the plant, meat or poultry is either discarded or reworked. Tracebacks from the consumer to the point of contamination, however, are not as simple. A consumer must first identify the store at which the purchase was made. The store must then identify the processor and the processor must identify either another processor or a slaughter plant.

In the consumer-slaughter plant traceback scenario, one traceback level depends on the previous traceback, suggesting that there is a higher probability of tracing meat to the store than to the processor and to the processor than the slaughter plant. For example, suppose the product label identifies the store at which a meat or poultry purchase was made. If the store purchases meat only from only one processor, then the processor can be identified. The processor, in turn, can identify its meat or poultry source only if it buys meat or poultry from one source.

It may be easier to identify a large plant as a source of contaminated meat than a smaller plant. Suppose there are 1,000 consumers of meat or poultry from plant "A" and only one consumer of meat or poultry from plant "B," then only 0.1 percent of the consumers of plant "A" production need to detect contamination for plant "A" to lose its reputation for selling meat or poultry free of contaminants, while 100 percent of the consumers of plant "B" production need to detect contamination for plant "B" to lose its reputation.

Large plants also have greater capacity to be single-source suppliers. Stores, restaurants, or processors may prefer to lower transaction costs by purchasing meat or poultry from only one source. For many of these buyers, only large plants have the capacity to meet this demand. One secondary impact of a single supplier is that it facilitates tracebacks. If meat or poultry containing contaminants is detected and the meat or poultry is determined to have come from a processor with a single carcass supplier, then the carcass supplier is also known.

Nelson (1970, 1974, and 1978) has argued and Milgrom and Roberts (1986) have shown that firms make long-term investments in advertising in order to earn a reputation that enables them to charge a price premium over competitors. Klein and Leffler (1981) argue that firms adhere to higher quality standards if the expected present value exceeds the expected short run gains from product deception. Long-run investments are necessary to be able to produce high quality products. If these investments have few alternative uses, then asset values drop dramatically with a loss of reputation. Thus, larger firms have a stronger incentive than smaller firms to avoid the perception of selling off-quality meat or poultry.

Alchian and Demsetz (1972) assert that more direct monitoring of employees by management reduces shirking. One supervisor can monitor several workers, but there is a limit to a manager's span of control (Keren and Levhari 1979), which, when reached, may result in greater shirking. Since workers maintain sanitation and process controls, a reduction in management effort that leads to an increase in shirking should result in an increase in sanitation and process control deficiencies.

Libecap (1992) argues that growth in product demand may encourage avoidance of selling meat or poultry with contaminants. If product demand is expected to grow, then expected long-term profits should be higher relative to the short-term gains from selling meat or poultry with contaminants.

## Plant Quality Control and the Role of FSIS Inspectors

By examining carcasses for unhealthy attributes and monitoring plant sanitation and process controls, FSIS inspectors act as quality control inspectors. The preamble to the final HACCP rule illustrates this role, stating that the inspection program (prior to HACCP) relied excessively on the detection and correction of problems rather than the assurance of prevention.

The analogy of FSIS inspectors to quality control agents is even more striking if one considers enforcement action. Prior to 1997, FSIS rarely withdrew inspection from any of the over 6,000 production plants it monitored because of sanitation or process control deficiencies.[3] Rather, FSIS temporarily interrupted affected production until deficiencies were corrected.

In the absence of stringent FSIS enforcement, public interest groups could also have monitored plants if sanitation and process control deficiency rates were publicly available. However, deficiency rates were not publicized. The combination of this inability of the public to monitor deficiency rates and the focus of FSIS on detection and correction makes it reasonable to assume that regulatory penalties have not been the controlling factor in product quality.

A far more powerful force than regulatory stringency encouraging plants to avoid deficiencies is the need to maintain product quality. Both rodent feces and harmful bacteria can adversely affect product shelf life and can strongly discourage consumer repurchases. The major fast food chains, such as McDonalds and Burger King, illustrate this view. All require stricter quality standards than those imposed by FSIS and punish noncompliance by changing suppliers (Ollinger 1996).

## Inspection Effort Per Plant

FSIS inspectors examine plants for five types of sanitation and five types of process control activities. Sanitation activities include pre-operation sanitation of facilities equipment -- both assembled and disassembled, operational sanitation, sanitation of product handling equipment, and contaminated/adulterated product handling. Process control activities are water supply/sewage disposal, facilities sanitation/personal hygiene, pest and rodent control, receipt and control of incoming material, and product handling and preparation. A plant is assigned a deficiency if an FSIS inspector identifies a sanitation or process control violation. FSIS then computes deficiencies as a percent of all sanitation and process control activities.

The number of sanitation and process control activities are the same for all plants. For example, if a plant is found to have one rat hair, then a deficiency is assigned. Since all standards must be met and FSIS code assures reasonable uniformity, deficiencies as a percent of sanitation and process control activities provide a uniform and accurate assessment of plant process quality. However it is far more likely that a large plant will be cited for a deficiency than a small plant. For example, if a large plant has 20 storage rooms and a small plant has one room, then rat hairs must be found in only five percent of the large plant storage space for it to be assigned a deficiency, whereas rat hairs must be found in 100 percent of the small plant space for it to be assigned a deficiency. This large plant bias makes it necessary to adjust the percent deficiencies.

Equation 1 describes adjusted deficiencies:

$$(1) \quad DEFICIENCY_i = \frac{(\% DEF_i * (1 + MORDEF_i))}{EFFORT_i}$$

where DEFICIENCY is the percent adjusted deficiencies; %DEF is the number of reported deficiencies divided by total possible deficiencies; MORDEF is zero for plants with one task per sanitation and process control activity and the number of deficient tasks in excess of one for plants with more than one task per sanitation and process control activity.

To estimate the number of deficient tasks in excess of one, it was assumed that the probability of having more than one deficient task for any one activity followed a Bernoulli process. For this Bernoulli process, the number of tasks monitored by inspectors was assumed to be equal to the number of trials, and the percentage of deficiencies was assumed to be equal to the probability of one task being deficient. It was also assumed that plants with 30 or less employees have one task per sanitation and process control activity and that plants with 30 or more employees have a number of tasks equal to the nearest whole number from the ratio of number of plant employees divided by 30 employees.

FSIS maintains data on the number of on-line, off-line, and dual-role inspectors. On-line inspectors monitor animal carcasses; off-line inspectors examine plant sanitation and process control; and, dual-role inspectors that perform both roles. Some plants have one or more inspectors. Other plants are monitored on a patrol basis by dual-role inspectors in which inspectors may spend only two hours in a plant performing several roles, including monitoring carcasses, examining sanitation and process controls, and performing other activities.

Each FSIS inspector submits one sanitation and process control task sheet per day, giving plants with more than one off-line inspectors more reported sanitation and process control activities than patrol plants and plants with one inspector. Since a percentage of deficiencies per activity is employed, EFFORT is defined as the number of off-line inspectors divided by the number of sanitation and process control sheets submitted, yielding a maximum weight of one.

The precise amount of off-line inspection time at patrol plants is not known. However, employees usually are assigned sanitation and process control tasks as part of their work assignments, and both the amount of off-line inspection time required at nonpatrol plants and the number of employees are known. Thus, the weight for off-line inspection at patrol plants is assumed to equal the ratio of employees at the patrol plant to the average number of employees per off-line inspector at nonpatrol plants. Since patrol inspectors may spend two hours at a plant performing several functions, the minimum amount of inspector effort was assumed to be 30 minutes.

## Empirical Model

In equation 2, the percent of sanitation and process control deficiencies per inspection task (deficiencies) is regressed on several independent variables:

$$DEFICIENCY_i = \beta_0 + \beta_1 MIX_i + \beta_2 LOG(PSIZE)_i + \beta_3 LOG(FCAPITAL)_i$$

(2)
$$+ \beta_4 MGMT_i + \beta_5 GROWTH_i + \beta_6 SHAN_i + \beta_7 REGION\ 3_i$$

$$+ \beta_8 REGION\ 4_i + \beta_9 REGION\ 5_i + \beta_{10} REGION\ 6_i + \varepsilon_{i,t}$$

where DEFICIENCY is adjusted deficiencies; MIX is product mix; PSIZE is pounds of plant output; FCAPITAL is the book value of a firm's building and machinery assets; MGMT is management effort; GROWTH is growth in value of shipments of firm; SHAN is animal inputs as a share of total pounds of inputs; REGION3, 4, 5, and 6 are dummy variables for inspection region of the country in which the plant exists. The log of PSIZE and FCAPITAL is used in order to give less weight to these size variables.

Identification of plants producing finished and semi-finished products, such as boxed or ground beef, in a traceback process is less difficult than the identification of carcass producers. Processors and semi-processors supply meat and poultry to retailers who sell it directly to consumers, making it difficult for the processor to avoid consumer detection of selling meat or poultry with contaminants. Similarly, large plants serve more customers and are often the only supplier to a customer, thereby facilitating identification.

In equation 2, MIX is one minus the percent of carcass outputs for cattle slaughter; processed pork products, such as hams and sausages, as a share of total sales for hog slaughter; and, chicken traypacks as a share of total pounds of output for chicken slaughter and chicken processing. MIX and PSIZE (plant output in pounds of meat) should negatively affect plant deficiencies.

A loss of plant reputation affects an entire firm, not just a plant. Firms with substantial investments in highly specialized meat and poultry production plants and equipment can incur a much greater financial loss from a loss of reputation than do firms with smaller investments, suggesting that firm fixed capital (FCAPITAL) should negatively affect deficiencies. Plants that use greater management effort have more ability to maintain product quality than do plants extending less management effort, suggesting that MGMT should negatively affect deficiencies.

Libecap (1982) argues that high sales growth encourages higher quality output, but growth can also cause adjustment costs, suggesting an uncertain sign for GROWTH. SHAN is a control variable for the liveweight of cattle as a share of all meat inputs for beef slaughter, liveweight of hogs as share of all meat inputs for pork slaughter, liveweight of cattle and hogs as a share of all meat inputs for miscellaneous meat slaughter, and the liveweight of chickens as a share of all poultry inputs for chicken slaughter and chicken processing. REGION3, 4, 5, and 6 are dummy variables for plants located in various FSIS geographic management areas. Data for the management area came from Mike Michelli of FSIS in a personal interview. GROWTH, SHAN and the REGION3, 4, 5, and 6 variables are included as control variables.

## Data and Estimation

**Data**
Data come from an FSIS dataset and the Longitudinal Research Database (LRD) of the Center for Economic Studies at the Census Bureau. The FSIS dataset includes over 6,000 manufacturing and nonmanufacturing establishments inspected by FSIS inspectors in 1992. The LRD data includes over 3,200 manufacturers of red meat or poultry products in 1992. The two datasets were matched by zip code and name and verified by plant output and product type. Using this procedure, 2,579 Census plants were matched with FSIS plants. The unmatched Census plants included manufacturing plants inspected by state inspectors, egg

products establishments (SIC 20159), and plants that could not be matched.[4]

The dataset of 2,579 plants was further reduced by deleting all plants with less than 50 percent of their value of shipments from meat or poultry slaughter or processing. Additionally, the FSIS dataset lacked deficiency data for many slaughter-only operations. After eliminating these plants, the final dataset contained 2,276 plants. From this dataset, only those plants generating at least 50 percent of their revenue from beef (SIC 20111), pork (SIC 20114, 20116, and 20117), miscellaneous meat slaughter (SIC 20110), miscellaneous poultry (SIC 20150), chicken slaughter (SIC 20151), and poultry processing (SIC 20155) were used. From this dataset, plants that lacked essential data were also dropped.

Sanitation and process control deficiency data and inspector time data came from FSIS. Data on product mix, pounds of output, and share of live animal inputs came from both FSIS and Census. The number of employees, value of machinery and buildings, nonproduction labor costs, and total value of shipments came from Census. Since the LRD is, in general, more complete than FSIS data, pounds of output and the number of animals from this dataset were used unless it was either missing or appeared to be reported incorrectly, in which case FSIS data were used.

Besides the data outlined above, FSIS data includes an estimated four digit SIC code, plant location and name, categorical data on production process types, and estimated financial data. Neither the categorical data nor the financial data are of sufficient quality to permit use in this study. The LRD, in addition to the data, described above, has various categories of labor costs, various material inputs and outputs, and numerous other data.

**Estimation**

Sanitation deficiencies as a percent of all cleaning tasks and process control deficiencies as a percent of all process control tasks were added to yield a value for total plant deficiencies as a percent of sanitation and process control tasks. This value of total plant deficiencies is a continuous number ranging from 0 to 1. In practice, no plant was found to have 100 percent of their cleaning or process control tasks as deficient, but several had either no critical sanitation deficiencies, no process control deficiencies, or no deficiencies of any type. As a result, the dataset is bounded from below by zero, but not from above. Table 1 provides mean deficiency rates by four digit SIC codes. Note, the SIC codes are based on estimates provided to FSIS by the Research Triangle Institute.

Equation 2 is estimated with maximum likelihood techniques using the Newton-Raphson algorithm. The estimates are computed from the inverse of the observed information matrix. Since the response variable can be zero, the model, similar to that examined by Tobin (1958), is left censored. It is assumed to take the form of y = max (Xβ + σε), where X is an n by k matrix of covariate values, y is a vector of responses, and ε is a vector of errors with survival function S, cumulative distribution function, and probability density function f. The log likelihood can be written as:

(3) $$L = \Sigma \log (f(w_i/\sigma)) + \Sigma \log (S(w_i))$$

where $w_i = y_i - x_i'\beta/\sigma$ with the first sum over the uncensored values and the second sum over the left censored values.

## Results

Table 1 gives the percent of sanitation and process control deficiencies (not adjusted for off-line inspection time), number of plants, and mean meat and poultry output for 4,847 manufacturing and nonmanufacturing plants inspected by FSIS. Manufacturing plants have SIC codes less than 4000. Poultry slaughter/processing and meat slaughter exceeded the all industry average percent sanitation and process control deficiencies by about 200 and 25 percent, respectively. Canned specialties exceeded the

### TABLE 1
### Mean Cleaning and Process Deficiencies for Various Industries

| SIC | Industry | % Deficiencies Sanitation | Process | Plants | Mean Output (mill. lb.) |
|---|---|---|---|---|---|
| 2011 | Meat Slaughter | 2.19 | 1.01 | 794 | 27.1 |
| 2013 | Sausage/Prepared Meats | 1.59 | 0.67 | 1604 | 10.7 |
| 2015 | Poultry Slaughter/Process | 4.50 | 2.31 | 139 | 97.9 |
| 2032 | Canned Specialties | 2.07 | 0.68 | 52 | 22.3 |
| 2038 | Frozen Specialties | 1.28 | 0.54 | 121 | 13.1 |
| 2099 | Food Preparations | 1.44 | 0.62 | 51 | 3.7 |
| 4222 | Refrig. Warehouse/Storage | 1.16 | 0.49 | 13 | 8.3 |
| 5141 | Groceries | 0.93 | 0.48 | 188 | 2.5 |
| 5142 | Frozen Foods | 1.03 | 0.49 | 108 | 3.0 |
| 5144 | Dist. Poultry Products | 1.30 | 0.69 | 151 | 9.4 |
| 5147 | Dist. Meat Products | 1.32 | 0.69 | 970 | 2.7 |
| 5149 | Groceries/Related Products | 0.60 | 0.31 | 59 | 0.8 |
| 5411 | Grocery Stores | 0.81 | 0.41 | 83 | 2.7 |
| 5421 | Meat / Fish Markets | 0.89 | 0.43 | 204 | 0.6 |
| 5461 | Retail Bakeries | 0.80 | 0.50 | 11 | 0.1 |
| 5499 | Misc. Food Stores | 1.11 | 0.63 | 32 | 0.7 |
| 5812 | Eating Places | 0.91 | 0.46 | 138 | 1.0 |
| 9999 | Nonclassifiable | 0.54 | 0.27 | 12 | 0.9 |
|  | All Industries | 1.57 | 0.73 | 4847 | 15.8 |

Data Source: USDA Food Safety Inspection Service.

all industry average percent sanitation deficiencies by about 25 percent. No other industry had substantially higher and several had less than one half the all industry average deficiencies.

It is far more meaningful to compare deficiencies within industries than across industries because the type of work performed varies substantially. Equally important, deficiencies must be adjusted for off-line inspector time. Since Census disclosure requirements prevent the use of off-line inspector time, percent of deficiencies per pound of output for four plant size classes is presented in Tables 2 and 3 for four plant size classes (by quadrants of output produced) for various manufacturing and nonmanufacturing industries. The tables show that, in all cases, deficiencies per pound of meat or poultry output are much higher for small plants, and that, within each cell, deficiencies per pound drop dramatically

TABLE 2

Mean Cleaning and Process Deficiencies Per Million Pounds of Output for Various Manufacturing Industries by Plant Size

| SIC | Industry | Size | % Deficiencies | | Plants | Mean Output (mill. lb.) |
|---|---|---|---|---|---|---|
| | | | Sanitation | Process | | |
| 2011 | Meat Slaughter | Very Small | 46.0 | 25.0 | 184 | 0.06 |
| | | Small | 6.0 | 3.0 | 194 | 0.28 |
| | | Medium | 2.0 | 0.90 | 210 | 1.43 |
| | | Large | 0.20 | 0.10 | 206 | 106.76 |
| 2013 | Sausage/ Prepared Meats | Very Small | 177.0 | 81.0 | 401 | 0.08 |
| | | Small | 2.0 | 1.0 | 400 | 0.64 |
| | | Medium | 0.70 | 0.30 | 402 | 3.04 |
| | | Large | 0.10 | 0.050 | 401 | 39.18 |
| 2015 | Poultry Slaughter/ Process | Very Small | 39.0 | 23.0 | 38 | 3.57 |
| | | Small | 0.10 | 0.06 | 45 | 49.3 |
| | | Medium | 0.05 | 0.02 | 35 | 119.80 |
| | | Large | 0.03 | 0.01 | 21 | 217.60 |
| 2032 | Canned Specialties | Very Small | 18.0 | 13.0 | 13 | 0.09 |
| | | Small | 4.0 | 1.0 | 13 | 0.91 |
| | | Medium | 0.70 | 0.03 | 13 | 5.72 |
| | | Large | 0.10 | 0.03 | 13 | 83.9 |
| 2038 | Frozen Specialties | Very Small | 23.0 | 12.0 | 30 | 0.05 |
| | | Small | 4.0 | 2.0 | 30 | 0.48 |
| | | Medium | 4.0 | 0.20 | 31 | 3.57 |
| | | Large | 1.0 | 0.02 | 30 | 48.96 |
| 2099 | Food Preparations | Very Small | 48.0 | 37.0 | 12 | 0.04 |
| | | Small | 14.0 | 6.0 | 13 | 0.14 |
| | | Medium | 2.0 | 1.0 | 13 | 0.77 |
| | | Large | 0.2 | 0.01 | 13 | 13.50 |

Data Source: USDA Food Safety Inspection Service.

as plant size rises.

Table 4 contains deficiencies per animal slaughtered for three size classifications (by number of animals slaughtered) for various slaughter plants. Similar to Tables 2 and 3, it shows that deficiencies per animal are much higher for small plants.

The log likelihood results for the estimation of equation 2 are shown in Table 5 for six meat and poultry industries. Product mix (MIX) is negative in all cases and significant in four. Plant size (PSIZE) is negative in five of the six cases and negative and significant in two cases. Recall that MIX -- share of output from further processed products -- and PSIZE -- pounds of meat output -- are linked to the ease with which a supplier of meat or

### TABLE 3
### Mean Cleaning and Process Deficiencies Per Million Pounds of Output for Various Nonmanufacturing Industries by Size

| SIC | Industry | Size | % Deficiencies | | Plants | Mean Output (mill. lb.) |
|---|---|---|---|---|---|---|
| | | | Sanitation | Process | | |
| 5141/ 5149/ 5411 | Groceries/ Groceries and Related Products/ Grocery Stores | Very Small | 650.0 | 302.0 | 82 | 0.01 |
| | | Small | 10.0 | 5.0 | 83 | 0.09 |
| | | Medium | 2.0 | 1.0 | 83 | 0.38 |
| | | Large | 0.70 | 0.30 | 82 | 8.42 |
| 5142 | Frozen Foods | Very Small | 17.0 | 11.0 | 27 | 0.05 |
| | | Small | 3.50 | 2.0 | 27 | 0.30 |
| | | Medium | 1.30 | 0.60 | 27 | 0.95 |
| | | Large | 0.30 | 0.10 | 27 | 10.72 |
| 5144 | Dist. Poultry Products | Very Small | 12.1 | 77.0 | 37 | 0.14 |
| | | Small | 2.0 | 0.90 | 38 | 0.57 |
| | | Medium | 0.90 | 0.50 | 38 | 1.55 |
| | | Large | 0.20 | 0.10 | 38 | 35.19 |
| 5147 | Dist. Meat Products | Very Small | 97.0 | 56.0 | 242 | 0.07 |
| | | Small | 5.0 | 3.0 | 243 | 0.29 |
| | | Medium | 2.0 | 0.90 | 243 | 0.89 |
| | | Large | 0.40 | 0.20 | 242 | 9.09 |
| 5421 | Meat/Fish Markets | Very Small | 45.0 | 21.0 | 51 | 0.02 |
| | | Small | 9.0 | 4.0 | 51 | 0.11 |
| | | Medium | 4.0 | 2.0 | 51 | 0.28 |
| | | Large | 1.0 | 0.50 | 51 | 2.13 |
| 5812 | Eating Places | Very Small | 152.0 | 93.0 | 34 | 0.01 |
| | | Small | 18.0 | 10.0 | 35 | 0.04 |
| | | Medium | 7.0 | 4.0 | 35 | 0.16 |
| | | Large | 2.0 | 0.70 | 34 | 3.70 |

Data Source: USDA Food Safety Inspection Service.

poultry with contaminants can be identified. The negative parameter suggests that more easily identifiable plants incur fewer deficiencies. This result is consistent with Holmstrom (1979 and 1982) and other economists who assert that moral hazard arises when complete information about the actions of others is not available.

Fixed firm capital investment (FCAPITAL) is negative in all cases and significant in four cases, suggesting that greater fixed investment encourages firms to reduce sanitation and process control deficiencies. In agreement with Klein and Leffler (1981), this negative sign suggests that firms reduce deficiencies because greater fixed investment exposes them to greater financial setbacks in the event of a loss of reputation arising from the sale of meat or poultry with contaminants.

Management effort (MGMT) is not significant. The control variables --

TABLE 4

Percentage of Total Deficiencies Per Animal for Various Animal Species by Plant Size[a]

| Animal | Size[b] | % Total Deficiencies | Number Plants | Mean Animals |
|---|---|---|---|---|
| Cattle | Small (0-10) | 0.004 | 749 | 900 |
| | Medium (10-50) | 0.0001 | 69 | 24,300 |
| | Large (over 50) | 0.00004 | 86 | 332,000 |
| | Average | 0.003 | 904 | 34,200 |
| Hogs | Small (0-50) | 0.004 | 751 | 3,300 |
| | Medium (50-200) | 0.00005 | 28 | 85,500 |
| | Large (over 200) | 0.000008 | 44 | 1,720,000 |
| | Average | 0.004 | 823 | 97,900 |
| Sheep/Goats | Small (0-50) | 0.0026 | 15 | 1,100 |
| | Medium (over 50) | 0.00002 | 650 | 306,500 |
| | Average | 0.0026 | 665 | 8,000 |
| Chicken | Small (0-10) | 0.52 | 60 | 2,400,000 |
| | Medium (10-50) | 0.0003 | 134 | 30,400,000 |
| | Large (over 50) | 0.00013 | 34 | 68,000,000 |
| | Average | 0.13 | 228 | 28,700,000 |
| Turkey | Small (0-1) | 0.065 | 40 | 82,800 |
| | Medium (1-5) | 0.002 | 19 | 3,162,000 |
| | Large (over 5) | 0.005 | 25 | 8,317,000 |
| | Average | 0.026 | 84 | 3,200,000 |

Data Source: USDA Food Safety Inspection Service.

[a]Total deficiencies include cleaning and process deficiencies. Total deficiencies are weighted by the number of animals in order to adjust for a natural bias for larger plants to incur more deficiencies. Bias occurs because the number of deficiencies is constant for all plants, regardless of size, and large plants have a greater processing area in which a violation can be incurred.

[b]Number in parentheses refers to range of number of animals in size category. In thousands for cattle, hog and sheep and goats and in millions for chickens and turkeys.

plant sales growth (GROWTH) and the share of animal inputs (SHAN) -- have mixed results.

## Conclusion

Using proxies linked to incentives for behavior with moral hazard and the costs of a lost reputation, this paper examines the impact of plant characteristics on sanitation and process control deficiencies. Results suggest that the share of output from further processed products and plant and firm size discourages sanitation and process control deficiencies. These results are consistent with Holmstrom (1979 and 1982) in that plants that are more easily identified by consumers have fewer sanitation and process control deficiencies. Results are also consistent with Klein and Leffler (1981) because firms with a greater financial investment in the production of meat or poultry products and, thus, much more to lose for failing to maintain a positive reputation for selling high-quality products were found to have fewer sanitation and process control deficiencies.

TABLE 5

Likelihood of Meat or Poultry Plants Having More Plant Sanitation and Process Control Deficiencies[a]

| Variable | Beef Slaughter | Pork Slaughter | Other Meat Slaughter | Chicken Slaughter | Other Poultry Slaughter | Poultry Process |
|---|---|---|---|---|---|---|
| Intercept | -4.28*** | -4.70*** | -3.60*** | -3.24*** | -3.92*** | -4.65*** |
|  | (0.57) | (0.47) | (0.26) | (1.06) | (1.51) | (0.74) |
| MIX | -0.73*** | -0.76* | -0.30* | -0.70 | -0.48 | -0.56** |
|  | (0.06) | (0.42) | (0.19) | (1.92) | (1.16) | (0.22) |
| PSIZE | -0.21** | -0.15** | -0.01 | -0.18 | -0.09 | -0.26** |
|  | (0.10) | (0.07) | (0.05) | (0.13) | (0.14) | (0.12) |
| FCAPITAL | -0.05 | -0.11** | -0.03 | -0.23*** | -0.12* | -0.04 |
|  | (0.50) | (0.05) | (0.03) | (0.001) | (0.07) | (0.05) |
| MGMT | 0.03 | 0.04 | -0.001 | -0.175 | -0.001 | -0.40 |
|  | (0.45) | (0.10) | (0.01) | (0.22) | (0.002) | (0.67) |
| GROWTH | 0.01 | 0.17** | 0.01 | -0.15 | -0.03 | -0.05 |
|  | (0.03) | (0.08) | (0.04) | (0.10) | (0.06) | (0.05) |
| SHAN | -0.22 | -0.03 | 0.03 | 0.07 | 0.05 | 0.04 |
|  | (0.26) | (0.23) | (0.72) | (0.05) | (0.12) | (0.03) |
| Log Likelihood | -203.3 | -161.1 | -837.8 | -94.9 | -133.6 | -100.4 |
| Obs. | 129 | 109 | 513 | 77 | 83 | 86 |

Standard errors are in parentheses.

*Significant at 90 percent level; **significant at 95 percent level; ***significant at 99 percent level.

[a]The dummy variables for FSIS inspection region are suppressed because of disclosure restrictions.

Results have implications for food safety regulation. Private market influences do encourage adherence to better food safety management practices. In terms of public policy, these results suggest that improvements in being able to trace food to the supplier could improve food safety. More generally, since additional publicly available information can encourage actions to reduce contaminants, product labeling and making pathogen count and deficiency rating data publicly available could encourage better food safety management practices.

## Notes

[1] The author is an economist at the United States Department of Agriculture, Economic Research Service, 1800 M Street, NW, Washington, D.C., 20036. The analysis was conducted as a Research Associate at the Center for Economic Studies, U.S. Bureau of the Census. Any findings, opinions, or conclusions expressed here are those of the authors and do not necessarily reflect the views of either the Census Bureau or the U.S. Department of Agriculture. The author gratefully acknowledges the assistance of Ashu Paul for his diligent data work and the help and assistance given by the Center for Economic Studies in their research effort. Special thanks go to Arnie Rezneck.

[2] Processors also can detect obvious contamination, but, since they do not eat the meat or poultry, they have less knowledge than consumers about product quality.

[3] Since regulatory requirements mandate that plants must have inspection in order to produce meat or poultry, the withdrawal of inspection effectively closes a plant.

[4] State inspection agencies, under the oversight of FSIS, inspect meat and poultry in some states. In 1992 there were approximately 3,000 state-inspected establishments, including both manufacturing and nonmanufacturing plants.

## References

Alchian A., and H. Demsetz. 1972. "Production, Information Costs, and Economic Organization." *American Economic Review* 62 (December): 777-95.

Akerlof, G.A. 1970. "The Market for 'Lemons': Quality Uncertainty and the Market Mechanism."
*Quarterly Journal of Economics.* 84 (November): 488-500.

Barzel, Yoram. 1982. "Measurement Cost and the Organization of Markets." *Journal of Law and Economics.* 25(April): 27-48.

Bureau of the Census. 1992. *Longitudinal Research Data Base.* Washington, D.C.: Bureau of the Census.

Caswell, Julie A. and Eliza Mojduszka. 1996. "Using Informational Labeling to Influence the Market for Quality in Food Products." *American Journal of Agricultural Economics* 78 (December): 1248-53.

Food Safety Inspection Service. 1995. *RTI Dataset.* Washington, D.C.: Food Safety Inspection Service.

Food Safety Inspection Service. 1995. *Deficiency Dataset.* Washington, D.C.: Food Safety Inspection Service.

Holmstrom, Bengt. 1979. "Moral Hazard and Observability." *Bell Journal of Economics* 10(1): 74-91.

Holmstrom, Bengt. 1982. "Moral Hazard in Teams." *Bell Journal of Economics* 13 (1):324-40.

Keren, M., and D. Levhari. 1979. "The optimum span of control in a pure hierarchy."
*Management Science* 25 (November): 1162-72.

Klein, Benjamin and Keith b. Leffler. 1981. "The Role of Market Forces in Assuring Contractual Performance." *Journal of Political Economy* 89 (4): 615-1.

Libecap, Gary D. 1992. "The Rise of the Chicago Packers and the Origins of Meat Inspection and Antitrust." *Economic Inquiry* 30 (April): 242-62.

Milgrom, Paul and John Roberts. "Price and Advertising Signals of Product Quality." *Journal of Political Economy* 96 (4): 796-821.

Nelson, Phillip. 1970. "Information and Consumer Behavior." *Journal of Political Economy.* 78 (2): 311-29.

Nelson, Phillip. 1974. "Advertising as Information." *Journal of Political Economy.* 81 (4): 729-54.

Nelson, Phillip. 1978. "Advertising as Information Once More." In *Issues in Advertising : The Economics of Persuasion*, ed. David G. Tuerck. Washington: American Enterprise Institute.

Ollinger, Michael. 1997. "Implementing HACCP in U.S. Meat Plants: Role of Incentives in Changing the Behavior of Firms, Managers, and Employees." Speech given at the 1996 American Agricultural Economic Association meetings in San Antonio, Texas.

Ollinger, Michael, Jim MacDonald, Ken Nelson, and Charles Handy. 1997. "Structural Change in the U.S. Meat and Poultry Industries." In *Strategy and Policy in the Food System: Emerging Issues*, eds.., Julie Caswell and Ron Cotterill, 3-22. Storrs CT: Food Marketing Policy Center.

Tobin, James. 1958. "Estimation of Relationships for Limited Dependent Variables." *Econometrica* 26 (1): 24-36.

Weiss, Roger W. 1964. "The Case For Federal Meat Inspection Examined." *Journal of Law and Economics* 7 (October): 107-20.

Chapter 12

# The Cost of an Outbreak in the Fresh Strawberry Market

Thomas W. Worth[1]

## Introduction

Foodborne illnesses caused by produce is a growing problem. According to the Food and Drug Administration (FDA), the percentage of foodborne illnesses attributed to produce rose from 2 percent in the 1973-87 period to 8 percent in the 1988-91 period. This is likely due to increased consumption of fresh and therefore uncooked produce. Another problem is the evolution of more virulent varieties of microbial pathogens such as the E. Coli O157:H7 strain that appeared in the early 1980s.

The problem, though, should not be overstated. Most foodborne disease outbreaks occur from meats and seafoods. Only a small proportion of outbreaks are related to fresh produce. According to the Center for Disease Control and Prevention (CDC), 21 out of 3,277 reported outbreaks were linked to fresh produce. Of the 21 outbreaks linked to fresh produce, ten were due to improper food preparation. One of the most common sources of contamination is preparing fresh produce on the same surfaces used for meat.

Consumers judge fresh produce according to visual characteristics, a poor indicator for the presence of microbial pathogens. Currently, the most common method of detection occurs when a consumer becomes ill after eating contaminated food. Besides the serious costs to human health, this method of microbial detection is imperfect in that most foodborne illnesses are mistaken for other illnesses and are never reported.

Once detected, tracing the source of the contamination is a difficult task. The CDC estimates that less than 1 percent of foodborne illnesses are both reported and traced back to the source of contamination.[2] Another complication related to traceback is the growing presence of imports. In 1997, 38 percent of fruit and 12 percent of vegetable consumption in the U.S. was supplied by imports. When an outbreak occurs the source may be any one of several exporting countries, domestic producers, or other handlers along the distribution channels.

According to a recent report by the General Accounting Office (1998), the increase in imports are overwhelming the limited resources allocated to foodborne illness detection. In 1992, the FDA, the agency responsible for inspecting imports of produce, tested 8 percent of the fresh produce shipments entering the United States. In 1997, the rate fell to 1.7 percent. In contrast, the Food Safety and Inspection Service (FSIS), an agency in the Department of Agriculture responsible for inspecting meat and poultry products, inspected 20 percent of meat and poultry imports and visited 336 foreign processing plants to check safety practices.

In response to the growth of foodborne illness outbreaks, both the U.S. government and domestic producer associations are developing guidelines for the production and handling of fresh fruits and vegetables. The President has proposed legislation to ensure the safety of imported and domestic fruits and vegetables throughout the production and distribution chain, including the development of guidelines for good agricultural practices and good manufacturing practices. To prevent disease outbreaks from imported produce, the President has proposed increased monitoring of food safety programs abroad and more frequent food inspections at the border. The legislation authorizes the FDA to block fruit and vegetable imports from any foreign country or facility that does not meet U.S. food safety requirements, a comparable responsibility that FSIS has for meat and poultry imports.

One large difference between the food safety guidelines for produce versus meat, poultry, and seafood standards is that the guidelines are voluntary. The main incentive for the produce industry to adopt food safety guidelines is to maintain and further the healthful image of the produce sector. If an outbreak occurs, the reputation of the produce sector may be diminished and consumer demand for produce may decline. The appearance of E. Coli O157:H7 in apple juice produced by Odwalla lead to a drop in demand and steep financial losses for the company. Odwalla responded by launching a new quality improvement program and by pasteurizing their juice products. In this example, the decrease in demand was sufficiently large to motivate Odwalla to voluntarily raise its quality. Other producers started pasteurizing their juice as well.[3]

News reports of hepatitis-tainted strawberries reduced sales of California strawberries by an estimated $20 to $40 million in 1997, even though the tainted strawberries came from Mexico (although the actual source of the contamination was never found).[4] Testing is made difficult because strawberries have a short shelf-life and strawberries often do not bear company brand labels. Since an individual producer is often difficult to identify as a source of a microbial pathogen, the industry, through its associations, has developed various food safety guidelines. For example, the International Fresh-Cut Produce Association has published a lengthy document of voluntary recommended guidelines on sanitary procedures for the fresh-cut industry.

A shortcoming of voluntary guidelines is that some firms may avoid improving food safety precautions because of the anonymity of the marketplace, that is, tracing back to find the source of a microbial pathogen may be very difficult. Fresh fruits and vegetables generally do not have labels indicating brand names. Since consumers cannot distinguish brand names, they may implicitly assign an average quality to each good. Thus, all firms in the industry share in developing and maintaining reputation. In more concentrated industries, where there may be less reliance on many domestic and foreign sources of supply, each firm would have a large stake in maintaining and improving the industry's reputation, that is, there is less of a free-rider problem. The willingness of firms to voluntarily bear the expense of food safety improvements also depends on how much consumers value quality. If the price premium for higher quality is small, then firms will not be willing to undertake expensive measures.

In the first section of this paper a conceptual framework is developed for how consumers respond to a change in quality and how the response affects price. For an indication of how much additional cost firms may be willing to absorb to improve the quality of their output, I estimate the demand response to foodborne disease outbreaks in the fresh strawberries market. Other costs which firms may consider in determining their choice of the level of food safety include: expected costs associated with tort liability, administrative fines, potential future (export) supply restrictions, costs of food safety testing when violations occur, and concern for stricter government regulations. While I do not consider these other factors, they may be less relevant when it is difficult to trace back the source of an outbreak to an individual firm.

## The Modeling Framework

This paper adapts previous models on reputation and quality to a situation where consumers do not observe the product quality until after purchase and there is no brand name. In his famous study, Akerlof (1970) demonstrates how asymmetric information between buyers and sellers on the quality of a good in a one-shot game can lead to adverse selection and market failure. One solution to the problem of adverse selection is a minimum quality standard (Leland 1979). This increases the average quality of goods on the market and can increase overall welfare. In a repeated game, firms may have an incentive to provide higher quality due to reputation. Shapiro (1983) shows that firms in a perfectly competitive industry can earn positive profits (a "quality premium") through consistently providing high-quality goods. The quality premium depends on consumers knowing which firm produced each good. I modify the demand structure in Shapiro (1983) to consider the case where consumers cannot identify the brand name.

## The Model

I will first examine consumer behavior. Consumers are heterogeneous in their willingness to pay for quality, denoted by $\theta$. The utility gained from purchasing the product depends on their quality preference and the industry's reputation for providing quality, or $R_t$. If the willingness to pay for quality plus the industry's reputation for providing quality is greater than the price, the agent buys one unit of the good.[5] The utility function for consumer $i$ at time $t$ is of the form:

(1)
$$U_{i,t} = \begin{cases} \theta_i + R_t - p_t & \text{If Agent Buys} \\ 0 & \text{Otherwise} \end{cases}$$

The willingness to pay for quality follows a frequency distribution $f(\theta)$. The cumulative density function $F(\theta)$ is the fraction of agents with a quality preference of $\theta$ or lower and falls within the bounds of $F(0)=0$ and $F(B)=1$. Agent $i$ will buy only if $\theta_i R_t - p_t \geq 0$. Thus the values of $\theta$ where the agent buys is:

(2)
$$\theta_i \geq p_t - R_t.$$

The critical value, $\hat{\theta}$, is equal to $p_t - R_t$. Assuming that there are $N$ agents leads to a demand function of the form:

(3)
$$D_t = N[1 - F(\hat{\theta})],$$
$$D_t = N[1 - F(p_t - R_t)].$$

Total demand at time $t$, or $D_t$, equals the number of agents times the proportion of agents willing to buy given a particular price and level of industry reputation. As for supply, there are $M$ firms producing one unit each. Thus the supply is fixed at $M$ units and $D_t = M$. This simplification allows the model to focus on the effect of reputation on demand. This may also be justified by the fact that, at least in the short-run, agricultural commodities are inelastically supplied. Applying this to equation (3) and solving for $p_t$ yields:

(4)
$$p_t = R_t + F^{-1}(1 - M/N).$$

The price is an increasing function in reputation ($R_t$). The price is also an increasing function in the amount of demand ($N$) and a decreasing function in market supply ($M$).

Consumers, being unable to observe quality before purchase and not knowing which firm produced the good, calculate an expected quality, or reputation, based on the quality of goods offered in previous time periods. This is reasonable since supply is fixed, the same set of firms are supplying the goods each time period. Quality in this situation is the presence of microbial contamination. Other quality characteristics that are observable, such as size or coloration, are already factored into the price. Consumers calculate reputation according to a quality function:

(5)
$$R_t = Q(q_{t-i}^m) \quad \text{where } i = \{1,2,...\} \text{ and } m = \{1,2,...,M\}.$$

The reputation of the industry at time $t$ depends on the quality of the output of the individual firms that consumers observed in earlier time periods ($Q(q^m_{t\text{-}ii})$). If consumers were able to distinguish between the output from various firms, the above function would become a vector of functions -- one reputation function per firm. Since firms would then bear directly and completely the costs of a reduction in reputation from lower quality output, they would have an incentive to provide a higher level of quality.[6]

Substituting the reputation function into the inverse demand function (Equation (4)) leads to:

(6)
$$p_t = Q(q_{t-i}^m) + F^{-1}(1 - M/N)$$
$$\text{where } i = \{1,2,...\} \quad \text{and } m = \{1,2,...,M\}.$$

This is the inverse demand function that used for the estimation in the next section of the paper.

## Demand Response to a Change in Quality

According to the model, the factors influencing price are quantity supplied, amount of demand, and industry reputation. The extent to which prices react to quality (via reputation) gives an indication of the industry's willingness to undertake the costs of improving quality, such as developing a HACCP plan or instituting GAPs. The price response to a change in reputation is measured in this section by examining price changes following news of an outbreak in the fresh strawberry and lettuce markets. I then estimate how much the industry, as a whole, loses per outbreak through lost sales.

Microbial contamination is not a broad problem for fresh produce. Only a few fruits and vegetables account for all of the foodborne disease outbreaks recorded by the CDC.[7] Table 1 shows which fruits and vegetables have been implicated in disease outbreaks in the last ten years. These commodities all grow close to the ground and either have a shape that easily catches debris (lettuce, alfalfa sprouts, basil, and scallions) or have a rough surface that is difficult to clean (raspberries, strawberries, and carrots). The source of the contamination, if known, is about evenly split between the U.S. and the rest of the world. A recent study by Zepp et al. (1998) finds no evidence that the risk of microbial contamination is greater with domestic versus imported fresh produce.

Although there are not many outbreaks attributable to fresh produce, the rate of outbreaks appears to be growing. Seven of the 18 outbreaks

TABLE 1
Outbreaks, Source, and Dates for Fresh Produce[a]

| Commodity | Disease | Source | Year |
|---|---|---|---|
| Alfalfa Sprouts | Salmonella | Unknown | 1995 |
| | Salmonella | Netherlands | 1997 |
| Cantaloups | Salmonella | Latin America | 1989 |
| | Salmonella | Latin America | 1990 |
| | Salmonella | Latin America | 1991 |
| Carrots | E. Coli | Unknown | 1993 |
| Fresh Basil | Cyclospora | Uknown | 1997 |
| Lettuce | Hepatitis A | Latin America | 1988 |
| | E. Coli | United States | 1995 |
| | E. Coli | United States | 1996 |
| Raspberries | Cyclospora | Latin America | 1996 |
| | Cyclospora | Latin America | 1997 |
| Scallions | Shigella | Latin America | 1994 |
| Strawberries | Hepatitis A | United States | 1990 |
| | Hepatitis A | Latin America | 1997 |
| Tomatoes | Salmonella | United States | 1990 |
| | Salmonella | United States | 1993 |

[a]Source: CDC and Tauxe (1997).

listed occurred in the last three years.

Since the presence of microbial contamination is not detectable by the consumer before purchase, fresh produce is an experience good in this respect. According to Tauxe (1997) most of the outbreaks listed in Table 1 have fewer than 100 recorded cases distributed accross three or fewer states. If consumers relied only on personal experience to determine the quality of a commodity, the effect of a disease outbreak would be minimal. Instead we assume that consumers rely not just on their own experience, but also the experiences of others as reported in the media.

The paper focuses on strawberries because they were involved in disease outbreaks that received widespread media attention and also because price and shipment data is available for these commodities. The disease outbreaks with carrots, fresh basil, cantaloupes, lettuce, and tomatoes did not affect many people and did not receive widespread media attention. Disease outbreaks in alfalfa sprouts and raspberries did receive significant media attention but there is no reliable shipment and price data available for these commodities. The requirement for significant media attention is important because of the use of national data. Small outbreaks that get only localized publicity will not have a measurable effect on national price and shipment statistics.

**Empirical Model**

I use a version of Muth's (1964) equilibrium displacement model developed by Richards and Patterson (1998). Richards and Patterson use this model to estimate how negative and positive news stories affect strawberry grower prices. They assume a linear relationship between number of news stories and price. The model in this section does not have that constraint. The model will also allow for lagged effects and will use retail price data as well as the producer price data used before. It is not clear how news of an outbreak will affect retail prices as compared to the effect on grower prices.

An indicator variable for positive news stories is also omitted. Smith et al. (1988), in examining the effect of news stories on the demand for milk in Hawaii, compare two specifications of their empirical model -- one that includes positive and negative news stories and one that includes only negative news stories. A $J$-test revealed the former to be mis-specified when compared to the later. Richards and Patterson find positive news stories to have only a small and statistically insignificant effect on strawberry prices. It seems that consumers give more weight to negative news. They may not view positive news as credible.

**Data**

Price data for strawberries and lettuce (as well as other fruits and vegetables) and consumer price indexes are provided by the Bureau of

Labor Statistics of the U.S. Department of Labor. Shipment and import data come from the Agricultural Marketing Service of the U.S. Department of Agriculture. Measures of personal disposable income are from the Bureau of Economic Analysis at the U.S. Department of Commerce. Since the price and shipment data is only sporadically available before 1989, the time period that will be tested is January 1989 to April 1998.

The determination of whether an outbreak received widespread media attention was based on a Lexus/Nexus search of the two major news wires (Associated Press and United Press International) and three major daily newspapers (New York Time, Los Angeles Time, and the Washington Post). Articles from these sources are often reprinted in smaller local papers. A search of major television network news stories was also done through Vanderbilt University's Network News Archive. In the case of strawberries, there was a disease outbreak in April 1997 that generated 78 articles and 11 television network news stories. In June 1996, an outbreak of cylcospora was blamed on strawberries. Several months later it was determined that raspberries were the source. This "non-outbreak" generated 20 articles and five network news stories. Since we assume that consumers rate the quality of output based on media reports, we include the cyclospora scare in the estimation.

## The Estimation

As specified in Equation (6) of the model in the previous section, price is a function of reputation ($Q_t(q^m_{t-j})$), as well as demand and supply characteristics ($F^{-1}(1-M/N)$). This yields an estimation equation of:

(7)
$$p_t = \alpha + \beta N_t + \delta m_t + \gamma Q_t + \varepsilon_t.$$

Two separate estimations use either retail price or the grower price, both in dollars per pound for month $t$, as the dependent variable. The intercept term $\alpha$ is modified by a set of dummy variables for each year and for each month. The matrix $N_t$ contains a set of demand parameters consisting of per-capita disposable income and the retail price of substitute commodities. The supply variable $m_t$ contains total monthly shipments of the commodity. The variable $Q_t$ contains a dummy variable indicating the month an outbreak is first reported ($q_t$) as well as lags of the variable to measure the effect of the outbreak over time ($q_{t-1}$, $q_{t-2}$, ....). The residual is $\varepsilon_t$. The theoretical model suggests that the coefficients for the outbreak variables and the supply variable should be negative.

The regression using grower price as the dependent variable include data for all twelve months of the year. In the case of retail prices there are no retail price data for the months of November, December, January, and February because volume is so small. The related commodities included in the estimation are bananas, grapefruit, and grapes.[8] The outbreak dummy variable ($q_t$) is lagged up to five months. This allows the effects of an outbreak to be examined well after the news reports of the outbreak have subsided. Estimations using a range of lags from three months to eight months, not listed here, produce similar results.

The results of the estimations for strawberries are shown in Table 2. For brevity, the coefficients for the dummy variables are not listed. The estimation with grower prices yields results that are very different than with retail prices. In the case of grower prices, news of an outbreak has a negative and significant effect on prices in the month it occurs ($q_t$). The grower price falls by $.41 per pound. This effect, though severe, subsides quickly. The coefficients for the following months are half as large and are not statistically significant. The coefficient for shipments is negative and significant as expected.

The grower price is not responsive to changes in the demand parameters. None of the coefficients for the substitute commodity prices are significant. The coefficient for per-capita personal income is not

## TABLE 2
### Regression Results

| Variable | Grower Price Coefficient | t-Value | Retail Price Coefficient | t-Value |
|---|---|---|---|---|
| Constant | 0.090 | 0.186 | -1.132 | 3.111*** |
| $P_{bananas}$ | 1.114 | 1.359 | 2.261 | 3.281*** |
| $P_{grapes}$ | 0.203 | 1.064 | 0.469 | 2.979*** |
| $P_{grapefruit}$ | -0.011 | 0.419 | -0.021 | 0.810 |
| Personal Income | 0.0000268 | 1.373 | 0.0000998 | 4.993** |
| Shipments | -0.000213 | 1.836* | -0.000169 | 1.654* |
| $q_t$ | -0.412 | 2.050** | -0.059 | 0.376 |
| $q_{t-1}$ | -0.209 | 1.007 | -0.006 | 0.032 |
| $q_{t-2}$ | -0.218 | 1.071 | -0.006 | 0.031 |
| $q_{t-3}$ | -0.192 | 0.946 | 0.273 | 1.474 |
| $q_{t-4}$ | -0.262 | 1.305 | 0.203 | 1.110 |
| $q_{t-5}$ | -0.147 | 0.736 | 0.137 | 0.746 |
| Adjusted $R^2$ | 0.73 | | 0.83 | |
| No. of Obs. | 97 | | 76 | |

significant either. This contrasts with the second regression using retail price as the dependent variable. The prices of bananas and grapes have a positive and significant effect on the retail price of strawberries. The coefficient for per-capita personal income is positive and significant as well. The determinants for retail and grower prices are different.

Returning to the main variables of interest, the disease outbreak indicator variables, it appears that news of a disease outbreak has a negative but insignificant effect on retail strawberry prices. The initial effect is a price drop of about $.06 per pound during the month of the outbreak. The price effect decreases to less than $.01 by the next month. By the third month after the outbreak, the price drop is gone.

It appears that news of an outbreak mainly impacts grower prices. Retail prices do not fall in tandem with producer prices. Apparently grocery stores do not respond to outbreaks by lowering prices. Perhaps they assume that consumers would be suspicious of commodities that get deep discounts at the time of a disease outbreak. Maintaining the price could be an attempt to signal to the consumer that the commodity is safe.

The only statistically significant effect of the news of an outbreak is an initial drop in grower price. Assuming, as in the theoretical model, that strawberries are inelastically supplied, this translates into a sales loss to growers of $47.5 million for the 1996 outbreak and $75.4 million for the 1997 outbreak. These amounts are comparable to results of Richards and Patterson (1998). The annual grower value of all strawberries produced in the U.S. was $770 million in 1996 and $907 million in 1997 (USDA 1998). The sales loss figures represent a maximum impact because of the assumption that supply is inelastic.

The results in this paper show an interesting contrast to the study by van Ravenswaay and Hoehn (1991) on how media reports on the cancer risk of Alar affect apple sales in the New York City area. The announcement linking Alar to cancer in 1984 caused a drop in demand that persisted at least five years to 1989. Most of the drop in demand is attributed to the initial 1984 announcement. Later media articles had a smaller effect. In this paper, disease outbreaks have a more acute, and less chronic, effect on demand. The demand effect of the disease outbreaks were large but short-lived whereas the effect of the Alar announcement lasted for years generating a much larger total loss. This difference in demand behavior reflects the different types of risks posed by foodborne disease. The presence of Alar is difficult to detect and the effect on consumers takes many years to become apparent. In the case of cyclospora and hepatitis, the effects are quickly apparent. If a consumer does not hear any news reports of disease outbreaks, he or she may conclude that current shipment of produce is likely to be free of microbial contamination. A lack of negative news is not so reassuring in the case of long-term cancer risk from Alar.

## Conclusion

When determining the level of quality of production, producers must balance the costs of increasing quality against the benefits in terms of a higher price. In this paper quality is measured in terms of news reports of microbial contamination. The estimations of sales loss due to outbreaks presented in the second section of this paper give an indication of the benefits a firm enjoys from taking measures to avoid contamination of its output. If the cost of a food safety measure in the strawberry industry is significantly greater than the cost of an outbreak, then there may not be an incentive for firms to undertake that measure.

Another complication is that when a firm undertakes a food safety measure it bears all of the cost, but the benefits (fewer outbreaks) are distributed across the entire industry. This free rider problem can inhibit individual firms from taking food safety measures even when the industrywide cost of the measures is less than the benefit of outbreaks avoided. One way, then, to increase the average quality of the industry's output is to lower the cost of improving quality. This could be accomplished through improvements in food safety techniques or through greater dissemination of current food safety knowledge. An increase in the price response of demand to quality also improves the average quality of production in the industry. Greater awareness of food safety issues among consumers would accomplish this.

### Notes

[1] T. W. Worth is an Economist at the Economic Research Service, U.S. Department of Agriculture. The author would like to acknowledge Barry Krissoff and Fannye Lockley-Holly for their contributions in developing this paper.

[2] "President Wants to Tighten Rules on Imported Produce," New York Times, October 3, 1997.

[3] "E-Coli Scare Forces N.C. Apple Growers to Reconsider Selling Cider," Associated Press, November 19, 1997. Article also points out that growers are getting the expensive pasteurization equipment even though the FDA is not requiring it. In other words the pasteurization is mainly consumer-driven. Some small firms are faced with the choice of selling unpasteurized cider or none at all.

[4] "California Officials Tout Food-Safety Initiative," The Sacramento Bee, California, August, 20, 1997.

[5] This utility function is also used by Shapiro (1983) and Falvey (1989). Restricting the consumption of agents to one unit allows the model to focus on their willingness to pay for quality.

[6] This is the setting of the model by Shapiro (1983).

[7] Outbreaks of foodborne illness were compiled from various issues of the CDC's Morbidity Mortality and Weekly Report.

[8]The choice of substitute commodities is based on a study by Thompson et al. (1990) that measures own and cross price elasticities for various commodities using a nonlinear Almost Ideal Demand System.

**References**

Akerlof, George A. August 1970. The Market for 'Lemons': Quality Uncertainty and the Market Mechanism. *Quarterly Journal of Economics* 84:488-500.

Caswell, Julie A., and Gary V. Johnson. 1991. Firm Strategic Response to Food Safety and Nutrition Regulation. In *Economics of Food Safety*, ed. Julie A. Caswell, 273-297. New York: Elsevier Science Publishing Co., Inc.

Falvey, Rodney E. Trade. August 1989. Quality Reputations and Commercial Policy. *International Economic Review* 30(3):607-622.

International Fresh-cut Produce Association. 1996. *Food Safety Guidelines for the Fresh-cut Produce Industry*. Third Edition.

Leland, Hayne E. 1979. Quacks, Lemons, and Licensing: A Theory of Minimum Quality Standards. *Journal of Political Economy* 87(6):1328-1346.

Muth, Richard F. 1964. Derived Demand Curve for a Productive Factor and the Industry Supply Curve. *Oxford Economic Papers* 16:221-234.

van Ravenswaay, Eileen O. and John P. Hoehn. 1991. The Impact of Health Risk Information on Food Demand: A Case Study of Alar and Apples. In *Economics of Food Safety*, ed. Julie A. Caswell, 155-174. New York: Elsevier Science Publishing Co., Inc.

Richards, Timothy J. and Paul M. Patterson. February 1998. The Economic Value of Spin Control: Food Safety and the Strawberry Case. *National Food and Agricultural Policy Project Discussion Paper* #98-1.

Shapiro, Carl. November 1983. Premiums for High Quality Products as Returns to Reputations. *Quarterly Journal of Economics* 98:659-679.

Smith, Mark E., Eileen O. van Ravenswaay, and Stanely R. Thompson. August 1988. Sales Loss Determination in Food Contamination Incidents: An Application to Milk Bans in Hawaii. *American Journal of Agricultural Economics* 70:513-520.

Tauxe, Robert V. December 1997. Emerging Foodborne Diseases: An Evolving Public Health Challenge. *Emerging Infectious Diseases* 3(4):425-434.

Thompson, Gary D., Neilson C. Conklin, and Gabriele Dono. November 1990. The Demand for Fresh Fruit. *Fruit and Tree Nuts Situation and Outlook Report* THS-256:39-44.

United Fresh Fruit & Vegetable Association. 1997. *Industrywide Guidance to Minimize Microbiological Food Safety Risks for Produce*.

United States Department of Agriculture, Economic Research Service. October 1998. *Fruit and Tree Nuts Situation and Outlook Report: Yearbook Issue* FTS-284.

United States General Accounting Office. April 1998. Food Safety: Federal Eforts to Ensure the Safety of Imported Foods Are Inconsistent and Unreliable. GAO/RCED-98-103.

Zepp, Glenn, Fred Kuchler, and Gary Lucier. April 1998. Food Safety and Fresh Fruits and Vegetables: Is There a Difference Between Imported and Domesitcally Produced Products? *Vegetables and Specialties Situation and Outlook Report* VGS 274:23-28.

Chapter 13

# Cost-Effective Hazard Control in Food Handling

John A. Fox and David A. Hennessy[1]

Assurance of food safety is problematic because of uncertainties in detection, untraceability of source, and scientific ignorance about the relationships between cause and effect. Thus, market failures abound in the economics of food quality (Antle). In recent years, food safety has assumed greater prominence in the policy arena. Advances in biological and engineering disciplines have increased the technical and economic feasibility of interventions that can protect food quality as it passes through production, processing, and distribution. Regulators seeking to remedy market failures should with some confidence a) quantify the net social benefit of quality food and how it relates to actions taken; b) quantify the net private benefit of quality food to food industry entrepreneurs and how it relates to actions taken; and c) design regulations to increase net social benefits by encouraging alterations in private actions. Downstream processors faced with upstream externalities will seek to a) relate their profitability to actions taken upstream; b) understand the relationship between upstream actions and upstream profitability; and c) design incentive mechanisms to facilitate congruence between downstream profitability and the goals of upstream food suppliers.

The aim of this paper is to determine privately optimal intervention strategies to maintain product quality in the presence of both regulatory and market incentives, and so provide a better understanding of the problems facing the regulator and downstream processor in directing actions to maximize overall economic surplus. The recent regulatory focus on food safety, which provides incentives for additional quality control actions on the part of processors, has been the impetus for our study. Unnevehr and Jensen provide a discussion of issues concerning recent regulations. To meet our objective, we develop a biology-driven microeconomic model of the private cost of food contamination. The model bears analogy with the literature on economic pest incidence thresholds below which control inputs are not applied (e.g., Hall and Norgaard; Feder; Szmedra et al.). However, unlike that literature, which focuses on a single intervention and where pest incidence must be estimated, our model studies repeated interventions to control pathogens and the thresholds are functions of pathogen biological parameters so that actual incidence need not be observed in practice.

Although the model is developed with reference to specific food quality examples, it is robust to a large number of disparate production environments. During the time taken for food to pass through to the point of consumption it is exposed to pathogen contamination. Contaminations occur in a random manner, and uncontrolled pathogens often grow at an exponential rate. Applying simple statistical methods to accommodate these stylized biological facts, we develop expressions for the private economic cost of uncontrolled contamination as a function of time. A costly means of control is then introduced into the model. The effects of the control on private economic cost depend upon the production environment. For example, pathogen damage can sometimes be eradicated by application of the control whereas at other times the control can just eliminate the potential for further damage. For different production environments, the privately optimal number of interventions (i.e., controlling actions) is studied as a function of biological and control cost parameters.

The model is then applied to policy design problems. Regulatory agencies may seek to influence private actions by checking quality just before end use, or at other points in the transformation process. These quality checks may be conducted in a deterministic or a random manner, and several plausible penalty structures for substandard quality can be postulated. We complete our formal model presentation by studying the effects of some candidate regulations on private actions.

To validate the analysis, we apply the model to the control of the lesser grain borer in stored wheat. We adapt our model to determine the optimal number of interventions given representative cost and biological parameters. Our conclusions concerning the optimal number of control treatments given existing price penalties for insect damaged kernels are consistent with decisions made in practice. We also examine the effects of random inspections when there are regulatory and market price penalties for violations of threshold quality levels.

## Model

Consider an ongoing production process over a sample period of duration $T$. There are discrete stochastic quality contaminations that follow a Poisson process with mean rate $\lambda$. This mean rate of contamination might be measured in microbial infections per day, insect infestations per day, and so on. The total (random and possibly infinite) number of contaminations is denoted by $x(T)$ with the contaminations occurring at random time points $t_1, t_2, ...,$ . The Poisson process can be motivated as either the true distribution of actual outcomes or the limiting distribution associated with the law of rare events in combination with other true distributions of outcomes (Taylor and Karlin, pp. 179-185). Regardless of its motivation, it is commonly used, simple, and very versatile.

Each quality contamination is initially of unit size, but is assumed to increase exponentially with time at rate $\beta$. For example, suppose that a single pathogen contaminates a food product or process at time $t_j$. Then the level of contamination at time $T$ arising from that single event is $e^{\beta(T-t_j)}$. The time $T$ expected level of contamination, $G(T)$, arising from **all** occurrences over $(0, T]$, when production was uncontaminated at time 0, is then:

$$(1) \quad G(T) = E\left[\sum_{k=1}^{x(T)} e^{\beta(T-t_k)}\right] = e^{\beta T} E\left[\sum_{k=1}^{x(T)} e^{-\beta t_k}\right] = e^{\beta T} M,$$

where $E[\cdot]$ is the expectation operator over the Poisson process and $M \equiv E\left[\sum_{k=1}^{x(T)} e^{-\beta t_k}\right]$. Taylor and Karlin (p. 193) have shown that $M$ may be rewritten as $\lambda[1 - e^{-\beta T}]/\beta$, and so we may state:

$$(2) \quad G(T) = \frac{\lambda}{\beta}\left[e^{\beta T} - 1\right].$$

Clearly, $G(T)$ is increasing in $\lambda$, the mean rate of contamination. It is also increasing in $\beta$, the growth rate, an inference that can be verified from the Taylor series expansion of (2) about $T = 0$.

To convert this contamination level into economic loss over the interval $(0, T]$, we set:

$$(3) \quad \mathcal{L} = H[G(T)]$$

where $H[\cdot]$ can be an increasing linear or convex function if economic loss is hypothesized to increase at a nondecreasing rate as contamination spreads. Economic loss can be controlled by interventions at time points before $T$.

Interventions could be of two types: one in which all traces of contamination are removed, and the other in which the source of damage is removed but the damage accumulates with time. An example of the latter is the fumigation of stored grain. This will kill all insects, but insect damage is irreversible and will accumulate. An example of the former is the cleaning of a restaurant bathroom. An important characteristic of these situations is that contamination, whether it accumulates or not, always imposes a cost. Thus, while cleaning the restaurant bathroom completely eliminates contamination, an irreversible cost, perhaps in the form of consumer dissatisfaction and loss of repeat business, has been incurred over the period that the bathroom was dirty.[2] We will address each in separate sections.

## Nonaccumulating Damage

For the situation in which intervention completely eliminates contamination, but where an irreversible cost has resulted from the contamination, denote the time points where interventions occur by $\{u_0, u_1, u_2, \ldots, u_n\}$. For the sake of exposition, we will assume that the problem at hand involves cleaning the bathroom in a restaurant. We focus on the time interval $(0, T]$ from the indefinitely long duration for which the restaurant is an ongoing business. For the context in question, $T$ is no more than a numeraire against which to measure the number of interventions. As we will discuss later, in other contexts it may be a decision variable. The number of interventions, $n + 1$, is chosen by the manager. It is convenient to initiate this intervention policy at a time point where an intervention occurs, i.e., $u_0 = 0$. In this way, we do not have to worry about under or over counting the number of interventions that occur. The assumption also allows us to start with a clean slate in that there are no contaminations in the system at initiation.

Let an intervention have fixed costs $F$, and variable costs proportional to the level of contamination at the time of intervention. Then the cost minimization objective is to choose an intervention strategy $u = (u_1 - u_0, u_2 - u_1, \ldots, u_n - u_{n-1})$ to minimize:

(4)
$$\mathbb{C}(\lambda, \beta, F, T, \phi, u) = H[(G(T-u_n)] + \sum_{i=1}^{n} H[G(u_i - u_{i-1})] + \phi \sum_{i=0}^{n} G(u_i - u_{i-1}) + (n+1)F,$$

where $\phi$ is the positive constant of proportionality associated with the variable costs of mitigating contamination. Notice that we have included the cost of the initial intervention at time $u_0 = 0$, which is dependent upon the time of the last intervention before the period under consideration, $u_{-1}$. Total cost is the sum of the direct costs of intervening to clean the equipment and the opportunity costs of nonintervention. Cost function (4) could be considered in the narrow sense of conventional costs, but might best be viewed as a loss function. For example, $H[\cdot]$ may include revenue losses due to reduced product quality or loss of reputation.

The intervention strategy involves choosing a sequence of intervention times over the period $(0, T]$ to minimize the value of (4). Because contaminations are assumed to increase exponentially with time, the cost function is convex in $(u_1 - u_0, u_2 - u_1, ..., u_n - u_{n-1}, T - u_n)$ if $H[\cdot]$ is either linearly increasing or increasing and strictly convex in $G$. Given convexity of $C(\cdot)$ in strategy $u$, it follows from Jensen's inequality (Royden, p. 115) that one should partition time into uniform intervals.[3] Therefore, if there are $n$ interventions (in addition to the initial intervention), then each occurs at $T/(n+1)$ time units apart. Accounting for the cost of the initial intervention at $u_0 = 0$, we may rewrite problem (4) as that of choosing $n$ to minimize the primal cost function:

(5)
$$C(\lambda, \beta, F, T, \phi, n) = (n+1)H[G(T/(n+1))] + (n+1)\phi G(T/(n+1)) + (n+1)F$$

over $n$.[4] In this equation it is helpful to remember that $G(T/(n+1)) = \lambda(e^{\beta T/(n+1)} - 1)/\beta$.

The associated first-order condition with respect to $n$ is:

(6)
$$H[G(T/(n+1))] + \phi G(T/(n+1)) + F - \frac{\lambda T e^{\beta T/(n+1)}}{n+1} H'[G(T/(n+1))]$$

$$-\frac{\phi \lambda T e^{\beta T/(n+1)}}{n+1} = 0.$$

Here, the prime on a function signifies a differentiation. The second, third, and fifth terms on the left-hand side sum to the direct marginal cost of increasing the rate of intervention, while the first and fourth terms sum to the marginal opportunity cost of intervention. The problem is to choose $n$, as represented by $n^*(\lambda, \beta, F, T, \phi)$, to achieve minimum cost $C^*(\lambda, \beta, F, T, \phi)$. Thus, the primal-dual problem is to choose $n$ to maximize $C^*(\lambda, \beta, F, T, \phi) - C(\lambda, \beta, F, T, \phi, n)$. Silberberg and also Chavas and Pope have studied the fundamental structure, such as symmetry, of general optimization problems such as this. A second differentiation of (5) with respect to $n$ renders:

(7)
$$\delta = \frac{\lambda T^2 e^{\beta T/(n+1)}}{(n+1)^3} \{\beta H'[G(T/(n+1))] + \lambda e^{\beta T/(n+1)} H''[G(T/(n+1))] + \phi\beta\} > 0,$$

where $\delta = \partial^2 C(\cdot)/\partial n^2$. Therefore, cost is strictly convex in $n$.

The problem facing the regulator may be quite different, however. Setting $D(n)$ as the external social damage associated with intervention choice $n$, the costs embodied in this function might represent food safety and information asymmetry problems facing the consumer. Alternatively, they might represent free-riding on brand or country reputation. It is most likely that $dD(n)/dn<0$, suggesting that $n^*$ is socially suboptimal.

We will now investigate how the optimal intervention strategy is affected by alterations in the economic and biological environment. Taking comparative statics with respect to $n^*$ and $F$, we have:

(8)
$$\frac{\partial n^*}{\partial F} = -\frac{1}{\delta} < 0.$$

As one would expect, an increase in the fixed costs of intervention reduces the incentive to intervene. It would further the goals of the regulator to subsidize $F$. The envelope derivative of (5) reveals that $\partial C^*(\lambda,\beta,F,T,\phi)/\partial F = \partial C(\lambda,\beta,F,T,\phi,n)/\partial F\big|_{n=n^*(\lambda,\beta,F,T,\phi)} = n^*(\lambda,\beta,F,T,\phi)+1$, so cost rises in strict proportion to the number of interventions.

A similarly intuitive result is developed when the effect of increasing the length of the production process is considered, i.e., $\partial Ln(n^*+1)/\partial Ln(T) = 1$. The elasticity of interventions with respect to the length of the process is unity, which arises from the fact that (6) is homogeneous of degree zero in $T$ and $n^* + 1$. This stands to reason because the process being modeled is, in a sense, stationary, and this fact motivates the use of $T$ as numeraire. In reality, one might expect the optimal number of interventions per unit time to vary as the product moves through the processing channel or as biological parameters change.

Now let us consider the biological environment. Differentiating equation (6) with respect to $\lambda$, the rate of contamination, it is shown in Appendix A that $\partial n^*/\partial \lambda \geq 0$. Therefore, the privately optimal number of interventions increases as the contamination rate increases. The effect of an increase in the biological rate of growth, $\beta$, on interventions is ambiguous. Even in the simple case where $H[G]=AG$, when the first-order condition is:

(9)
$$\frac{(A+\phi)\lambda}{\beta}[e^{\beta T/(n^*+1)}-1]+F-\frac{(A+\phi)\lambda T}{n^*+1}e^{\beta T/(n^*+1)}=0,$$

the comparative static is $\partial n^*/\partial \beta = -[F-(A+\phi)\lambda\beta(T/(n^*+1))^2 e^{\beta T/(n^*+1)}]/[\beta\delta]$ which seems to be indeterminate in sign. The ambiguity arises because while a higher $\beta$ means that interventions must be more regular if contamination is to be controlled, it also means that the control is less effective because an increase in the rate of growth acts to reduce the time interval until the effects of the control are undone. The expression for $\partial n^*/\partial \beta$ suggests that one would expect an increase in $F$ or a decrease in $\lambda$ to reduce the incentive to control when $\beta$ increases.

Even with the ambiguity of sign, costs certainly rise with the rate of damage:

(10)
$$\frac{\partial C^*(\lambda,\beta,F,T,\phi)}{\partial \beta}=\frac{(n^*+1)\{H'[G(T/(n+1))]+\phi\}}{\beta}\frac{\partial G(T/(n+1))}{\partial \beta}>0,$$

where it is shown in Appendix A that $\partial G(\cdot)/\partial \beta > 0$. For cost specification (5), it is shown in Appendix A that $\partial n^*/\partial \phi > 0$. Early intervention is preferred in order to control a more serious problem later because variable costs are proportional to the exponentially growing contamination. Essentially, a stitch in time saves nine.

Setting social cost equal to $W(\lambda,\beta,F,T,\phi,n^*) = C^*(\lambda,\beta,F,T,\phi)+D(n^*)$, we have $\partial W(\cdot)/\partial \lambda = \partial C^*(\cdot)/\partial \lambda + [dD(n^*)/dn^*][\partial n^*/\partial \lambda]$ which is ambiguous in sign. As is usual in the world of second bests, mitigation through optimal choices could conceivably overwhelm a deterioration in the biological environment caused by an increase in $\lambda$. The impact of $\beta$ on the regulator's objective function, $\partial W(\cdot)/\partial \beta = \partial C^*(\cdot)/\partial \beta + [\partial D(n^*)/\partial n^*][\partial n^*/\partial \beta]$, is also ambiguous, as is the impact of $\phi$.

## Accumulating Damage

An alternative context involves an ongoing production process in which the intervention completely eliminates the contamination, but the damage accumulates. That is, economic loss occurs just once and it is a function of the aggregate contamination. Such is often the case with disease or insect infestations on stored grain or meat. Here, the intervention may kill the source of contamination; but the product remains damaged, and damage accumulates with subsequent contaminations over the remaining time. In this situation, the problem is to choose $u$ to minimize:

(11)
$$\mathbb{C}(\lambda,\beta,F,T,\phi,u) = H\left[G(T-u_n)+\sum_{i=1}^{n}G(u_i-u_{i-1})\right]+\phi\sum_{i=0}^{n}G(u_i-u_{i-1})+(n+1)F,$$

where the summation of contaminations is interior to the $H[\cdot]$ function instead of being exterior. We set $\phi = 0$ and invoke Jensen's inequality again to simplify the problem of identifying the optimal intervention strategy, and the primal cost function becomes

(12)
$$C(\lambda,\beta,F,T,n) = H[(n+1)G(T/(n+1))]+(n+1)F.$$

A comparative static analysis of this equation provides results similar to those in the previous section, and is reported in Appendix B.

## Application: Insects in Stored Grain

A number of insect pest species cause damage to stored grain. In the United States, one of the most common and most destructive is the lesser grain borer. Damage is caused when adult females deposit eggs on the outside of kernels, and the hatched larvae bore into the kernel to complete development inside (USDA). The resulting insect damaged kernels (*idk*) are recognized by the Federal Grain Inspection Service as a separate quality factor. Levels of *idk* above 32 per 100 grams result in designation of grain as "sample grade" -- suitable only for animal feed -- whereas lower levels of *idk* will usually result in price discounts. Similarly, the presence of live insects in grain will, above a threshold of 1 per 1,000 grams, result in grain being designated infested and subject to price discounts. Reed et al. found evidence of current or previous insect infestation in 60 percent of wheat samples collected in Kansas in 1986-87. In that study the mean number of *idk* was 7.3 per 100g and the presence of live insects was significantly associated with the probability of receiving a price discount.

Because infestation occurs after storage, prestorage control practices are ineffective. Before the development of insect resistance, effective control was achieved by using single applications of residual chemical protectants such as malathion. Today, largely due to insect resistance but also as a result of increased concerns about organophosphate and other chemical residues in food, control relies more on fumigants such as aluminum phosphide or methyl bromide, on atmospheric modifications that create oxygen deficiency, and on other nonchemical methods such as drying or aeration (Beckett et al.; Arthur). Fumigants kill insects and eggs present at the time of treatment, but do not provide any residual protection against reinfestation. Fumigation is widely used -- for example Reed and Worman reported that 50 percent of on-farm storage bins containing grain in January 1987 had been fumigated at least once. The average cost of phosphide fumigation was estimated at $0.009/bu.

In contrast to the indefinite time process considered when developing the model, in the present context $T$ is a decision variable. Being more than just an arbitrary numeraire, we must now consider the optimized dual to equation (4) in a different light. While demand for consumption may be heavier when prices are low just post-harvest, the demand for many grains is very consistent through the crop year. As mentioned in Williams and Wright (p. 28), one can conclude from the model of Holthausen, among others, that when a sufficiently rich menu of maturities on futures contracts are available to a storer, then futures market prices and storage costs completely determine storer behavior. Thus, $T$ will be chosen to equate the marginal cost of storage to time $T$ with the futures price at time $T$.[5,6] The marginal cost of storage is dominated by the opportunity cost of capital. From results that we will present in Table 1 it can be concluded that the marginal cost per day of controlling the lesser grain borer in wheat is about 0.005c/bu while the daily capital cost at an 8 percent annual interest rate on wheat valued at $3.50/bu is about 0.077c/bu. Therefore, the effect of pest protection activities on optimal $T$ is minimal, and we assume that $T$ is exogenous in our illustration below.

Given $T$, grain storage is a finite production process for which, as noted earlier, the primal cost function represented by (5), and its counterpart in the accumulating damage situation, (12), are not necessarily valid unless the variable costs of intervention are zero. Equal spacing of earlier interventions remains the optimal strategy but, if the cost of intervention is proportional to contamination, there is an incentive to time the last intervention earlier. Thus, the interval between the last intervention and $T$ will exceed $T/(n + 1)$. To illustrate, consider a simple example where $T = 10$ and it is optimal to have one intervention. Damage at $T$ is minimized by placing the intervention at $t = 5$. But total cost may be reduced by placing the intervention at $t = 4.5$; damage will be higher but the cost of the single intervention is reduced and there is no subsequent intervention for which cost will be increased. The same applies if the optimal number of interventions were nine. Instead of placing the final intervention at $t = 9$, it may be cheaper to place it at 8.5. The earlier interventions would then be equally spaced over [0, 8.5).

The application is based on grain storage in a 20,000 bushel bin for a period, $T$, of 250 days. We assumed that development of an insect population can be described by a simple exponential model of the form $N = Ae^{\beta T}$, where $A$ represents the initial population and $\beta$ is the population growth rate (Trematerra, Fontana, and Mancini). In the simulation, we used values of 0.05, 0.06, and 0.07 for $\beta$. Consultation with entomologists provided an estimate of l=100 for the insect reinfestation rate, and in the simulation we use values of 80, 100, and 120. The penalty function for *idk*, $H[\cdot]$, is given in Figure 1. This function is piecewise linear and overall convex on the domain [0, 32] where grain is eligible to enter human feed markets. Grain is rejected above the threshold of 32 *idk*/100g. In the simulation, we arbitrarily used a penalty of $2 per bushel for rejected grain but found that, with one or more interventions, the final level of *idk* never approached this threshold. Thus, the results are not sensitive to variation in that upper penalty, provided of course that it is at least as large as the maximum penalty for accepted grain. The discount schedule presented in Figure 1 is representative of schedules typically applied.

Conveniently, because of the manner in which grain borers damage grain, there is roughly a one-to-one relation between the number of emerged insects and the number of damaged kernels. Damage (i.e., *idk*) is cumulative and its economic consequence is realized only at the end of the period. To complete the cost function we used values of $100 and $120 for the fixed cost, $F$, and values of 0.0032, 0.0056, and 0.008 cents/insect for $\phi$, the variable cost parameter. These were based on commercial fumigation rates.

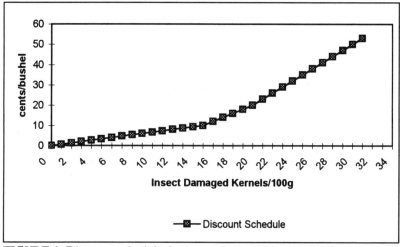

FIGURE 1. Discount schedule for insect damaged kernels (idk).

Optimal *n* was identified by line search minimization of the cost objective function, modified to account for earlier placement of the final intervention when costs are proportional to contamination. Results are presented in Table 1. The entries show both the *hypothetical* minimum total cost and the total costs associated with the relevant, integer-valued, cost-minimizing numbers of interventions. Overall, for the chosen parameter values, costs are in line with those found in practice. Also, the number of interventions is in accord with commercial practice where either one or two interventions occur per 250 days. Entries in the lower section of the table show that the optimum number of interventions falls

TABLE 1

Number of Interventions and Costs ($) as Cost and Biological Parameters Change

| Reinfestation Cost Parameters | Growth | $\lambda=120$ | | | $\lambda=100$ | | | $\lambda=80$ | | |
|---|---|---|---|---|---|---|---|---|---|---|
| | | $\beta=0.07$ | $\beta=0.06$ | $\beta=0.05$ | $\beta=0.07$ | $\beta=0.06$ | $\beta=0.05$ | $\beta=0.07$ | $\beta=0.06$ | $\beta=0.05$ |
| $\phi=0.008$ | Min.Cost | 309 | 257 | 203 | 295 | 244 | 191 | 278 | 229 | 178 |
| F=$100 | Opt. n | 2.27 | 1.83 | 1.38 | 2.18 | 1.74 | 1.31 | 2.06 | 1.65 | 1.23 |
| Cost | n=1 | 1254 | 486 | 232 | 1061 | 421 | 210 | 869 | 357 | 188 |
| | n=2 | 318 | 259 | 231 | 298 | 249 | 226 | 278 | 240 | 220 |
| | n=3 | 340 | | | 333 | | | 327 | | |
| $\phi=0.0056$ | Min.Cost | 295 | 245 | 193 | 282 | 233 | 183 | 266 | 219 | 171 |
| F=$100 | Opt. n | 2.18 | 1.77 | 1.34 | 2.09 | 1.69 | 1.27 | 2.00 | 1.60 | 1.19 |
| Cost | n=1 | 1116 | 439 | 216 | 946 | 383 | 197 | 777 | 326 | 178 |
| | n=2 | 229 | 250 | 226 | 283 | 242 | 222 | 266 | 233 | 217 |
| | n=3 | 333 | | | 328 | | | 322 | | |
| $\phi=0.0032$ | Min.Cost | 278 | 231 | 182 | 266 | 220 | 173 | 252 | 207 | 162 |
| F=$100 | Opt. n | 2.08 | 1.69 | 1.27 | 2.00 | 1.61 | 1.20 | 1.91 | 1.53 | 1.14 |
| Cost | n=1 | 956 | 386 | 198 | 813 | 338 | 182 | 671 | 291 | 165 |
| | n=2 | 279 | 240 | 221 | 266 | 233 | 217 | 253 | 227 | 214 |
| | n=3 | 326 | | | 322 | | | | | |
| $\phi=0$ | Min.Cost | 245 | 204 | 161 | 235 | 195 | 153 | 224 | 184 | 144 |
| F=$100 | Opt. n | 1.88 | 1.52 | 1.16 | 1.82 | 1.46 | 1.10 | 1.74 | 1.38 | 1.04 |
| Cost | n=1 | 677 | 293 | 166 | 581 | 261 | 155 | 485 | 229 | 144 |
| | n=2 | 247 | 224 | 212 | 239 | 220 | 210 | 231 | 216 | 208 |
| $\phi=0$ | Min.Cost | 282 | 233 | 184 | 271 | 223 | 174 | 257 | 211 | 164 |
| F=$120 | Opt. n | 1.82 | 1.46 | 1.10 | 1.74 | 1.40 | 1.06 | 1.66 | 1.34 | 1.00 |
| Cost | n=1 | 697 | 313 | 186 | 601 | 281 | 175 | 505 | 249 | 164 |
| | n=2 | 287 | 264 | 252 | 279 | 260 | 250 | 271 | 256 | 248 |

$\lambda$ = Poisson parameter for mean rate of grain borer infestation; $\beta$ = growth rate of damage caused by a given infestation; $\phi$ = constant of proportionality for variable cost of mitigating a given level of damage; F = fixed cost of an intervention.

with increases in the level of fixed costs. Comparisons among the three main columns show that for all values of $F$, $\phi$, and $\beta$, an increase in $\lambda$, the reinfestation rate, is associated with increases in the optimum number of interventions.

The contamination growth rate, $\beta$, is shown here to have a positive effect on $n$, although this effect could not be positively identified in the model (see Appendix B). But an analysis of derivatives, i.e., equation (9) and the discussion of it together with results in Appendix B, leads to the conjecture that an increase in $F$ or a decrease in $\lambda$ should reduce the incentive to intervene when $\beta$ increases. The simulation results support this conjecture. For example, with $\phi = 0$, and $F = \$100$, and $\lambda = 100$, the optimum number of interventions increases by 0.72 as $\beta$ increases from 0.05 to 0.07. If $F$ is raised to $\$120$, the increase in optimum $n$ falls to 0.68, and if $\lambda$ then falls to 80, the increase in optimum $n$ falls to 0.66. Clearly, both minimized cost and cost for a given number of interventions rise with increases in $\beta$. Also, increases in the variable cost parameter, $\phi$, do indeed induce an increase in the optimal number of interventions.[7]

## Regulations Through Fines

A popular form of regulation is to impose fines for being detected to have exceeded a maximum limit on contamination. Suppose that hygiene guidelines require that the contamination level not exceed a level $k$ at any point in time. Given the difficulty in establishing comparative statics for the more general case, let us consider the contamination proportional cost specification, $H[G]=AG$, in more detail. If an establishment is inspected and:

(13) $$\frac{\lambda}{\beta}(e^{\beta T/(n+1)} - 1) \leq k,$$

then the proprietor is not fined. But if relation (13) does not hold, then a fine of magnitude $P$ is imposed. The probability of a random inspection per unit time is the constant $s$, so there will be an average of $sT$ inspections over the time period $T$. Isolating the choice variable, $n$, we have the condition under which no fine occurs as $\beta T / [Ln(\lambda + \beta k) - Ln(\lambda)] \leq n+1$.

Setting the value of $n$ for which (13) holds with equality as $\bar{n}$, there is not a regulatory infringement if $n \geq \bar{n}$. However, we assume that violations occur, i.e., that in the absence of a penalty it is optimal to set $n < \bar{n}$. To establish the probability of a violation, note that time in a contamination build-up equals $T/(n + 1)$ whereas the maximum time in a build-up that would give zero probability of being fined is $T/(\bar{n}+1)$. Then the ratio of these times, $(n+1)/(\bar{n}+1)$, is the probability of not being in violation of the regulatory standard, and so the probability of being in violation is $1-(n+1)/(\bar{n}+1)$. Substituting in for $\bar{n}$, the expected money value of fines over time $T$ that a firm choosing $n < \bar{n}$ will face is:

(14)
$$R(\lambda, \beta, T, P, s, k, n) = \frac{Ps}{\beta}\{\beta T - (n+1)[Ln(\lambda + \beta k) - Ln(\lambda)]\}.$$

If $n \geq \bar{n}$, then the expected money value of the fine equals 0. Augmenting the firm's primal cost function by this expected penalty, the firm chooses $n$ to minimize:

(15)
$$C(\cdot) = \frac{(A+\phi)(n+1)\lambda}{\beta}[e^{\beta T/(n+1)} - 1] + (n+1)F + LR(\lambda, \beta, T, P, s, k, n),$$

where $L = 1$ when $n < \bar{n}$ and $L = 0$ otherwise. The associated first-order condition is:

(16)
$$\frac{(A+\phi)\lambda}{\beta}\left[e^{\beta T/(n+1)} - 1 - \frac{\beta T}{n+1}e^{\beta T/(n+1)}\right] + F - L\frac{Ps}{\beta}[Ln(\lambda + \beta k) - Ln(\lambda)] = 0.$$

the second derivative with respect to $n$ is $\delta$ as before. We denote the optimal choice by $n^*$ as before. Because we assume $n < \bar{n}$, we may set $L = 1$. This specification will lead to the social first-best solution in the event that $\partial D(n)/\partial n = -Ps\, Ln(1 + \beta k/\lambda)/\beta$. There always exist some $k$, given $P$ and $s$, such that this condition is true. But the level of $k$ in question may be such that the fine is not binding. However, if $Ps$ is altered it is likely that a $k$ can be found such that the fine is binding and the marginal condition holds.

Notice that optimal $n$ is larger under the regulation than absent the regulation. To see this, observe that the effect of the regulation on the first-order condition comes through $-Ps[Ln(\lambda + \beta k) - Ln(\lambda)]/\beta$, which is negative and which would equal zero if either $P = 0$ or $s = 0$. Therefore, if $Ps \neq 0$, then the first parts of the first-order condition must be positive. But these parts constitute the first-order condition when $Ps = 0$, i.e.,

(17)
$$\frac{(A+\phi)\lambda}{\beta}\left[e^{\beta T/(n^*+1)}-1-\frac{\beta T}{n^*+1}e^{\beta T/(n^*+1)}\right]+F=0.$$

As the left-hand side of (17) is positive under the fining regulation, and $\delta > 0$, therefore $n^*$ must be larger under the regulation than absent the regulation.

Similar to equation (8), $\partial n^*/\partial F = -(n^*+1)^3/[(A+\phi)\lambda\beta T^2 e^{\beta T/(n^*+1)}] < 0$. Although this is the same formula as in (8), the value of $n^*$ is now larger because of the fine. Therefore, given that the numerator, $(n^*+1)^3$, is larger and that the denominator is smaller (since $e^{\beta T/(n^*+1)}$ is decreasing in $n$), it follows that $n^*$ is more sensitive to $F$ under the regulation than absent the regulation, i.e., $\partial n^*/\partial F|_{reg} < \partial n^*/\partial F|_{no\,reg}$ where *reg* and *no reg* identify the regulated and unregulated situations, respectively.

Viewing (16), the first-order condition under the regulation continues to be homogeneous of degree zero in $T$ and $n + 1$. Therefore, $\partial Ln(n^*+1)/\partial Ln(T) = 1$ continues to hold regardless of the regulation; however, regulation does affect other comparative static results. For example, consider the impact of $\lambda$ on $n^*$. In the unregulated situation it is positive, but when a firm can be fined for excess contamination then the effect is:

(18)
$$\frac{\partial n^*}{\partial \lambda} = \frac{1}{\delta}\left[\frac{F}{\lambda}-\frac{Ps}{\lambda\beta}[Ln(\lambda+\beta k)-Ln(\lambda)]\right]-\frac{Psk}{\lambda(\lambda+\beta k)\delta}.$$

This cannot be signed, although it is likely to be positive if any of $k$ or $P$ or $s$ are very small, or if $F$ is very large. As under the unregulated case, the sign of $\partial n^*/\partial \beta$ remains ambiguous. This was to be expected because the regulation just adds nonlinearity to the problem.

Because $P$ and $s$ enter objective function (15) symmetrically, it suffices to analyze the comparative statics of $P$. Differentiating (16) with respect to $n$ and $P$ establishes $\partial n^*/\partial P = s[Ln(\lambda+\beta k)-Ln(\lambda)]/[\delta\beta] > 0$, and the number of interventions increases with an increase in either $P$ or $s$. Given $dD(\cdot)/dn < 0$, social cost will decrease due to the regulatory change.

A third way of toughening penalties might be to reduce tolerance for contamination, $k$. The relevant comparative static is

$\partial n^* / \partial k = Ps /[\delta(\lambda + \beta k)] > 0$, so a decrease in tolerance decreases optimal
$n$, and increases social cost. This counterintuitive (to the authors at least)
result arises because the distortion in the first-order condition due to the
fine regulation becomes even more distorting as $k$ increases. Effectively,
the regulation increases cost but reduces the marginal cost of $n$ because the
expected fine is decreasing in $n$. An increase in $k$ reduces the marginal
cost even further by decreasing the $n$ at which no fine ever occurs
$(\partial \overline{n} / \partial k < 0)$, which of course decreases the probability of a fine. The
critical, and counterintuitive, point is that the fining system in (14) can be
viewed as the sum of fixed cost, $PsT$, and subsidy per unit intervention
$Ps\, Ln[(\lambda + \beta k)/\lambda]/\beta$; and an increase in $k$ increases the unit subsidy.

Table 2 shows how costs and the optimum *hypothetical* and integer
numbers of interventions respond to changes in regulatory parameters in
the grain storage example. The standard, $k$, might here refer to live insects
in grain.[8] Relative to Table 1, costs in Table 2 are calculated for mid-
range values of the insect infestation and insect growth parameters, and,
with the exception of the bottom row, for a fixed cost level of $100.
Absent the regulation (i.e., with $P = 0$), the optimal *hypothetical* number
of interventions given those parameters is $n^* = 1.46$ at a cost of $195.
This increases under the regulation -- the entries show that, for all positive
values of the fine, $n^*$ exceeds 1.46.

For higher fines and higher, i.e., less stringent, thresholds, we find a
number of "corner solutions," that is, the optimal number of interventions
being that which just eliminates the possibility of a fine $(n = \overline{n})$. These
corner solutions may mask, in some instances, the effects of changes in the
regulatory parameter. For example, increases in fixed costs or in the size
of the fine often have no effect on the optimal number of interventions at
the highest ($k = 15$) contamination threshold. But at lower thresholds ($k =
5$), increases in the level of the fine have the predicted positive effect.
Finally, comparisons across the columns of Table 2 illustrate the
counterintuitive result -- that tightening the regulatory standard (reductions
in $k$) leads to a reduction in the number of interventions when corner
solutions do not pertain. This effect is more pronounced for higher levels
of the fine (compare $P = \$10$ with $P = \$25$).

## Terminal Penalties

In practice, grain is inspected only when sold out of storage, i.e., at time $T$. At this time, in addition to losses incurred due to past contaminations, penalties may also be applied as a function of the level of current contamination. For insects that attack grain, the penalty would be a function of the level of insect infestation at the time of sale. This penalty might be a condition in a commercial contract, or it might be imposed by the state. If the penalty function is $J(\cdot)$, then the magnitude

### TABLE 2
### Number of Interventions and Costs ($) as Regulatory System Parameters Change[a]

| s, P | Standard: Insects/bu $\overline{n}$ | k=15 1.88 | k=10 2.12 | k=5 2.65 |
|---|---|---|---|---|
| s=0.02 | Min.Cost | 195 | 195 | 195 |
| P=$0 | Opt. n | 1.46 | 1.46 | 1.46 |
| F=$100 | cost, n=1 | 261 | 261 | 261 |
|  | cost, n=2 | <u>220</u> | <u>220</u> | <u>220</u> |
| s=0.02 | Min.Cost | 201 | 205 | 210 |
| P=$10 | Opt. n | 1.53 | 1.52 | 1.51 |
| F=$100 | cost, n=1 | 276 | 279 | 283 |
|  | cost, n=2 | <u>220</u> | 222 | 229 |
| s=0.02 | Min.Cost | 209 | 218 | 233 |
| P=$25 | Opt. n | 1.67 | 1.65 | 1.61 |
| F=$100 | cost, n=1 | 299 | 306 | 317 |
|  | cost, n=2 | <u>220</u> | 225 | 242 |
| s=0.02 | Min.Cost | <u>212</u> | <u>229</u> | 264 |
| P=$50 | Opt. n | <u>1.88</u> | <u>2.12</u> | 1.92 |
| F=$100 | cost, n=1 | 337 | 351 | 374 |
|  | cost, n=2 | <u>220</u> | 230 | 264 |
| s=0.02 | Min.Cost | <u>212</u> | <u>229</u> | <u>275</u> |
| P=$75 | Opt. n | <u>1.88</u> | <u>2.12</u> | <u>2.65</u> |
| F=$100 | cost, n=1 | 376 | 396 | 430 |
|  | cost, n=2 | <u>220</u> | 235 | 286 |
| s=0.02 | Min.Cost | <u>250</u> | <u>272</u> | 325 |
| P=$75 | Opt. n | <u>1.88</u> | <u>2.12</u> | 2.22 |
| F=$120 | cost, n=1 | 396 | 416 | 450 |
|  | cost, n=2 | <u>260</u> | 275 | 326 |

[a]Notes: 1) All table entries calculated for $\lambda=100$, $\beta=0.06$, $\phi=0$; 2) underlined figures signify "corner solutions," i.e., where the number of interventions results in the expected value of the fine being equal to zero; 3) at n=3 and k=5, 10, or 15; cost =$307 when F=$100 and cost = $367 when F= $120.

$J[\lambda(e^{\beta T/(n+1)} - 1)/\beta]$ would have to be added to the objective function, e.g., added to equations (4) and (11). A particularly interesting penalty function is one in which the penalty is zero up to a threshold level of contamination (infestation), say $k$ as before, and is a large fixed penalty, $\overline{R}$, thereafter. It should first be noted that this penalty treats the time period subsequent to the final intervention in a manner different to that for the prior periods. Therefore, it breaks the inherent symmetry that had led us to the conclusion that time intervals between interventions would be of equal durations. So it is no longer clear that this very helpful simplification can be applied even in the case where $\phi = 0$.

To discern whether it can, first optimize without including the terminal penalty function and let the solution be $n = n^*$. Now, if $k > \lambda(e^{\beta T/(n^*+1)} - 1)/\beta$, then the terminal penalty will not occur and can be omitted from analysis except to check that subsequent parameter shifts do not induce it to be binding. If $k \leq \lambda(e^{\beta T/(n^*+1)} - 1)/\beta$, then the decision maker will either choose to accept the penalty as a fixed cost or will act to ensure that the penalty is not incurred.

With a terminal penalty, avoidance strategies can be classified into two broad categories: a) adding interventions, or b) rescheduling existing interventions. At most, one additional intervention will be required because it can be placed immediately before inspection so as to ensure avoidance of the penalty. Therefore, if in a period of length $T$ the optimal number of interventions in the absence of a terminal penalty is $n^* = 2$, then any terminal penalty can be avoided by placing a third intervention at $T - \varepsilon$, where $\varepsilon$ is arbitrarily small. The cost of the additional intervention will be equal to the cost of the two existing interventions, which occur at points $T/3$ and $2T/3$, because each occurs following an equally long period of contamination build-up.

The above strategy is unlikely to be optimal. When costs are convex in time, cost can be decreased by earlier timing of the additional intervention. In particular, it can be placed at $T - t^*$ where $t^*$ is the length of time in a contamination build-up following which the level of contamination is exactly equal to the penalty threshold, $k$. Thus, timing the final intervention at $T - t^*$ will just avoid the penalty, and will reduce both the cost related to product contamination (given exponential growth) and the costs of the final intervention if that cost is proportional to the level of contamination (because contamination will be lower at $T - t^*$ than at $T - \varepsilon$).[9] Choice $t^*$ depends only on parameters $\beta$, $\lambda$, and $k$ and is given by $t^* = [Ln(\lambda + k\beta) - Ln(\lambda)]/\beta$. Values of $t^*$ for the grain storage example are presented in Table 3 in which the threshold, $k$, is at a level that ensures that a penalty would occur in the absence of a response. That is, $t^*$ is less than $T/(n^*+1)$, the length of the final period with $n^*$ equally spaced interventions.

With an additional intervention at $T-t^*$, convexity of the cost function ensures that cost may be further reduced by rescheduling the earlier interventions in the interval $[0, T-t^*)$ to achieve equal spacing. Thus, with this response strategy, which we label A1, the $n^*+1$ interventions would take place at times $(T-t^*)/(n^*+1), 2(T-t^*)/(n^*+1)$, and so on to $T-t^*$.

An alternative strategy, A2, involving equal spacing of the $n^*+1$ interventions may then provide a further reduction in cost. Strategy A1 does not have equal spacing of all interventions unless, by chance, $t^* = T/(n^*+2)$. If $t^* < T/(n^*+2)$, then equal spacing of the interventions will **not** avoid the penalty and strategy A1 will be preferred. However, if $t^* > T/(n^*+2)$, then equal spacing will push the final intervention closer to the inspection date and so reduce costs while continuing to avoid the penalty.

The terminal penalty may also be avoided without an additional intervention by changing the timing of existing interventions. In particular, the final intervention can be delayed until time $T-t^*$ to just eliminate the penalty. To further reduce costs, the earlier interventions could then be rescheduled to achieve equal spacing in the period up to $T-t^*$. We label this strategy B1. It is, of course, very similar to strategy A1 above, both involving an intervention at $T-t^*$ and equal spacing of the earlier interventions. The benefits of the additional intervention in A1 are that it not only reduces *idk* contamination, but also reduces the cost of

TABLE 3

Values of t*, the Maximum Interval between Final Intervention and T to Avoid a Terminal Penalty at a Threshold of $k = 1$ Insect Per 5 Kg[a]

| Growth: | Reinfestation | | |
| --- | --- | --- | --- |
| | $\lambda=120$ | $\lambda=100$ | $\lambda=80$ |
| $\beta=0.07$ | 58.3 | 60.9 | 64.0 |
| $\beta=0.06$ | 65.5 | 68.5 | 72.2 |
| $\beta=0.05$ | 75.1 | 78.6 | 83.0 |

[a]This threshold of $k = 1$ insect per 5 Kg was chosen to ensure that the penalty will occur for all intervention strategies given in Table 1. Thus, $t^* < T/(n^*+1)$ where $n^*$ is the optimal (integer) number of interventions before the penalty is introduced.

each intervention when cost is proportional to contamination. Thus, increases in $\beta$ or $\lambda$ will reduce the cost of A1 relative to B1, whereas increases in the fixed cost of intervention, $F$, will increase it.

Table 4 shows the costs associated with strategies A1, A2, and B1 for the stored grain situation with the introduction of a terminal penalty. For simplicity, costs of intervention are restricted to a fixed cost of $100. Also shown in Table 4 is the optimal number of interventions and the associated cost in the absence of the terminal penalty. All intervention strategies result in higher costs than under the no penalty situation. For six of the nine parameter combinations examined, strategy B1, rescheduling the existing number of interventions, is the optimum response. For the others, strategy A1, adding an intervention at $T-t^*$, is optimal relative to B1. Notice that for situations in which $t^* > T/(n^*+2)$, strategy A2, with

**TABLE 4**

Intervention Strategies and Costs with a Terminal Penalty

|  | Reinfestation | | |
|---|---|---|---|
| Growth: $\lambda=120$ | $\lambda=100$ | $\lambda=80$ |
| $\beta=0.07$ | B1: $277.48[a] | B1: $259.65 | B1: $243.49 |
|  | A1: $314.54 | A1: $311.98 | A1: $309.58 |
|  | A2: $314.34 + Penalty[b] | A2: $311.95 + Penalty | A2: $309.56 |
|  | n*=2 costs $247[c] | n*=2 costs $239 | n*=2 costs $231, |
|  |  |  | "$t^*$>" |
| $\beta=0.06$ | B1: $229.57 | B1: $223.15 | B1: $217.34 |
|  | A1: $308.91 | A1: $307.56 | A1: $306.30 |
|  | A2: $308.86 | A2: $307.38 | A2: $305.91 |
|  | n*=2 costs $224, | n*=2 costs $220, | n*=2 costs $216, |
|  | "$t^*$>" [d] | "$t^*$>" | "$t^*$>" |
| $\beta=0.05$ | B1: $505.11 | B1: $383.23 | B1: $282.92 |
|  | A1: $212.69 | A1: $210.30 | A1: $208.13 |
|  | A2: $212.19 + Penalty | A2: $210.16 + Penalty | A2: $208.12 + Penalty |
|  | n*=1 costs $166 | n*=1 costs $155 | n*=1 costs $144 |

[a]Strategies are: B1 – maintain $n^*$ interventions, adjust timing; A1 – add intervention at $T-t^*$; A2 – add intervention and maintain equal spacing. Cost parameters are $F = \$100$, $\phi = 0$.

[b]Simulation based on penalty of $0.50/bu, i.e., $10,000. Strategy A2 includes the penalty when $t^* < T/(n^*+2)$.

[c]Underlined figures show the optimal (integer valued) number of interventions and associated cost in the absence of a terminal penalty (from Table 1).

[d]Notation "$t^*$>" indicates a parameter combination for which $t^* > T/(n^*+2)$. Otherwise $t^* \leq T/(n^*+2)$.

equally spaced interventions, costs less than A1. But in each of these cases, and there are four such cases, B1 dominates A2 so strategy A2 is never optimal in the simulations. For situations with $t^* < T/(n^* + 2)$, A2 does not avoid the terminal penalty. Increasing $\beta$ from 0.06 to 0.07 is shown to increase the cost of strategy B1 to a greater extent than it does that of A1, and the same finding applies to increases in l, the reinfestation rate.

## Conclusions

If food safety regulations are intended to redress externalities, then suboptimal incentives for guiding actions must be present at the outset. Thus, to understand the welfare effects of food safety regulations it is necessary to understand the nature of incentives structures. In this paper we develop a model that describes a processor's strategy for a particular type of hazard -- one that occurs randomly in a homogeneous system, whose damage may or may not accumulate over time and for which control is costly. All contamination incidents are assumed to impose cost -- even if intervention restores the product to its initial condition. In this situation we find that the optimal strategy involves equal spacing of one or more interventions. With an application to the control of an insect pest in stored grain, we show how the frequency of intervention varies directly with the frequency of contamination and inversely with the fixed cost of an intervention. We also show the less obvious conclusion that an increase in the variable cost of intervening actually increases the incentive to intervene.

We then consider the effects of regulatory regimes consisting of either: a) random inspection with fines for violation of a threshold, or b) terminal inspections for violation of a threshold. Random inspections unambiguously result in more frequent interventions, and increases in either the frequency of inspection or the penalty enhance the processor's incentive to control the hazard. However, we find that variation in the threshold contamination level in this regime has a counterintuitive effect. Because a lowering of the standard (i.e., an increase in the tolerable contamination level) has the effect of lowering the number of interventions at which the fine will be avoided, it increases the benefit of (i.e., subsidizes) intervention, and the frequency of intervention will increase. With terminal penalties, the optimal response involves either an additional intervention or a rescheduling of existing interventions, and higher rates of infestation or higher growth rates by infecting pathogens will favor the former.

While we focus only on costs that are directly borne by a processor, we recognize that regulatory interventions are usually aimed at external costs. We show how a social damage function can create a wedge between private and social cost, and thus may affect policy design. However, we leave the specification of that social damage function to others.

**Notes**

[1]Fox is an assistant professor at the Department of Agricultural Economics, Kansas State University, Manhattan. Hennessy is an assistant professor at the Department of Economics, Iowa State University, Ames. Senior authorship is not assigned. Journal paper No. JB17679 of the Iowa Agriculture and Home Economics Experiment Station, Ames, Iowa, Project No. 3463, and supported by Hatch Act and State of Iowa Funds. The authors would like to thank Ron Mittelhammer, Rich Sexton, three anonymous reviewers, and participants at the NE-165 Conference, Washington, D.C., June 1998. Republished with permission of The American Agricultural Economics Association, Iowa State University, 415 S. Duff St., Suite 3, Ames, IA 50010. "Cost Effective Hazard Control in Food Handling," D.A. Hennessy & J.A. Fox, *American Journal of Agricultural Economics*, May, 1999. Reproduced by permission of the publisher via Copyright Clearance Center, Inc.

[2]We thank an anonymous reviewer for seeking clarifications on the issue of irreversible costs.

[3]Noting that the process is ongoing, we will verify that partitioning into equal time periods acts to minimize cost. Consider the case of two adjacent time periods, $[u_{j-1}, u_j]$ and $[u_j, u_{j+1}]$, which are not of equal length. Primal cost over these two periods is the function:

$$C(\lambda, \beta, F, u_{j+1} - u_{j-1}, \phi, n = 2) = \sum_{i=0}^{i=1} H[G(u_{j+i} - u_{j+i-1})] + 2F.$$

Because this function is convex in the argument of $G(\cdot)$, the cost is minimized by setting $u_j - u_{j-1} = u_{j+1} - u_j = (u_{j+1} - u_{j-1})/2$. Now consider the entire period, $[0, T]$. Then the process of cost reduction, through repeated averaging, will give rise to equation (5).

[4]Technically, this equation is valid only for an infinite process because it assumes that the cost of the initial application equals the cost of all other applications. The watering of vegetables on supermarket shelves and the cleaning of food preparation areas at a restaurant might be viewed as infinite production processes. The equation might not be true for finite production processes because the first intervention may differ, e.g., product entering may have strictly positive contamination. This is not a problem if $\phi = 0$, i.e., if variable costs of application are zero. Finite production processes can be modeled when $\phi \neq 0$, but they do not lend themselves as readily to analysis.

[5]Of course, there is the discreteness problem that only a limited number of futures contracts are traded each year. If forward markets are readily available, then this may not be as much of a problem. Also, we ignore basis risk.

[6]Heterogeneity in storage costs across firms, or opportunity costs that vary as other crops are harvested, will ensure heterogeneity in optimal $T$.

[7]A supplement to Table 1 showing the effect of the variable intervention cost on the placement of the final intervention when the process is not of infinite duration is available from the authors. As expected, for a fixed number of interventions, increases in $\beta$ and decreases in $\phi$ reduce the incentive to intervene earlier. For example, when $\phi = 0$ then a single intervention will be placed at day 125. With $\phi = 0.008$, $\beta = 0.05$ and $\lambda = 80$, placing the intervention at day 111 will reduce cost by \$22 compared with acting on day 125. When $\beta$ increases to 0.07, cost is minimized by intervening on day 115.

[8]Note that while insect damage is cumulative, the number of insects is not.

[9]In fact, minimum intervals for pesticide use before worker re-entry (e.g., to harvest) have sometimes been mandated for fruit orchards, and Lichtenberg et al. have reported that Oregon plum growers tended to use parathion insecticide at the minimum required interval to protect against coddling moth damage even if there were no signs of a problem at that time.

**References**

Antle, J.M. "Efficient Food Safety Regulation in the Food Manufacturing Sector." *Amer. J. Agric. Econ.* 78(December 1996):1242–47.

Arthur, F.H. "Grain Protectants: Current Status and Prospects for the Future." *J. Stor. Prod. Res.* 32(1996):293–302.

Beckett, S.J., B.C. Longstaff, and D.E. Evans. "A Comparison of the Demography of Four Major Stored Grain Coleopteran Pest Species and their Implications for Pest Management." Proceedings of the 6th International Working Conference on Stored-product Protection, Canberra, Australia, April 1994.

Chavas, J.P., and R. Pope. "Nullity Restrictions and Comparative Statics Analysis." *Int. Econ. Rev.* 33(February 1992):15–31.

Feder, G. "Pesticides, Information, and Pest Management under Uncertainty." *Amer. J. Agric. Econ.* 61(February 1979):97–103.

Hall, D.C., and R.B. Norgaard. "On the Timing and Application of Pesticides." *Amer. J. Agric. Econ.* 78(May 1973):198–201.

Holthausen, D.M. "Hedging and the Competitive Firm under Price Uncertainty." *Amer. Econ. Rev.* 69(December 1979):989–95.

Lichtenberg, E., R.C. Spear, and D. Zilberman. "The Economics of Reentry Regulations of Pesticides." *Amer. J. Agric. Econ.* 75(November 1993):946–58.

Reed, C., and F. Worman. *Quality Maintenance and Marketing of Wheat Stored on Farms and in Elevators in Kansas: Description, Techniques, and Innovations.* Kansas Agricultural Experiment Station, Bulletin #660, Manhattan KS, 1993.

Reed, C., V.F. Wright, J.R. Pedersen, and K. Anderson. "Effects of Insect Infestation of Farm-Stored Wheat on its Sale Price at Country and Terminal Elevators." *J. Econ. Entomology* 82(1989):1254–61.

Royden, H.L. *Real Analysis*, 3rd ed. New York, NY: Macmillan Publishing Co., 1988.

Silberberg, E. "A Revision of Comparative Statics Methodology in Economics, or, How to Do Comparative Statics on the Back of an Envelope." *J. Econ. Theory* 7(February 1974):159-72.

Szmedra, P.I., M.E. Wetzstein, and R.W. McClendon. "Economic Threshold under Risk: A Case Study of Soybean Production." *J. Econ. Entomology* 83(June 1990):641-46.

Taylor, H.M., and S. Karlin. *An Introduction to Stochastic Modeling*. San Diego: Academic Press, Inc., 1984.

Trematerra, P., F. Fontana, and M. Mancini. "Analysis of Development Rates of *Sitophilus oryzae* (L.) in Five Cereals of the Genus *Triticum*." *J. Stor. Prod. Res.* 32(1996):315-22.

United States Department of Agriculture (USDA). "Stored Grain Advisor: An Expert System for Stored Grain Management." Agricultural Research Service, U.S. Grain Marketing Research Lab., Manhattan Kansas, April 1995.

Unnevehr, L.J., and H.H. Jensen. "HACCP as a Regulatory Innovation to Improve Food Safety in the Meat Industry." *Amer. J. Agric. Econ.* 78(August 1996):764-69.

Williams, J.C., and B.D. Wright. *Storage and Commodity Markets*. Cambridge, U.K.: Cambridge University Press, 1991.

# Appendix A

Using the first-order condition yields:

(A1)
$$\frac{\partial n^*}{\partial \lambda} = -\frac{(H'[\cdot]+\phi)}{\delta\beta}\left(e^{\beta T/(n^*+1)} - 1 - \frac{\beta T}{n^*+1}e^{\beta T/(n^*+1)}\right)$$

$$+ \frac{H''[\cdot]\lambda T e^{\beta T/(n^*+1)}(e^{\beta T/(n^*+1)} - 1)}{\delta\beta(n^*+1)}$$

The lower term is assuredly positive. Setting $x = \beta T/(n+1)$, the upper right-hand term in parentheses becomes $y(x) = e^x - 1 - xe^x$. Noting that $y(0) = 0$ and $dy(x)/dx = -xe^x \leq 0$ for $x \geq 0$, we have $y(x) \leq 0$ for $x \geq 0$, and so $\partial n^*/\partial \lambda \geq 0$. Also, using the same substitution, it can be shown that $\partial G(T/(n+1))/\partial \beta = \lambda(xe^x - e^x + 1)/[\beta^2] > 0$. Turning to $\phi$,

(A2)
$$\frac{\partial n^*}{\partial \phi} = -\frac{\lambda}{\delta\beta}\left[e^{\beta T/(n^*+1)} - 1 - \frac{\beta T}{n^*+1}e^{\beta T/(n^*+1)}\right] > 0.$$

# Appendix B

For accumulating damage, the first-order condition and second-order conditions for (12) are:

(B1)
$$H'[\cdot]\left[G(T/(n+1)) - \frac{\lambda T}{n+1}e^{\beta T/(n+1)}\right] + F = 0,$$

(B2)
$$\delta_1 = \frac{\partial^2 C(\cdot)}{\partial n^2} = H''[\cdot]\left\{G(T/(n+1)) - \frac{\lambda T}{n+1}e^{\beta T/(n+1)}\right\}^2 + \frac{H'[\cdot]\lambda T^2 \beta}{(n+1)^3}e^{\beta T/(n+1)} > 0.$$

Clearly, $\partial n^* / \partial F = -1/[\delta_1] < 0$ and $\partial Ln(n^*+1)/\partial Ln(T) \equiv 1$. After some work, we obtain:

(B3)
$$\frac{\partial n^*}{\partial \lambda} = \frac{F}{\delta_1 \lambda} - \frac{1}{\delta_1}\frac{H''[G(T/(n+1))]\lambda}{\beta^2}(e^{\beta T/(n+1)} - 1)\left[e^{\beta T/(n+1)} - 1 - \frac{\beta T}{n+1}e^{\beta T/(n+1)}\right],$$

which is positive given the analysis in Appendix A. As for $\partial n^* / \partial \beta$, we could not sign it even for the linear damage specification in the nonaccumulating damage case. Because the accumulating damage objective function degenerates into the nonaccumulating damage objective function for the linear specification, the sign is ambiguous for accumulating contamination also.

Chapter 14

# A Real Option Approach to Valuing Food Safety Risks

Victoria Salin[1]

## Introduction

Business risks related to price and/or cost variability can affect investment decisions, according to research in the area known as "real options" (Dixit and Pindyck 1994, among others). Uncertainty about future returns from an investment project generates an incentive for investors to postpone the initiation of a project, even if standard capital budgeting rules would call for immediate investment. Delays in investments can have significant implications for modernization of a firm or an industry, and also can affect industry structure over the long-run.

This research incorporates the risks of a food safety event into an agribusiness investor's decision problem and uses real option valuation techniques to assess how investments would be affected by a food safety incident. Concerns about food safety affect investors' perceptions of future returns from a food processing project, and would likely be important factors affecting processing firms' decisions of whether to invest in maintenance of equipment or installation of upgraded capital. HACCP implementation may include significant new capital expenditures, perhaps in equipment or in information systems to provide trace back capability. To the extent that HACCP changes the probability distribution of food safety risks, investment decisions might be affected.

Risks related to food safety do not follow well-known, bell-shaped distributions. Generally, the probability of contamination reaching consumers or products being recalled is very low. But if the event occurs, the returns for the firm can be expected to drop in a discrete fashion, perhaps to disastrously low levels. For example, Odwalla, Inc., lost more than $12 million in the quarter following the October 1996 outbreak of E. coli O157:H7 in its apple juice (*New York Times*, Jan. 4, 1998).

The objective of this research is to assess the impact of food safety risks on capital investment by agricultural and food firms. The real option framework is used because it is a way to explicitly value how uncertainty affects financial decisions. The option value is a measure of the opportunity cost that the firm incurs by making an investment immediately, rather than holding the option on a potential project and waiting to see how uncertainty is resolved. Significant option values can explain why firms delay investments even when the present value of expected cash flows exceeds the trigger levels that are found from deterministic capital budgeting models.

Without HACCP, the uncertainty associated with food safety risks might provide an incentive for firms to delay entering the industry, or to delay renovating existing plants and equipment. To the extent that HACCP reduces food safety risks, the opportunity costs of immediate investment would fall, and HACCP requirements could add value to the industry. If HACCP eliminates all food safety risk, then part of the option value disappears and new investment would occur faster than in the system without HACCP. More investment, or more rapid investment, could lead to a more competitive and efficient food processing and distribution sector, increasing the total value of the food industry.

Experience with the real option model in valuing uncertainty indicates that results are sensitive to parameters assumed for the risks the firm faces. Critical parameters in the option pricing function relate to project characteristics, market prices, and interest rates. In applying the model to the context of food safety, the parameters that describe the risk of a discrete drop in the firm's returns are introduced as well. Another objective of this research is to examine the sensitivity of the option value to the food safety related parameters, and to consider whether the option value is more strongly affected by other risks in the market. If the results prove to be strongly affected by the food safety parameters, more emphasis on researching the appropriate elements of the distributions of food safety risks will be warranted.

The remainder of this paper is organized as follows: (1) Presentation of the real option model that represents food safety risks. The underlying stochastic processes are explained and the expression for option valuation is derived. (2) Data needs are described and the suitability of the Food Safety and Inspection Service recall database is evaluated. (3) Results in two areas are presented: first, option values are calculated for a hypothetical investment project, and second, the effects of various model parameters are assessed. (4) Conclusions.

## The Model

Consider the following example: In order to satisfy HACCP requirements, suppose a firm must install new processing equipment. This equipment is specific to the firm and thus is a sunk cost. Uncertainty about the returns from that investment come from two major sources: (1) market risk (prices, input costs, and consumer choices); and (2) some remaining small food safety risk. These uncertainties can generate a large positive opportunity cost to investing immediately, rather than waiting to see how the uncertainty is resolved (Dixit and Pindyck 1994). The option value, or opportunity cost, constitutes an additional hurdle that the expected net present value of cash flows must cover, above and beyond the sunk cost of the project.

The basic concept of real options begins with an investor who holds an opportunity to undertake a project. That investment opportunity has value, whether or not the project is in place. The value of the project over time is not known for certain at the time of investment. But an opportunity to invest in a project that might provide positive returns is an asset that can be valued, using financial option valuation techniques. The effect on investment behavior of the uncertainty related to food safety can be valued precisely in an option-based framework, given certain critical parameters and assumptions about the stochastic process for future returns from the investment.

In this model, future returns are assumed to follow a mixed Brownian motion and Poisson process. This mixed stochastic process moves continuously (perhaps due to price or production variability), and is also subject to discrete jumps (due to food safety events). The characteristics of these processes are described in the following sections.

**Stochastic Process: Poisson Component**

Assume that the length of time a firm operates before a food safety problem occurs follows an exponential distribution. The time before the jump to a lower level of returns is a random variable with range on the interval $[0,\infty)$. The probability of the jump occurring at time T is:

$$(1) \qquad P(a < T \le b) = \int_a^b \lambda e^{-\lambda t} \, dt,$$

where t is the current time, $\lambda$ is the positive exponential hazard rate parameter, and $e$ is the natural exponential function (Pitman 1993). The probability that a food safety problem sufficient to affect firm revenue occurs is defined as $exp(-\lambda t)$. The hazard rate, or probability per unit time of the event occurring just after time t, is $\lambda$. Because the expectation of T is inversely related to $\lambda$ ($E(T) = 1/\lambda$), a subjective determination of the size of $\lambda$ can be made based on investors' prior beliefs. This feature makes the model adaptable to business decision makers' perceptions of risks. For example, suppose the investor expects that no problem will occur during the next two years. One approximation is to define expected lifetime of the project operating without a safety problem as $E(T) = 2$, which yields $\lambda = \frac{1}{2}$. There is a chance of a food safety event during any period, but the probabilities of safe operation are higher in periods before $t = E(T)$ (Table 1).

Three properties of the exponential distribution should be noted: (1) the upper bound is infinite, (2) constant failure rate or hazard rate, and (3) memoryless property (Pitman 1993). A constant failure rate generates probabilities of the event occurring that generally increase over time (Table 1). Such an assumption might be consistent with a view that bacterial contamination in the food supply is increasing in frequency and in severity, so that firms should expect more costly food safety problems to occur in the future.

## Stochastic Process: Brownian Motion Component

In addition to the discrete movements in cash flows, the future returns from the investment project are assumed to move in a continuous fashion. The typical model for the continuous movement is a Brownian motion process. The origins of Brownian motion processes are in physics, specifically the characteristics of a heavy particle being bombarded by lighter particles. It has been adopted in financial economics as a mathematical description of the probability distribution of the future price of an asset. The movements of the asset price are assumed to be normally distributed, with mean and standard deviation that depend only on the amount of time that has passed (Chriss 1997: 96-97). A typical Brownian motion process would be written as:

(2) $$dV = \alpha \, V \, dt + \sigma \, V \, dz,$$

where $dz$ is the increment of a Brownian motion process with drift parameter $\alpha$ and variance rate $\sigma$. The first term on the right side is the expected growth rate of V, or trend parameter. The second term is sporadic variability, or volatility in the value of the asset.

## Model of Returns to Investment Under Uncertainty

The model used to estimate a value for the option to postpone the investment begins with standard expected value concepts in finance and extends them to include uncertainty. Define the value of an investment opportunity, F(V), as the expected present value from investing at the optimal time (Dixit and Pindyck 1994):

**TABLE 1**
**Illustrative Probabilities From Exponential Survival Function**

| t | Probability of No Food Safety Problem, if $E(T) = 2$ $P(T>t) = e^{-\lambda t}$ |
|---|---|
| 1 | .60653 |
| 1.5 | .47237 |
| 2 | .36788 |

(3) $$F(V) = \max_T E_0 \left[ (V_T - K) e^{-\rho T} \right],$$

where E denotes the expectation operator, T is the optimal time to invest,[2] $V_T$ is the expected present value of the investment made at time T, K is the sunk cost of the project, $e$ is the natural exponential function, and r is the discount rate. This is a Bellman equation for the value of an asset, F(V), which is the investment opportunity or option. V is a function of state variables ($x$) and control variables ($u$) in the usual style of Bellman's equations. The state variable that defines the level of profit might be whether the firm has invested or not. Control variables, or choice variables, might be the quantity of labor hired. V is also a function of the current time, $t$. The functional relationships are written in explicitly in the next equation.

The key factor in valuation of the option on the investment is the future cash flows from the operating project, should it be undertaken. Thus the maximization problem is subject to a stochastic process specified for the evolution of V, which is the value of the operating project.

Both sides of equation (3) are multiplied by $(1+\rho \Delta t)$ and the limit as $\Delta t \to 0$ is taken, in order to convert to continuous time. The Bellman equation in continuous time is:

(4)
$$\rho F(V(x, u, t)) = \max_u \left[ \pi(x, u, t) + \frac{1}{dt} E[dF(V)] \right]$$

The left side of this equation is the normal return per unit time that is required to hold the asset F(V). The right side of equation (4) is the immediate profit, if the investment is made ($\pi(x,u,t)$), plus the capital gain or loss expected from holding the option ($E[dF]$). Profit is zero if the investor is still holding the option. Capital gain depends on the instantaneous change in the value of the investment opportunity and is tied to the stochastic process specified for V.

In order to find the value of the option, consider periods before the investment is made.[3] Profit flow from the project is zero, so the first term on the right drops out of the Bellman equation. Multiplying through by $dt$ yields:

(5) $$\rho F(V(x, u, t)) dt = E[dF(V)]$$

This is the basic relationship between the value of the investment opportunity ($F(V)$, the option) and the expected value of the project in place ($V$). On the left side is the total return on the investment opportunity.

It equals the expected gain from holding the option and depends on the evolution of the asset returns. Suppose V evolves according to a combined geometric Brownian motion and jump process so that:

(6) $$dV = \alpha V dt + \sigma V dz - V dq$$

The last term, $Vdq$, represents the Poisson process for the random timing of discrete jumps in V. It defines the probability that the disaster occurs during an infinitesimal interval of time, $dt$. Assume that $dz$ and $dq$ are independent and let

(7) $$dq = \begin{cases} 0 & \text{with probability } 1 - \lambda\, dt \\ \phi & \text{with probability } \lambda\, dt, \end{cases}$$

where $\phi$ is the percentage by which $q$ will change if the Poisson event occurs ($0 \leq \phi \leq 1$). If $\phi = 1$, the firm shuts down and stays closed forever.

Returning to the Bellman equation (5), expand $dF$ using Ito's lemma[4] for differentiation of stochastic processes and obtain an expression in terms of $dV$. With the right side in terms of V, the option valuation function can be seen to be a function of the evolution of the underlying asset=s value. That is, the value of the investment opportunity depends on the expected value of the project. At this point, the choice of stochastic process for V enters the derivation. Expanding the right side of equation (5), substituting, and rearranging:

(8)
$$\rho F(V) dt = E\left[ F'(V) dV + \frac{1}{2} F''(V) dV^2 \right]$$
$$\rho F(V) dt = \alpha V F'(V) dt + \frac{1}{2}\sigma^2 V^2 F''(V) dt - \lambda\left[ F(V) - F((1-\phi)V) \right] dt$$
$$0 = -(\rho + \lambda) F(V) + \alpha V F'(V) + \frac{1}{2}\sigma^2 V^2 F''(V) + \lambda F((1-\phi)V)$$

This is a second order homogeneous differential equation in F(V), the value of the investment opportunity. It is solved for F(V) subject to the following boundary conditions:

(9)
$$F(0) = 0$$
$$F(V^*) = V^* - K$$
$$F'(V^*) = 1$$

where V* is the expected net present value of the project at which immediate investment is optimal (trigger value). This equation has a solution of the form:

(10)
$$F(V) = AV^\beta, \text{ where } \beta = f(\sigma, \rho, \alpha, \lambda, \phi), \beta > 1$$

The parameter $\beta$ is found by obtaining a solution to a nonlinear equation. Numerical methods are typically required to solve for $\beta$; however, if $\phi = 1$, an algebraic solution for $\beta$ exists.

## Data

The data required to apply this model are the parameters of the probability distributions for (1) returns on food processing industry investments, and (2) occurrence and severity of food safety events, from the point of view of business cash flows. The Food Safety and Inspection Service (FSIS), U.S. Department of Agriculture, maintains a database of meat product recalls, which is used as a starting point for examining the probability distribution of a food safety event (Table 2). This database records the quantity of products recalled, by firm, but it is not ideal for using directly in the real option model. Ideally, the data would provide a direct estimate of the effect on expected cash flows from a food safety problem. The cash flow effects are a function of the value of sales lost from the recall itself, and perhaps some sales losses to the brand or product that last after the recall has ended. The recall database is in quantity terms, so more inquiry is needed to estimate value of the recall. Duration of market reaction to a recall is not provided in the FSIS records, and is likely to be more difficult to uncover than the value of recalled products.

The frequency distribution of product recalls during 1996 and 1997 indicates that most recalls for bacterial contamination were for relatively small quantities of meat (Figure 1). Thirteen of the 23 product recalls involved fewer than 10,000 pounds of meat products. One large recall in 1997 (25 million pounds of ground beef produced by Hudson Foods) raised the average size of a recall incident to one million pounds over the two years.

The examination of the FSIS database indicates that frequency of food safety problems and quantity recalled are inversely related. These are the two key parameters needed to operationalize the jump process option valuation model. To consider estimating the frequency of occurrence, one must look at the 23 recalls, totaling 26.48 million pounds, against the total pounds shipped in the $94 billion meat products industry (U.S. Census Bureau 1992).

## Results

Real option values were calculated for a hypothetical investment project. Depending on the specifics of the model, these experiments indicate that option values are between 39 and 59 percent of sunk costs of the investment project (Table 3). These estimates suggest that the opportunity cost of investing immediately is between $390,000 and $590,000. The uncertainty associated with future cash flows generates a significant incentive to delay investments until uncertainty is resolved.

The limited data on food safety related product recalls indicates an inverse relationship between the probability of occurrence and the severity of the problem, so it is interesting to consider option values for

TABLE 2

Meat Products Recalled by Food Safety and Inspection Service for Bacterial Contamination, 1996 and 1997

| Case No. | Company | Pounds Recalled | Product |
|---|---|---|---|
| *1997* | | | |
| 027-97 | John Volpi & Co. | 500 | salami |
| 025-97 | Beef America Production Co. | 168,725 | ground beef |
| 022-97 | Kayseri Basterma Inc. | 340 | sausage |
| 021-97 | Beef America Operation Co. Inc. | 443,656 | ground beef |
| 020-97 | Herman's Quality Meat Shop | 100 | ground beef |
| 018-97 | Great Value Supermarket | 70 | ground beef |
| 017-97 | Bavarian Meat Products | 450 | sausage |
| 015-97 | Hudson Foods | 25,000,000 | ground beef |
| 010-97 | Angus Meats | 1,427 | ground beef |
| 009-97 | Tyson Foods Inc. | 14,000 | grilled chicken |
| 008-97 | Texas American Foodservice | 576,000 | beef patties |
| 007-97 | Hans Kissle Co. | 5,400 | chicken salad |
| 006-97 | Hester Industries | 50,000 | cooked chicken |
| 004-97 | Iowa Ham Canning | 33,000 | ham |
| 001-97 | Smokey Hollow Foods | 18,848 | sausage |
| *1996* | | | |
| 024-96 | Goulds Country Store | 52 | ham |
| 021-96 | Pioneer Packing Co. | 400 | ham |
| 017-96 | Rochester Meats | 152,000 | ground beef patties, frozen |
| 013-96 | Freedman Food Service | 2,608 | cooked beef |
| 012-96 | Snowden's Sausage Co. | 2,000 | sausage |
| 009-96 | Corfu Foods Inc. | 720 | roast beef, sliced |
| 007-96 | ConAgra (Monfort) | 11,539 | ground beef |
| 006-96 | Mann's Intl. Specialties | 487 | roast beef |

combinations of parameters that roughly match this picture. The numerical experiments generate option values that are 40 percent or less of sunk costs when size of the jump ($\phi$) is high and probability of the jump ($\lambda$) is low. These are at the lowest end of the range of option values calculated. A 10 percent reduction in $\lambda$ results in a 1 to 2 percent decrease in option values in this set of experiments. If $\phi$ is reduced in ten percent increments, option values decline by 0.5 percent or less, close to zero if $\phi$ begins at larger levels. These relationships imply a relatively small anticipated effect of HACCP on investment decisions, to the extent that HACCP would reduce the jump process parameters.

Suppose instead that the probability of a jump is relatively high but the size of the jump is expected to be small. In these experiments, where $\lambda$ is set in the range of 0.4 to 0.5, option values are larger (approximately 45 percent of sunk costs). A 10 percent reduction in $\lambda$ has an impact on option values of around -3 percent. If these parameters more closely describe the investment scenario, then the conclusion could be reached that food safety risks have a more significant effect on the decision.

The results above, from the model that includes food safety risks, were compared to the outcomes from a hypothetical project of the same size, facing the same market risks, but excluding food safety risks. The calculated option value on the $1 million project is greater than if food safety risks are included -- equal to 48 percent of sunk costs of the project. It is apparent that the jump process modification to the model is not a straightforward risk-increasing component in the valuation. The food

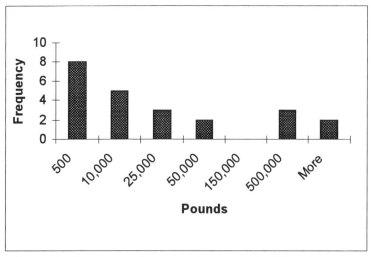

FIGURE 1. Histogram of recalls of meat, 1996-1997. Source: Food Safety and Inspection Service, U.S. Department of Agriculture.

## TABLE 3
### Effects of Food Safety Parameters on Option Values

| Parameter Values | | | % Change in Option Value | |
|---|---|---|---|---|
| Phi | Lamda | Option Value F(V) | From 10% Change in Phi | From 10% Change in Lamda |
| 1.000 | 0.900 | 0.586 | | |
| 0.900 | | 0.586 | -0.000 | |
| 0.729 | | 0.586 | -0.004 | |
| 0.656 | | 0.586 | -0.019 | |
| 0.531 | | 0.585 | -0.139 | |
| 0.478 | | 0.583 | -0.270 | |
| 0.387 | | 0.576 | -0.700 | |
| 1.000 | 0.810 | 0.553 | | -5.703 |
| 0.900 | | 0.553 | -0.000 | -5.703 |
| 0.729 | | 0.553 | -0.005 | -5.705 |
| 0.656 | | 0.552 | -0.024 | -5.710 |
| 0.531 | | 0.551 | -0.159 | -5.739 |
| 0.478 | | 0.550 | -0.297 | -5.764 |
| 0.387 | | 0.543 | -0.727 | -5.819 |
| 1.000 | 0.450 | 0.438 | | -3.081 |
| 0.900 | | 0.438 | -0.000 | -3.081 |
| 0.729 | | 0.438 | -0.019 | -3.084 |
| 0.656 | | 0.437 | -0.059 | -3.090 |
| 0.531 | | 0.436 | -0.236 | -3.105 |
| 0.478 | | 0.434 | -0.370 | -3.104 |
| 0.387 | | 0.429 | -0.691 | -3.060 |
| 1.000 | 0.328 | 0.408 | | -1.985 |
| 0.900 | | 0.408 | -0.000 | -1.985 |
| 0.729 | | 0.408 | -0.027 | -1.988 |
| 0.656 | | 0.407 | -0.070 | -1.990 |
| 0.531 | | 0.406 | -0.225 | -1.980 |
| 0.478 | | 0.404 | -0.324 | -1.957 |
| 0.387 | | 0.401 | -0.525 | -1.851 |
| 1.000 | 0.266 | 0.395 | | -1.367 |
| 0.900 | | 0.395 | -0.000 | -1.367 |
| 0.729 | | 0.395 | -0.028 | -1.367 |
| 0.656 | | 0.395 | -0.068 | -1.365 |
| 0.531 | | 0.394 | -0.187 | -1.334 |
| 0.478 | | 0.393 | -0.255 | -1.296 |
| 0.387 | | 0.390 | -0.367 | -1.156 |

Note: $\alpha = 0$, $\sigma = 0.2$, $I = 1$

safety parameters have an unexpected effect that merits further investigation (reported in the following section).

## Assessment of Relative Importance of Model Parameters

The effects of various parameters are considered separately for their influence on the jump process model results. The findings suggest that there are complex interactions among parameters and that simple comparative statics rules are not obtainable. Thus, the use of the option valuation model will require researchers to cautiously investigate the context of the decision problem before reaching conclusions about the importance of food safety related risks in the investment decision.

A key question is whether the option values are more sensitive to size of the jump or to frequency of the jump's occurrence. Previous research using a jump process option valuation model suggests that business decision makers are more concerned with the size of a jump than the frequency of the jump (Willner 1994: 232). Willner's discussions with venture capitalists were the basis for this conclusion about the relative importance of jump size. The venture capitalists stood to gain from a successful start-up enterprise, while their liability for losses was limited. In spite of the anecdotal evidence about the importance of jump size, Willner's quantitative model did not place a larger weight on the size parameter over the probability parameter; his results were that a 10 percent increase in either the size of the jump or the frequency of the jumps had similar effects on the option value of the venture.

In this application to food safety, the jumps are reductions in the firm's revenue and even a small incident may affect the reputation of a firm or the entire industry. Further inquiry is warranted to determine if jump size or frequency is more critical to food and agribusiness decision makers, or if both are equally important. Then the model structure should be such that the appropriate weights are given to the parameters.

### Influence of Jump Process Parameters on Real Option Values

The next section of this paper examines how the structure of this option valuation model treats changes in the two food safety parameters. The results are from numerical examples of model solutions and do not represent a particular firm or industry situation. The findings indicate that:
- the probability of a jump has ambiguous effects on the results,
- jump probability dominates size of the jump in determining option values, and
- size of the project matters to the influence of jump probability.

First consider the case of a large and fatal jump downward in the firm's revenue following a food safety problem. The firm's cash flows drop to zero and never recover. This case is illustrated when $\phi$, jump size, is set equal to one. This assumption provides the useful feature of an exact algebraic solution to the option valuation model. Figure 2 shows the results of the option valuation exercise when $\phi = 1$, $\lambda$ varies from 0.1 to 0.9, and the size of the project varies from $0.5 million to $1.38 million. Increasing $\lambda$, the likelihood of a jump in project returns, increases option values when the project is relatively large, as can be seen from the lines in Figure 2 beginning at higher option values with higher $\lambda$. But the lines cross, resulting in regions in which option values are decreasing in $\lambda$.

It seems an unhelpful and counter-intuitive result that rising probabilities of a jump could either increase or decrease the option value. The reason for this contradictory result is that $\lambda$ enters into the stochastic process for future returns in two places, affecting both variance and trend components of the stochastic process. These two components interact in determining the final option value. Larger values of $\lambda$ add to the variance in project returns. This added uncertainty tends to increase the value of an option to postpone the investment. The variance effects of $\lambda$ are counteracted by the impact of $\lambda$ on expected growth of the project, or capital gain (Dixit and Pindyck 1994: 168). Expected capital gain on a project that occurs while the option is held adds to the value from waiting. Lamda reduces the expected capital gain from holding the option, by the probability of a jump occurring ($E(dV\ dt/V) = \alpha - \lambda$). Thus, higher $\lambda$ reduces the capital gains and reduces the option value.

These experiments demonstrate that the capital gain effect dominates the variance effect when the project is relatively small, so that higher levels for $\lambda$ generate net reductions in the value of an option on the investment. This conclusion is opposite to what one would expect from interpreting $\lambda$ purely as a risk-increasing parameter. For a larger project, which carries with it a larger value from waiting to avoid a mistaken

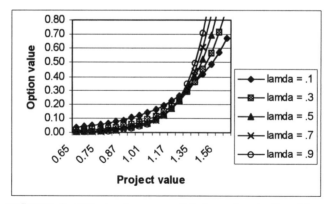

FIGURE 2. Effect of exponential hazard rate (lamda) on option values.

investment decision, the effect of λ is consistent with its interpretation as a risk-increasing parameter because the variance effect is more significant than the reduction in capital gain.

Now consider jumps that harm the firm but do not cause a complete shutdown.[5] The jump size and the probability of its occurrence interact in determining option values, leading to the possibility of option values increasing in jump size or decreasing in jump size. The key parameter in understanding this result is β. The option valuation function is a U-shaped curve when plotted against the parameter β (Figure 3). Beta is positively related to both the size of the jump (φ) and to the probability of a jump (λ), but the relationship proceeds at different rates.

When λ is very low,[6] increases in jump size lead to an increase in β, yet the option value falls. This is another counter-intuitive result, apparent in the downward sloping part of the curve in Figure 3. The finding that option values can decrease in jump size is unexpected, but might be considered irrelevant given that the event is highly unlikely.

If the decision maker is less confident about the prospects of avoiding a food safety event, then a larger value of λ would reflect that belief. A range of .5 to .9 for λ implies that the expected time before a problem occurs is from 1 to 2 years. For values of λ in this range, the option valuation function is along the upward portion of the U-shaped curve (Figure 3), implying that option values are increasing in λ and increasing in φ. Thus both the jump size and the probability of a jump tend to increase the value of waiting to invest. In this range, HACCP programs would provide investors with an incentive to proceed more rapidly than in a scenario without HACCP efforts in place.

The frequency of a jump (λ) has a larger effect on option values than does the size of the jump (φ) in this model. In experiments in which φ is reduced by increments of 10 percent, holding λ fixed, the percentage change in option values never exceeds 1 percent (Table 3). When λ is reduced by increments of 10 percent and φ held constant, the option values

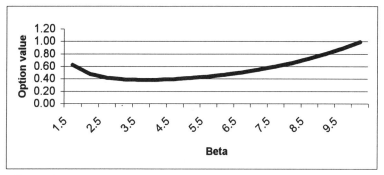

FIGURE 3. U-shaped relationship between option values and beta.

fall by 4.6 to 5.7 percent, beginning at large $\lambda$. For $\lambda$ at lower levels, the percentage change in option values is less (1.1 to 3 percent), but it is still clear that option values are more sensitive to changes in the frequency of jumps than to the size of jumps.

This model structure, in which the probability of a jump occurring is the most important parameter, would be consistent with business decision makers' concerns about "reputation effects" from a food safety problem, even if it were small in terms of economic harm. To eliminate or reduce the risk of even small food safety problems would have value in terms of investment decisions. Efforts to minimize the scale of problems without changing their probability would not affect investment financing decisions as much.

The finding that jump probability has a more significant influence than the size of the jump begs the question -- which parameter does HACCP affect? Does HACCP affect both parameters? The answer to this question will require further investigation of the context of the particular investment problem.

### Influence of Market and Project Characteristics on Real Option Values

In addition to the effects of changing jump process parameters, parameters that describe the investment project and the general business climate affect option values. The previous research in the finance area has uncovered some comparative statics rules (Dixit and Pindyck 1994). For example, option values increase in volatility of future cash flow ($\sigma$). But many of these parameters interact. Two parameters were considered in this chapter, the expected value of the project (V), and the market discount rate ($\rho$).

The expected present value of the investment project in place was allowed to vary in these experiments. Reductions in V reduce the option value, at a decreasing rate, as shown by the curves in Figure 1. This scale effect is the largest driving force in option values among all the parameters studied. A 10 percent increase in V causes an approximately 20 percent increase in the option value. This effect is best understood by considering V as the value at risk. If more is at stake, there is a larger opportunity cost of entering into the project too early.

The market risk represented by the discount rate parameter has a modest effect on option values. When the discount rate was increased (starting at 4 percent), option values fell by approximately 2 percent. The discount rate relationship is decreasing at a decreasing rate, with negligible change in option values occurring when interest rates moved above 12 percent. The effect of changing discount rates was comparable in size to the effect of changing the probability of the food safety problem occurring.

## Conclusions

Food safety risks are included in an investment decision model using a jump process. The model described in this paper allows consideration of two food safety parameters: (1) the probability of occurrence of a problem, and (2) the size of the discrete drop in cash flows should the problem occur. Option values range from 39 to 59 percent of sunk costs of the project, suggesting significant incentive to postpone investments due to uncertain future cash flows.

The probability of occurrence of a problem is the dominant food safety parameter in determining real option values in this model. This model structure is appropriate if business decision makers believe that the probability of a food safety event occurring is more important than the size of the event, in terms of the decrease in cash flows for the firm. If further inquiry reveals that decision makers are less concerned with frequency of problems than with scale of the problem, then an alternative model structure must be considered.

A decrease in the probability of occurrence of food safety problems can either raise or reduce real option values. This outcome is counter-intuitive, given that a decrease in risk would be expected to reduce option values. It implies that any risk-reducing effects of HACCP might not increase the incentives for rapid investment in food industry modernization. Researchers need to understand this ambiguity and be clear about the driving forces. The unexpected results occur when projects are smaller and when the probability of occurrence of a problem is very low. These general characteristics could easily describe many investments in the food processing industry.

The numerical experiments also revealed a range of parameters in which reductions in the probability of occurrence and the scale of the food safety problem reduced option values. If HACCP programs affect the probability of outbreaks, they have long-term financial implications for firms and for the food industry overall through the linkage to investment decisions. Risk reduction accomplished through HACCP would be an incentive for more rapid investment, which would enhance efficiency and could offset some costs of HACCP implementation.

Researchers have called for collection of scientific/medical data about disease incidence as well as information about economic tradeoffs related to food safety (Roberts and Smallwood 1991). This examination of financial decisions under uncertainty points out some specific data needs for evaluating business risks related to food safety. Understanding what factors determine the hazard rate parameter is the first target, according to this research. The probability of a drop in cash flows resulting from a food safety event may be defined at the firm level, and could perhaps be understood by surveys or interviews to elicit subjective probabilities or tolerances. It is also possible that food safety events in another firm affect cash flows industry-wide. Such spillover effects can be estimated from commodity market data or stock price data.

### Notes

[1] Victoria Salin is Assistant Professor, Department of Agricultural Economics, Texas A&M University, College Station, Texas.

[2] Note that the T at which the investment is made is not necessarily the same as the time T used to represent occurrence of the jump process.

[3] Once it is optimal to invest, the option value is irrelevant, so the only option pricing function of interest is the one for the continuation region, before investing.

[4] Ito's lemma is a rule for differentiation of functions of stochastic processes. Ito's lemma for combined Brownian motion and jump processes is shown by Dixit and Pindyck (1994:86).

[5] The model has no exact algebraic solution under this assumption. Numerical approximations were obtained by solving a nonlinear equation for $\beta$, given values for other parameters. The approximations, made using the Mathematica software, used the secant method of solution and were accurate on the order of $10^{-7}$ to $10^{-11}$.

[6] "Very low" is .1 or .2, implying that decision makers expect five to ten years to pass before a food safety problem occurs.

### References

Chriss, Neil A. 1997. *Black-Scholes and Beyond: Option Pricing Models.* Chicago, IL: Irwin Professional Publishing.

Dixit, Avinash K., and Robert S. Pindyck. 1994. *Investment Under Uncertainty.* Princeton, NJ: University Press.

Pitman, Jim. 1993. *Probability.* NY: Springer-Verlag.

Roberts, Tanya, and David Smallwood. 1991. Data Needs to Address Economic Issues in Food Safety, *Am. J. of Agric. Econ.,* Aug:933-942.

U.S. Census Bureau. 1992 Census of Manufactures Food and Kindred Products. *1992 Economic Census: Census of Manufactures.* 26 Nov.1996. <http://www.census.gov/epcd/www/mc92ht20.html> (22 May 1998).

U.S. Department of Agriculture, Food Safety and Inspection Service. FSIS Recall Cases: 1998. 20 Feb 1998. <http://www.usda.gov/fsis/ophs/recalls/rec1998.htm> (16 March 1998).

Willner, Ram. 1995. Valuing Start-up Venture Growth Options. In Real Options in Capital Investment: Models, Strategies, and Applications, ed. Lenos Trigeorgis, 221-239. Westport, CT: Praeger.

Chapter 15

# Economic Efficiency Analysis of HACCP in the U.S. Red Meat Industry

William E. Nganje and Michael A. Mazzocco[1]

## Introduction

Although the focus of Hazard Analysis of Critical Control Points (HACCP) in the meat industry is to increase safety of meat products, this study evaluates whether HACCP can be used by small processing and packing firms to meet the dual goals of increasing safety and improving efficiency in the U.S. meat industry. A major concern by industry participants is to quantify HACCP benefits that are specific to the firm. Although the likelihood and consequence of disease outbreak should be reduced with HACCP, several questions about the costs and benefits to small firms remain unanswered. The anticipated problem with HACCP is that firm managers do not perceive HACCP benefits that may compensate high implementation cost of HACCP systems, especially for smaller firms.

According to Crutchfield et al. (1997), the USDA's Economic Research Service (ERS) estimated the net present value of HACCP regulation for meat and poultry (over a 20-year period) to be between $6.4 and $23.9 billion to society as a whole while the costs of implementing HACCP is estimated to be only $1.9 billion. Comparing this HACCP cost to the benefits to society, one can easily conclude that HACCP will be good for consumers, firms in the meat industry, and regulatory authorities. There are substantial benefits from HACCP to society as a whole, but specific firm level benefits are yet to be quantified especially for small firms.

MacDonald et al. (1995) reported that for the first five years (long term projections) FSIS estimates that small establishments, defined as those with sales of less than $2.5 million, would bear about 45 percent of HACCP costs, or $330.6 million. This cost may be relatively high for small firms, which account for less than 2.0 percent of the industry shipments. Table 1 and Figure 1 support their assertion that smaller firms will bear even larger HACCP cost per pound than bigger firms. Consequently, small firms are concerned about possible HACCP incentives that may compensate HACCP cost. Other issues that support the fear by small firms who must implement HACCP are discussed under industry structure and upward price rigidity.

## Industry Structure

The majority of the firms in the meat packing and processing industry are small firms, defined as those with less than twenty employees or sales less than $2.5 million. Sixty-five percent of plants in the Midwest are in this category (Nganje et al. 1995). The Meat and Poultry Directory documents the number of firms that currently implement quality management tools like Total Quality Management (TQM). Firms operating TQM systems in the meat industry are larger firms. With the similarities of process control systems (e.g., TQM and HACCP), larger firms with TQM may incur lower costs with HACCP implementation than smaller firms. This is one probable reason why smaller firms are expected to incur a greater portion of the HACCP cost in terms of volume of products they produce (MacDonald et al. 1995). It is also an indication that the significant cost savings or increased margins, proposed by quality management experts like Deming (1986), Juran (1974), and Crosby (1979), that may result from a quality management system like HACCP (for smaller plants) may require in-depth analysis.

Though quality management systems are internally driven by the fact that the cost of poor quality exceeds the cost of developing processes

**TABLE 1**

**Actual and Anticipated Implementation and Operating HACCP Expenses**

| Cost Categories | HACCP Cost Categories | Minimum ($) | Mean ($) | Maximum ($) |
|---|---|---|---|---|
| Implementation Cost | Plan development | 1,000 | 5,588 | 35,400 |
| | Training | 900 | 5,074 | 36,000 |
| | Materials and building remodeling | 1,200 | 43,941 | 409,995 |
| Sub-Total ($) | | 3,100 | 54,603 | 481,395 |
| Sub-Total ($)/5 | | 620 | 10,921 | 96,279 |
| Operating Cost | Record keeping and USDA verification | 1,133 | 5,753 | 14,490 |
| | Bacteria testing | 1,560 | 18,510 | 39,000 |
| Sub-Total ($/yr) | | 2,693 | 24,263 | 53,490 |
| Total HACCP | Expenses/year | 3,313 | 35,184 | 149,769 |
| Implementation Amortization | Cost ($/Lb, five years) | 0.0001 | 0.0074 | 0.08886 |
| Operating Cost | ($/Lb) | 0.0003 | 0.0180 | 0.3462 |
| Total HACCP | Expenses ($/Lb) | 0.0004 | 0.0254 | 0.4351 |

Source: Nganje et al. (1995).

which produce high quality products, smaller firms may require different strategies to benefit from any reduction in average cost. This premise does not dispute the cost reduction that may accompany an HACCP system, but investigates whether cost savings are significant to serve as incentives for small firms. This is questionable because with the premise of scale economies, controlling a critical control point (CCP) for a large firm with a large volume of products may result in the firm saving significant amounts (e.g., millions of dollars), while controlling the same CCP for a small firm with a small volume of products may only save the firm an insignificant amount. This study therefore evaluates meaningful levels of cost savings with HACCP implementation, especially for small firms, using efficiency analysis. Such savings may provide firm specific benefits that will enable small firms to compensate high HACCP implementation expenses.

**Upward Price Rigidity with Increased Safety**

Seitz et al. (1994) identify major meat packers and processing firms as operating under oligopoly, with some distinguishing characteristics of their interdependence in pricing and marketing practices. Mutual interdependence in pricing behavior in this situation means that if a firm raise the price of its products, other firms in the industry may not raise prices to the same extent. McDonald et al. (1995) estimated that product prices would not rise with HACCP adoption because large plants produce most of the output and face trivial cost increases, but the costs of small plants could rise enough to place substantial pressure on them. This is true because only one of the HACCP proposal cost components varies with output (antimicrobial, chemical, and physical treatments), whereas the majority of cost components (plan development, pathogen testing, training, and record keeping) vary little as output increases. There was no significant increase in price received after HACCP implementation for

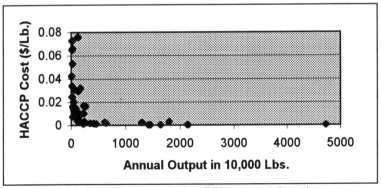

FIGURE 1. Variation of reported HACCP cost and firm size.

small firms (Nganje et al. 1995). It therefore becomes paramount to analyze cost cutting incentives or areas of higher leverage that may be associated with HACCP to provide added incentives other than increased prices. Evaluating strategic efficiency in this study may enable small firms to quantify HACCP incentives other than price increase.

This study develops a methodology to quantify and evaluate incentives (other than price increases) associated with HACCP systems for small meat processors and packers. The study evaluates HACCP cost and efficiency from the small firms' perspective. The objective of this study is to estimate strategic efficiency of firms with and without HACCP systems. Strategic efficiency evaluates whether it is advantageous for small firms to implement mandatory HACCP systems as leaders or as followers.[2] Strategic efficiency is evaluated in this study with economic efficiency and biased technical change. Economic efficiency in this study is analyzed with cost efficiency and pure technical efficiency. Cost efficiency is estimated with elasticities of size from a system of cost share equations. The existence of size elasticity will imply firms can decrease their average cost of production with HACCP systems and therefore significant cost savings can be achieved in the long run. Technical efficiency is estimated with corrected least squares procedure, using a Ray homothetic profit function. Analysis of biased technical change identifies areas of future technical change which have the greatest marginal impact on cost reduction, especially with respect to HACCP costs. This procedure uses parametric variation to quantify biases or cost cutting incentives.

This study addresses the issue of whether HACCP can improve the efficiency of small firms or reduce their marginal cost of production. The management literature confirms that HACCP can be used as an effective quality management tool. But whether this is true for small firms will be analyzed in this study. Though some of the advantages and incentives are often mentioned, it may take extra training to teach plant managers how they can improve their efficiency with HACCP. Training on implementing HACCP as a safety tool and as an efficiency tool may provide added incentives for firms to operate HACCP systems effectively.

**HACCP and Meat Safety**

Inspection for meat and poultry in the United States is about to enter a new era where meat and poultry products will be produced with a quality management system called HACCP. On July 4, 1996, a final proposed rule on pathogen reduction in meat processing and packaging (HACCP) systems was produced by the Food and Inspection Service, USDA (Federal Register/Vol. 60, No. 23). HACCP systems are anticipated to minimize current contamination problems and fulfill the intended inspection purpose of providing a product that is safe when properly handled and prepared for consumption. Under the regulation, processors have the primary responsibility for development and implementation of HACCP systems for meat animal slaughter, carcass fabrication, packaging, and distribution. The major role of the regulatory agency is to verify that a processor's HACCP system is effective and working as intended.

The USDA's goal to renovate the current meat inspection system stems from problems that are uncontrollable by the current inspection system. Domestically, disease outbreaks have been affecting consumers who desire maximum safety. Examples of bacteria outbreaks include: *E. coli* 0157:h7 in 1993, cited by Bjerklie (Meat and Poultry, April 1994) and Karr (Meat and Poultry, February 1994), and Hudson Foods shut down and purchased in 1997 by IBP. Consequences of continued microbial contamination may include a downward trend in meat consumption and may result in a serious economic impact on producers, packers, and processors. The United States Department of Agriculture estimated 6.5 to 81 million cases of foodborne illness occur each year. About 25 percent of these outbreaks have implicated meat or poultry as the food "vehicle" for the causal bacteria (USDA, 1990). In the United Kingdom, Japan, and other countries, mad cow disease or chicken flu are the major concerns. In the U.S., MacDonald et al. (1995), reported that *E. coli* outbreaks are the principal concern regarding meat safety today.

> In the past, the dual goals of public health and consumer protection of meat inspection were centered on hormonal problems, sick animals' inspection with organoleptic methods (using sight, touch, and smell), product adulteration with fillers, and deceptive labeling practices... Microbial contamination is the principal concern today, with pathogens that can cause human illness and are carried in the intestinal tracts of healthy animals. This problem arises because high throughput slaughter and processing methods often combine meat from many carcasses, raising the likelihood of pathogens spread through cross-contamination (MacDonald et al. 1995).

Consequently, inspection with organoleptic methods may be costly and inefficient. Process control inspection methods like HACCP are more effective in reducing pathogens in meat. This is one reason why HACCP may be the most appropriate inspection procedure. HACCP involves the identification of the likely hazards that can cause meat products to be unsafe to consumers; the identification of critical control points (CCPs) in the production process where failure could cause hazards to occur; the establishment of critical limits or operating parameters for each critical control point; and the establishment of procedures for monitoring the CCPs and recording the results (National Restaurant Association 1994). It is a continuous, comprehensive food safety monitoring system that is designed to prevent hazards from developing and thus ensures a high degree of food safety.

According to the WHO/ICMSF meeting (1982), HACCP can be applied to the handling of meat and poultry at home (Zottola and Wolfe 1980), food-service establishments (Bobeng and David 1977; Bryan 1981; Bryan and Mckinley 1974; and UnKlesbay et al. 1977), and to any type of food processing plant (Bauman 1974; Dhew 1972; Ito 1974; Kauffman and Scaffner 1974; Peterson and Gunnerson 1974). The concept completely refocuses attention on controlling the manufacturing process to produce end products that are safe and needed by consumers. However, increasing safety with HACCP systems may be costly for small firms. This raises the question of how small firms will compensate for HACCP expenses. The literature on HACCP as an effective business management tool gives some insights.

**HACCP as a Business Management Tool**

Other than safety, HACCP can be used as an effective business management tool to reduce production cost and improve the efficiency of the meat industry. The fact that HACCP is defined as a process control tool, makes it almost indistinguishable from total quality management (TQM) within the quality management literature in recent years. Like TQM, it is hypothesized that if firms effectively implement HACCP, they may not attain zero defect or bacteria tolerance levels, but they will meet a high degree of safety for meat products and simultaneously lower production cost. TQM and HACCP have several similarities as pointed out by Scott et al. (1992), who suggested that some elements of TQM should be considered as important factors in developing an effective HACCP program. These include 1) management and employment training and education; 2) operator control; 3) teamwork; 4) effective communication between management and workers; and 5) constancy of purpose by management. Consequently, if HACCP systems and TQM systems are properly operated, they can improve the overall management system, which may improve the long-run survival of the firm. This is because the philosophy of continuous improvement enables the firm to adapt easily to new technologies and changing customer needs. In general, firms may enjoy lower costs of production with HACCP because of reduced product reworks, sewer damaged products, and the efficient use of scare resources or resource reallocation.

Although the proposed use of HACCP in government regulation has been increasing in the 1990s, the concept of HACCP has been around for many decades. Mazzocco (1996) pointed out that HACCP systems are an integral part of quality management systems used by many firms in a variety of industries. The underlying principle of implementing HACCP is that the cost of monitoring processing systems and keeping them in control is cheaper (and safer) than the cost of inspecting finished products to determine whether they meet quality standards. In addition, monitoring efforts can be focused on critical control points in the process. If the process exceeds specific limits at a critical point in the process (for example in temperature measurement, pH, moisture, or bacteria counts), an early indication is given that the quality of the product may be compromised if the situation is not corrected.

HACCP systems like statistical process control systems (SPC) identify controllable and uncontrollable variables that are most likely to affect product quality. Critical limits are established to detect defects in the process. Errors are corrected as soon as they are detected by the critical limits. This procedure reduces process variability and produces wholesome end products. We are reminded here that although Pillsbury made the first HACCP food products for space astronauts in 1974, the concept of HACCP has long been used in many other industries.

Like HACCP, SPC has been applied in electronic plants, material factory management, environmental economics, and the military, all with a common goal of lowering costs, saving time, and improving the quality of end products. Kackar and Sheomaker (1986) demonstrated empirically the use of SPC as a cost-effective tool and as a tool to improve the processes of electronic plants with robust designed end products. Santana (1988) also showed that electronic plants can improve their yield forecasts with reduced production costs through SPC systems. Vaughan and Resell (1983) used SPC as a tool to monitor point source pollution. Sprow (1984) studied low cost SPC systems as an important first step for factory manufacturing. The U.S. NAVY (1988) showed in a procurement study how SPC can eliminate 30 percent of a $34.9 billion program by eliminating unnecessary manufacturing and testing procedures.

Though quality management systems can provide incentives to reconfigure production systems, several issues remain unanswered in the case of HACCP. First, regulating a mandatory quality management system like HACCP is complex. The fact that HACCP is based on continuous improvement makes it difficult for the regulatory agency to establish uniform (equivalent) guidelines across different firms. Second, incentives or benefits specific to small firms have not been quantified in the literature. A major reason for the delay of a mandatory HACCP system (since 1989) is the difficulty encountered in empirically analyzing HACCP incentives that may compensate the hurdle of high implementation cost, especially for small firms. HACCP expenses and ways with which small firms will compensate for these expenses are the major motivation for this study.

## Theoretical Development

Two forms of efficiency measures have been used in the literature to analyze cost cutting incentives. They are technical and allocative (economies of size) efficiencies.[3] The two types can be estimated with a cost or profit function. Technical efficiency is defined as the relative ability of a firm to produce maximum output. Allocative efficiency reflects the ability of a technically efficient firm to combine inputs in a cost minimizing way. Technical and allocative inefficiencies prevent a firm from achieving minimum cost.

The measurement of efficiency can be done via both frontier and non-frontier methods. Neff, Garcia, and Nelson (1993) did an extensive comparison of the production frontier methods for technical efficiency. In the frontier approach, inefficiency is captured through varying coefficients (Lau and Yotopoulos 1971) or via asymmetry (Toda 1976). This approach enables us to compare efficiency levels among groups. Garcia, Sonka, and Yoo (1980) use this approach to compare the efficiency of small versus large farms in Illinois. A graphical representation of the efficiency frontier using normalized input levels was discussed by Mazzocco and Cloutier. Toda's approach, though it measures just allocative efficiency, has the advantage that it can be superimposed on several functional forms or flexible functional forms (Chambers 1988). This is advantageous when technology is not homogenous and the elasticities from a Cobb-Douglas model are adversely affected. The frontier approach, in contrast, has the advantage of stemming from economic theory of maximizing behavior. Deviation from the frontier is interpreted as a measure of the efficiency with which economic units pursue their technical or behavioral objectives.

Neff, Garcia, and Nelson (1993), in their comparison of the production frontier methods, concluded that differences in the measure of efficiency may be related to the method employed (especially between non-parametric methods and parametric methods). This study uses both the cost and profit function for comparative purposes because efficiency measures vary widely with the method used.

Analysis in this study begins with cost estimates of HACCP safety regulation for different firm categories and sizes, and follows with concurrent cost cutting incentives, both from a view of efficiency with respect to economies of size, technical efficiency, and parametric variation of biased technical change. These three approaches are adopted in this study because of the focus of analyzing cost cutting incentives. Scale and size economies, and biased technical change will be estimated using a flexible cost function approach.

**Theoretical Model for Cost Functions and Model Assumptions**

The generalized translog cost function with m-outputs and n-inputs specified by Ray (1982) is presented in Equation 1. It incorporates a time or technological change component that is used to evaluate biased technical change (BTC). The procedure to estimate BTC and its interpretation is presented later in this section. Equation 1 will be modified or made empirical to estimate and analyze the following: 1) scale and size economies to evaluate cost cutting incentives for firms with and without HACCP systems and 2) parametric variation on biased technical change to evaluate future areas with HACCP where small firms can reduce production cost. The advantage of the translog (dual) functional form is that it is a quadratic approximation of the "true" cost function (Ray 1982). So, other than being a flexible functional form, a global minimum cost can be estimated. The translog is flexible because specific features of technology (like homotheticity) may be tested by examining the estimated model parameters.

Another theoretical cost function model used in the literature is the random coefficient regression model of Hildreth and Houck (1968). Hornbaker, Dixon, and Sonka (1989), adopted this model to estimate production activity costs. The model basically reduced to a heteroscedastic model, which was estimated using a generalized least squares (GLS) method. The shortcomings of this model are that it had no time series component and had the strong assumption about the convergence of the matrix (or sigma matrix, estimated from the square of the error terms). Good knowledge of the sigma matrix is usually unknown. The dual approach used in this study overcomes this problem and increases efficiency in estimation. The simplicity of this approach will be discussed explicitly in later sections.

(1)
$$lnC = lnK + \sum_{i=1}^{m} a_i lnq_i + 1/2 \sum_{i=1}^{m} \sum_{j=1}^{m} d_{ij} lnq_i lnq_j + \sum_{r=1}^{n} b_r lnw_r$$
$$+ 1/2 \sum_{r=1}^{n} \sum_{s=1}^{n} f_{rs} lnw_r lnw_s + \sum_{i=1}^{m} \sum_{r=1}^{n} g_{ir} lnq_i lnw_r + hT$$

The variables in this equation are: "C" is cost or the dependent variable, "$q_i$" is output of product $i$, "$w_r$" is the price of input r, "h" is the error term, "m" is the number of outputs produced, "n" is the number of inputs used, "T" is the annual or technological index of time, and "K" is the constant term. This model is a generalization of the Cobb-Douglas (C-D) model. The model differs from the Cobb-Douglas model in that it relaxes the C-D's assumption of a unitary elasticity of substitution. We can obtain the C-D model from this model by restricting $d_{ij} = f_{rs} = g_{ir} = 0$ (Greene 1993).

***Homogeneity restriction.*** A valid cost function must be homogenous of degree one in input prices. To ensure linear homogeneity conditions, the restriction below (adopted from Ray 1982) is imposed during estimation of the cost function.

$$\sum_{r=1}^{n} b_r = 1, \quad \sum_{s=1}^{n} f_{sr} = 0, \quad \sum_{i=1}^{m} g_{ir} = 0 \quad (r=1,2,\ldots\ldots,n)$$

***Slustky's symmetry restrictions and concavity.*** The fact that the translog cost function is a second-order approximation (Chambers 1988) implies Slustky's symmetry of the form: $d_{ij} = d_{ji}$ and $f_{rs} = f_{sr}$ for all i,j,r, and s. Concavity of the cost function is met by imposing the restriction that the parameter matrix $[f_{rs}]$ or the Hessian matrix of the cost function is negative semi-definite.

***Homotheticity restriction.*** If the technology is homothetic, the dual cost function is multiplicatively separable in output quantities and input prices (Ray 1982). The cost function $C = C(q,w)$ is of the form $h(q) * t(w)$, where "q" and "w" are vectors of output quantities and input prices. In Equation 1, this requires that $g_{ir} = 0$ (for all $i$ and r) so that the quadratic interaction term between output levels and input prices should disappear (Antle 1984; Ray 1982). The specified function will be tested for homotheticity to improve efficiency in estimation, that is, if the function is homothetic, $g_{ir} = 0$.

***Estimation problems.*** One problem with multi-output, multi-input cost functions is the large number of variables to be estimated. For an m-output, n-input model with matrices $(d_{ij})$ and $(f_{rs})$ symmetrical, one needs to estimate ½(m + n)(3 + m + n) parameters (Ray 1982). This does not include the intercept and the rate of Hicks-neutral technical progress. For example, in the case of two outputs and five inputs, we must estimate 37 parameters. In general, it is difficult to obtain a sample large enough to estimate the full cost function. Thus, estimating the full cost function, even with the restriction of homogeneity in input prices may result in a classic specification problem with negative degrees of freedom.

Estimating a full dual system of cost and cost shares leads to much higher efficiency (Garcia and Sonka 1984; Ray 1982) due to the decreased number of parameters estimated. This procedure resolves the problem of lost specification error due to the decreased degree of freedom required for the cost share system. Using Equation 1 and Shephard's lemma, the input share equations are derived.

(2) $\quad s_r = b_r + f_{r1}\ln w_{r,t} + \ldots + f_{rn}\ln w_{n,t} + g_{i1}\ln q_i + \ldots + g_{im}\ln q_m$

where $r=1, \ldots, n$, $s_r = w_r x_r / C$, and $x_r$ is the quantity of the rth input. The sum of these shares must be one. For this to be true for all prices and outputs, it requires:

$$\sum_{r=1}^{n} b_r = 1, \quad \sum_{s=1}^{n} f_{sr} = 0, \quad \sum_{i=1}^{m} g_{ir} = 0$$

(for $r = 1, \ldots, n$). This condition is same as the conditions for linear homogeneity of the cost function in input prices.[4]

**Theoretical Model to Analyze Efficiency**

Three methods are combined to analyze efficiency in this study. Economies of size efficiency analysis and corrected least squares (to estimate pure technical efficiency) with Ray homothetic profit are used as traditional efficiency estimates. These methods are combined with biased technical change to identify cost cutting incentives for firms with and without HACCP systems. This study analyzes economic efficiency from two different perspectives. First, using the system of cost function and cost shares, inefficiencies (the error term) in the cost shares system with the quadratic interaction terms of output and input prices is due to economies of size inefficiency and the random error only. This means that technical inefficiency only exists in the main translog cost function. Therefore, using the cost shares to evaluate efficiency implies assuming technical efficiency is constant. Second, allocative efficiency is constant in the Ray homothetic profit function (the profit function assumes cost minimization) and pure technical efficiency is analyzed. Combining economies of size efficiency and pure technical efficiency with the Ray homothetic profit function enables this study to analyze efficiency with HACCP systems from different perspectives so that the inferences can be robust. This measure of efficiency with HACCP systems goes a step further to identify areas where firms implementing HACCP systems may or may not be more efficient than firms not implementing HACCP systems. The efficiency results will determine whether it is strategic for firms to be leaders or followers in implementing HACCP systems.

*Scale and size elasticities.* Returns to scale refers to the change in output as inputs are multiplied by a scalar. The relative change in output can be represented by an elasticity of scale. Increasing returns to scale exist when the elasticity of scale is greater than one. Elasticity of size is the ratio of average cost to marginal cost. Chambers (1988) points out that these measures are very different. If one evaluates economies of size and finds them to be less than one, it implies that the firm involved can decrease average costs by decreasing production or implementing a different technology. This will be an interesting implication for small and large firms in the meat industry under HACCP.

The product specific economies of scale (PSES) gives information about changes in cost as individual activities within the firm expand (McClelland et al. 1988). Thus, Equations 3 and 4 assume marginal cost is equal to marginal revenue and equal to price.

(3)
$$PSES = \sum \partial lnC/\partial lnQ = (Q/C)\sum \partial C/\partial Q = PQ/C.$$
$$\text{where } Q = \sum_{i=1}^{m} q_i$$

Variables "C" and "q" are as specified previously. Economies of size from Equation 3 is given by;

(4)
$$ES = (\sum \partial lnC/\partial lnQ)^{-1}$$

**Pure technical efficiency with the Ray homothetic gross profit function.** Efficiency results often vary widely with the method used (Neff et al. 1993). For comparative purposes, traditional measures of allocative and technical efficiency are used in this study. Pure technical efficiency is estimated using corrected least squares. Because some of the firms had negative net profits, this makes it appropriate to use Ray homothetic gross profit function to estimate technical efficiency of firms with HACCP systems and firms without HACCP systems. The Ray homothetic profit function is specified in Equation 5. A corrected least squares procedure is used to estimate pure technical efficiency for the different groups with estimates of the residual from this model.

"P" is output price, "y" is output, and other notations are the same as outlined previously.

(5)
$$PY = (\frac{w_i x_i}{\sum_{i=1}^{m} w_i x_i})\sum_{i=1}^{m} \ln w_i x_i + \xi_i$$

***Technical change.*** Several methods have been used in the literature to estimate biased technical change. Antle (1984) presents a summary of these methods and their drawbacks. The Hicks Neutral technical change, based on the marginal rate of technical substitution, identifies biases between input pairs. However, it does not give a global picture of technical change. Therefore, the multi-factor measure proposed by Binswanger (1974) and adopted by Antle (1984) will be used in this study. In this study and as confirmed by Antle (1984), no difference is made between this method and the cost-share approach. Given the cost function $C(q,w)$, the ith cost elasticity for the ith input in the estimated cost function is given by,

(6)
$$\eta_i = \partial C/\partial X_i$$
$$\eta \equiv \sum_i^n \eta_i$$
*Where the cost share of input i is given by* $C_i = \eta_i/\eta$

Biased Technical change $B_i$ can now be defined using $C_i$ as in Equation 7 below. Technical change is biased against the use of input $i$ if $B_i$ is less than zero and it is biased toward input $i$ if $B_i$ is greater than zero. In the HACCP context, bias against input $i$ will indicate the possibility of cost cutting incentives or high leverage which the firm can enjoy by reducing cost with this input factor over time. Biased technical change is given by:

(7)
$$\beta_i = \partial \ln C_i / \partial \ln T$$

Variable $C_i$ is given as in Equation 6 and the specification of $T$ is before and after HACCP implementation. Technical change is neutral with respect to input $i$ when $B_i = 0$.

**Empirical model for HACCP Cost Structure and Efficiency Analysis**

To estimate a cost structure for firms with and without HACCP, Equation 1 reduces to a translog cost function with one aggregated output, one HACCP input variable aggregating all HACCP expenses, and three other inputs (carcass purchase, labor, and material) of the firm. The empirical cost function model is presented in Equation 8. This cost function model uses weighted input prices ($w_1$-$w_4$) of the variables listed above and output quantity (y). The price of labor ($w_3$) is in dollars per hour including benefits (e.g., health insurance and retirement benefits). The variable $w_2$ is the price per pound of fresh carcass or live animals purchased. The price for HACCP ($w_1$) is the price per pound of HACCP safety expenses since most of HACCP expenses consist of material and labor expenses. The variable $w_4$ is the weighted price for operating material expenses and utilities. The output y is the aggregated quantity of fresh cuts, ham, sausages, and others. In Equation 8, t = time index used for technical change, y = output quantity, and all other variables are as specified previously. To increase efficiency in the estimation, a system of four equations is estimated, including three cost shares. Using weighted prices and aggregated output should have no significant effect on the results since the systems of equations uses cost shares or expenditure shares in conjunction with prices.

(8)
$$\begin{aligned}
lnC = & \alpha_0 + \gamma_y \ln y_t + \alpha_1 \ln w_{1,t} + \alpha_2 \ln w_{2,t} + \alpha_3 \ln w_{3,t} + \alpha_4 \ln w_{4,t} + \\
& 1/2(\gamma_{11} \ln w_{1,t} \ln w_{1,t} + \gamma_{12} \ln w_{1,t} \ln w_{2,t} + \gamma_{13} \ln w_{1,t} \ln w_{3,t} + \\
& \gamma_{14} \ln w_{1,t} \ln w_{4,t} + \gamma_{22} \ln w_{2,t} \ln w_{2,t} + \gamma_{23} \ln w_{2,t} \ln w_{3,t} + \\
& \gamma_{24} \ln w_{2,t} \ln w_{4,t} + \gamma_{33} \ln w_{3,t} \ln w_{3,t} + \gamma_{34} \ln w_{3,t} \ln w_{4,t} + \\
& \gamma_{44} \ln w_{4,t} \ln w_{4,t}) + \gamma_{y1} \ln Y_t \ln w_{1,t} + \gamma_{y2} \ln Y_t \ln w_{2,t} + \\
& \gamma_{y3} \ln Y_t \ln w_{3,t} + \gamma_{y4} \ln Y_t \ln w_{4,t} + \gamma_{yy} (\ln y_t)^2 + \eta_t \\
s_1 = & \alpha_1 + \gamma_{11} \ln w_{1,t} + \gamma_{12} \ln w_{2,t} + \gamma_{13} \ln w_{3,t} + \gamma_{14} \ln w_{4,t} + \gamma_{y1} \ln Y_t \\
s_2 = & \alpha_2 + (\gamma_{22} \ln w_{2,t} + \gamma_{12} \ln w_{1,t} + \gamma_{23} \ln w_{3,t} + \gamma_{24} \ln w_{4,t} + \gamma_{y2} \ln Y_t \\
s_3 = & \alpha_3 + \gamma_{33} \ln w_{3,t} + \gamma_{23} \ln w_{2,t} + \gamma_{13} \ln w_{1,t} + \gamma_{34} \ln w_{4,t} + \gamma_{y3} \ln Y_t
\end{aligned}$$

This system of four equations is estimated using Shazam. We use three cost share equations because cost shares sum to one and using all cost shares will cause the matrix not to be full rank. Economies of size efficiency and biased technical change are derived from Equation 8 as outlined previously.

## Summary of Survey Data and Variable Definitions

Theoretically, data on inputs and outputs with their respective prices are required to estimate a cost function. However, Neff et al. (1993) point out that input quantity data often are unavailable. Firms often keep these data in order to preserve their technology from being adopted by other firms. Data are often available on farm output revenues and input expenditures. Therefore, a common approach is to use these revenue and expenditure data as proxies for output and input quantities (Neff et al. 1993; Aly et al. 1987; Grabowski et al.; Neff et al. 1991). Assuming a competitive input and homogenous output product market, prices are competitive and homogeneous across firms. Hence, differences in efficiency measures here are likely to reflect quantities and not price differences.

A field survey was conducted to collect specific HACCP data for the empirical models. Data were collected for all HACCP input variables and other firm data relating to labor, material, and carcass and live animals purchases. A survey was designed to obtain data from different categories of firms with an emphasis on small firms who focus on processing only, slaughtering only or firms that perform both slaughtering and processing in the red meat industry. This is because previous studies by McDonald et al. (1996) and Nganje et al. (1995) have pointed out that HACCP will have a different impact on firms of different sizes and categories. A random sample was selected across the entire industry of meat processors and packers in the United States.

A systematic random sampling technique was used to select a sample of 1,050 firms from a population of 13,572 firms across the U.S. provided by the American Association of Meat Processors. Based on the approach of Rea and Parke (1989), a planned sample size of 990 would provide for a minimum standard error of the sample distribution at 95 percent confidence level and provide a confidence interval (sample error) of 3 percent for the entire population. A mail survey approach was chosen due to cost considerations. After double mailing, follow-up post cards, and telephone reminders, only 98 total responses were received. Of the 98, only 68 were valid responses that could be used for this study. Although this response rate still maintains the level of confidence at 95 percent, the sampling error increases to 9.9 percent.

The questionnaire was designed to provide data for estimating Equations 4, 5, 7, and 8. The questionnaire was pre-tested three times to adjust the clarity of the questions. The questionnaire had three sections: general business characteristics, total production expenses, and HACCP performance and expenses. Section one of the survey collected information about general business characteristics. Respondents had to identify their firms in terms of their size (sales volume and number of employees), their categories (slaughter only, processing only, and both slaughter and processing), type and volume in pounds of products they sell, and their source of live animals or carcasses (to identify any link between source of carcass or live animals and *E. coli* testing). The next section provided data on total production and operating expenses, including the quantity and price of carcasses or live animals purchased, the hours and dollars per hour of labor used and labor incentives provided, and the units of material and utilities used and their unit prices. The last section provided HACCP expenses and performance. HACCP expenses include materials purchased, training of employees, plan development, record keeping and monitoring, and laboratory testing. The current and anticipated impact of HACCP on product rework, the extension of product shelf life, and labor use were grouped into levels (remain constant, increase by ..percent and reduce by ..percent).

## Estimation, Results, and Discussion

The section that follows contains estimates of the translog cost functions for all firms with and without HACCP systems and for a subset of small firms. Efficiency and technical change results using the cost and profit functions are discussed in subsequent sections.

**Cost Function Estimation**

A system of equations (Equations 8) was used to estimate and test the properties of a valid cost function. The system of equations is made up of a cost function equation and input cost share equations as described in the methodology section. Estimated cost functions were restricted with respect to properties of linear homogeneity in input prices and symmetry of the Hessian matrix. Estimated cost functions were also continuous in input prices since the translog cost function is twice differentiable in input prices. Though the literature writes a cost function as a function of input prices and output quantity, usual applications hold output fixed. If the cost function is intended to estimate economies of scale and size, then output quantity must be incorporated in the cost function. Estimation in this research incorporates the interaction term of output and input prices ($lnylnw_i$), output ($lny$), and the quadratic interaction term of output ($lnylny$) into the model. This is especially important because the test for non-homotheticity (Ho: Coefficient of $lnylnw_i = 0 \ \forall_i$) requires output in the cost function specification. Homotheticity tests are important because scale and size economies are the partial derivatives of the cost function with respect to output quantities. Results of regression estimates of cost structures are presented in Table 2. Blank spaces in Table 2 represent variables that are not relevant in the models. The variables HP is HACCP cost per pound, CP is price per pound of carcass purchased, LP is wage per hour of labor (this include hour wage, health insurance, and retirement benefits), MP = price per unit of operating material, and Y is output quantity.

In general, the cost structure for all firm size categories were non-homothetic translog cost functions.[5] The models have very good fit according to Table 2. The $R^2$ and raw moment $R^2$ are high. The standard errors of estimation are low compared to the estimated annual total costs, and all cost structure models have a good number of significant variables. The constant terms were positive and significant at the 1 percent level for all models. This implies that firms may lose some fixed cost expenses if zero output is produced. In general, the cost function results indicate that per unit production costs decrease as output increases. This is an indication that firms with larger output in the industry have some economies of scale or that small firms are underutilizing current capacity or that they are inefficient. In general input price increases are expected to increase production costs if firms are efficient. All models show

TABLE 2

Cost Structure of All Firms With HACCP, of Small Firms Before and After HACCP Implementation, and of Firms Without HACCP Systems

|  | Coefficients of All Firms With HACCP n=47 | Coefficients of Small Firms With HACCP n=34 | Coefficients of Small Firms Prior to HACCP n=34 | Coefficient of Firms Without HACCP n=21 |
|---|---|---|---|---|
| HACCP (HP) | 0.0125 | 0.0411 | - | - |
| Carcass (CP) | 0.0268 | -0.3197 | 1.3975*** | -2.4876 |
| Labor (LP) | 0.8193*** | 0.9890*** | 1.4825* | 0.8457* |
| Material (MP) | 0.1405 | 0.28971 | 1.0850 | 2.6419 |
| HP*HP | 0.0092*** | 0.0134*** | - | - |
| HP*CP | -0.0059 | -0.0073 | - | - |
| HP*LP | 0.0020 | 0.0013 | - | - |
| HP*MP | -0.0054** | -0.0073*** | - | - |
| CP*CP | 0.1879* | 0.2353*** | 0.06252*** | 0.0131* |
| CP*LP | -0.1177*** | -0.1389*** | -0.09270*** | -0.0251* |
| CP*MP | -0.0643 | -0.0891*** | 0.0303** | 0.0119 |
| LP*LP | 0.0893** | 0.0893 | 0.06261*** | 0.0131* |
| LP*MP | 0.0263 | 0.0483** | 0.03028** | 0.0120 |
| MP*MP | 0.0434 | 0.0482 | -0.0606** | -0.0239 |
| Output (Y) | -0.9288*** | -0.6081** | -1.4903*** | -0.0227* |
| Y*HP | 0.0081* | -0.0012 | - | - |
| Y*CP | 0.0689** | 0.1053*** | -0.0798* | 0.1355 |
| Y*LP | -0.0611*** | -0.0760*** | 0.0971 | 0.0029 |
| Y*MP | -0.0159 | -0.0281 | -0.0203 | -0.1419 |
| Y*Y | 0.0722*** | 0.0614*** | 0.0798*** | -0.0103 |
| Constant | 13.451*** | 10.752*** | 19.708*** | 16.220*** |
| $R^2$ | 0.8667 | 0.9653 | 0.8564 | 0.627 |
| Raw Moment $R^2$ | 0.9988 | 0.9999 | 0.9992 | 0.9979 |
| Mean of Dependent Variable (log of Cost) | 14.613 | 13.899 | 14.104 | 13.17 |
| σ (Std. Error) | 0.5191 | 0.1751 | 0.4030 | 0.6478 |
| Homotheticity | 6.3219*** | 22.8067 | 31.3692*** | 60.21*** |
| F-stats (P-value) | (0.0000) | (0.0000) | (0.0000) | (0.0000) |

\*\*\*, \*\*, \* represent significance at the 1 percent, 5 percent, and 10 percent levels of significance, respectively. "-" represents the variable was omitted. Because of symmetry restrictions the coefficients of the cost share equations are the same with the quadratic interaction term coefficients.

production cost will increase as prices of individual inputs increase, except for the price of carcass for the 21 firms without HACCP and the 34 small firms with HACCP (but the quadratic interaction term for these variables will positively increase production as they increase). The quadratic interaction price coefficients that are negative and significant need further investigation. The efficiency and technical change analysis provide more insight on the efficient use of input factors.

It would have been ideal to analyze cost structures for different categories and sizes (slaughter only, process only, and both slaughter and process) for detailed comparison. However, there was only sufficient data to analyze size differences and stage of HACCP implementation.

**Summary of Cost Function Results**

For firms with HACCP systems, the variables that contribute significantly to total firm cost are HACCP expenses, labor use, and carcass purchases (in their linear or quadratic forms). For firms without HACCP systems, carcass purchases and labor contributed the most to total firm cost. This indicates that the predominant technology used in the industry is labor intensive. This is probably the case because the majority of the firms are small firms (34 out of 47 firms with HACCP systems were small, each with about $2.5 million dollars in sales annually). Material only becomes significant for all the categories through the quadratic terms of material and order variables.

The cost structure across all categories suggests inefficiency in some input use, and suggest in-depth analysis of the differences in efficiency levels. Also, if these firms have to reallocate input use, especially with HACCP implementation, it is important to measure elasticities of substitution between HACCP and other input variables to determine the substitutability of other inputs with HACCP. From the cost structure results it is difficult to conclude that one category of firms is more efficient than the other. The efficiency results further investigate firms that may be less efficient overall or more efficient in their cost sets.

**Economic Efficiency Results**

Cost efficiency analysis and the corrected least squares estimates of pure technical efficiency using Ray homothetic profit are presented in this section. This study analyzes economic efficiency from two different perspectives. First, using the cost function and cost shares system, inefficiencies detected in the cost shares are due to allocative inefficiency and the random error only. This means that technical inefficiency is only detected in the main translog cost function. Therefore, using the cost shares to evaluate efficiency implies the assumption that technical efficiency is constant.

Second, because the profit function assumes cost minimization, pure technical efficiency is analyzed using the Ray-homothetic profit function. This holds allocative efficiency constant. Combining economies of size efficiency and pure technical efficiency with Ray homothetic profit function enables this study to analyze efficiency with HACCP systems from different angles so that the inferences can be robust. This study uses both allocative and technical efficiency measures to test the hypothesis that firms with HACCP systems are more efficient. This is tested in two steps.

Step one addresses whether firms with HACCP systems are technically more efficient than firms without HACCP systems. The corrected least squares technique is used to measure pure technical efficiency of firms with and without HACCP systems with the profit function frontier. Since efficiency results vary widely with the methods used, a second step tests whether differences in efficiency levels can be attributed to HACCP systems. To analyze step two, cost efficiency is estimated for small firms prior to and after HACCP implementation. Also the marginal cost of all firms with and without HACCP systems are estimated and the results are used to determine whether firms with HACCP systems enjoy lower marginal cost than firms without HACCP systems. If firm A produces output Y with marginal cost MC and firm B produces Y at marginal cost MC*, and if MC*<MC and both firms face the same output price P, it can be said that firm B is allocatively more efficient than firm A. Efficiency results of Equations 4 and 5 are presented in Table 3 Firms are technically and size efficient if the Pure Technical Efficiency Ratio or economies of size equals one. The procedure for testing the hypothesis of whether efficiency levels are different or equal is done using a two sample t-test. The t-test results indicate that although all firms have some allocative and technical inefficiencies, firms without HACCP are the most inefficient followed by small firms prior to HACCP implementation. Comparing technical efficiency ratios indicates that small firms with HACCP systems are technically more efficient than those without HACCP systems. Though

TABLE 3

Results of Pure Technical and Economics of Size Efficiency

|  | All Firms With HACCP | Small Firms Before HACCP | Small Firms With HACCP | Firms Without HACCP |
|---|---|---|---|---|
| **Pure Technical Efficiency Ratio** | 0.92201 | 0.82102 | 0.88965 | 0.79086 |
| **Cost Efficiency (economies of size)** | -1.2281 | -2.3405 | -1.6020 | -6.0248 |
| **Sample Size** | 47 | 34 | 34 | 21 |

all firms size categories are allocatively inefficient, results in Table 3 shows that all firms with HACCP systems enjoy lower marginal cost than firms without HACCP systems.

***Comparison of economic efficiency of firms with and without HACCP systems.*** The difficult question addressed in this section is to determine whether differences in efficiency levels can be attributed to HACCP implementation and not to other unknown variations. In an attempt to address this question, comparisons were made between the efficiency levels of firms prior to and after HACCP implementation. The t-test indicated that economies of size value for firms prior to HACCP implementation is (-2.3405) significantly lower than the value of firms after HACCP implementation (-1.6020). Therefore, firms enjoy lower marginal cost after HACCP implementation. Also, since the coefficient of determination of the cost structure explains most of the variation in cost (96.5 percent and 85.6 percent before and after HACCP implementation), we can conclude that at least 86 percent of the time HACCP systems will make firms more efficient by cutting down on production cost.

One efficiency issue is whether or not firms are disadvantaged by implementing HACCP as leaders or as followers. Leaders are those firms currently implementing HACCP systems, and followers are those firms which wait four years (the time for HACCP implementation given by the regulatory agency) before implementing HACCP systems. Traditional economic studies on efficiency (allocative and technical efficiency) fail to capture the strategic savings that followers of HACCP systems may loose in the long run. Biased technical change measures identifies cost cutting points that may be strategic for small firms in the meat industry to increase their efficiency as they implement HACCP systems. Biased technical change shows why firms with HACCP systems are more efficient and identifies input variables that can be used to increase efficiency or reduce production costs.

***HACCP as a technical change which biases input mix.*** Biased technical change is evaluated for two periods: before and after HACCP implementation in small firms. It would have been better to work with more time series data of HACCP implementation or several years worth of data for firms with HACCP systems. However, this was not possible because of the stage of HACCP implementation in the industry. Substituting coefficients from Table 2 into cost shares of Equation 8 gives biases for input factors in Table 4. From the literature, biases less than one imply firms can reduce marginal cost by efficiently reallocating that variable. From Table 2 it can be seen that before HACCP implementation firms could reduce marginal cost with the efficient use of labor at the 5 percent significance level. With HACCP implementation no significant inefficiencies exist, though labor use and carcass purchase are negative. These results show that firms can reduce the inefficiency of labor use that exists before they implement HACCP systems. HACCP systems help firms to efficiently reallocate scarce labor. Labor was parametrically varied to determine the point where biased technical change was neutral. Analysis of parametric variation of bias technical change indicate that firms with HACCP systems enjoy an 18-21 percent lower marginal cost of labor than their counterparts without HACCP systems.[6]

## Summary and Conclusion

Two major concerns by industry participants were investigated in this study. These are HACCP cost and benefits specific to small firms. This study quantifies HACCP costs and evaluates meaningful levels of cost savings with HACCP implementation, for small firms, using efficiency analysis.

Two efficiency tests are used to compare the performance between firms before and after HACCP implementation and firms with and without HACCP systems. Allocative efficiency maps the marginal cost set and the price set, technical efficiency maps the output and input sets, and economies of size efficiency maps the marginal cost and output set. This study uses economies of size efficiency and technical efficiency because the literature and survey data reveals that prices may not serve as incentives to compensate small firms for their HACCP expenses. This study uses both efficiency measures because efficiency results vary widely with the method used.

To run the empirical models for this study, data on input and output prices and quantities for all production activities are needed. Secondary data on HACCP were not available for the detailed analysis required for this study. Therefore, a field survey was conducted to collect specific HACCP data and other input data on prices and quantities. The response rate from the survey and survey results reveal that there is limited scope of HACCP implementation in the meat industry. HACCP cost estimates reported by survey respondents were relatively higher for smaller firms as compared to their larger counterparts.

### TABLE 4
#### Biased Technical Change of Input Factors

|  | Small Firms Before HACCP | Small Firms With HACCP |
|---|---|---|
| HACCP (HP) | - | -0.0213 |
| Carcass (CP) | 1.2357 | -0.7681 |
| Labor (LP) | -1.4408[a] | 1.2799 |
| Material (MP) | 1.2051 | 0.5095 |
| Sample Size | 34 | 34 |

[a]Implies negative and significantly different from one at the 5 percent level of significance (example, 1 ó -1.441 ± st. error(1.96)).

The methodology developed in this study was tested using the primary data collected from meat processing and packing firms in the industry. Cost structure analyses indicate the cost functions were non-homothetic. Estimation in this research therefore incorporates the interaction term of output and input prices ($lnylnw_i$), output ($lny$), and the quadratic interaction term of output ($lnylny$) into the model. This is especially important because the test for non-homotheticity (Ho: Coefficient of $lnylnw_i = 0 \; \forall_i$) requires output in the cost function specification. Homotheticity tests are important because scale and size economies are the partial derivatives of the cost function with respect to output quantities. This implies that technical change is due to both the Hicksian and scale effect.

Economies of size efficiency analysis of primary data collected across the U.S. indicate that meat and poultry processing firms enjoy lower marginal costs with HACCP systems as opposed to their marginal cost prior to HACCP implementation and firms without HACCP systems are less cost efficient than firms with HACCP systems. Using a Ray homothetic function and adjusted ordinary least square procedure, firms with HACCP systems have greater technical efficiency than firms without HACCP systems. This study confirms the proposition that HACCP can be used as an effective management tool to increase efficiency in the meat industry.

Analysis of biased technical change suggests reasons why small firms may be more efficient with HACCP systems. These results indicate that HACCP can improve the overall efficiency of the meat industry by efficient reallocation of labor use and carcass purchases. Small firms that may not have output price incentives or economies of scale incentives can enjoy cost cutting incentives with HACCP systems. HACCP will be economically beneficial to small and large firms.

Though (HACCP) is intended to increase the safety of meat and poultry products by decreasing harmful bacteria levels, it also has the potential as a quality management tool to reduce production cost and improve the efficiency of the industry.

**Notes**

[1] The authors are Assistant Professor, North Dakota State University, and Associate Professor, University of Illinois, Urbana-Champaign.

[2] Small firms may want to be followers in implementing HACCP because HACCP implementation is costly and their knowledge of HACCP systems is limited. The agency has not provided specific guidelines on HACCP for different firm categories and sizes. This is probably because of the complexity to establish guidelines since HACCP is based on continuous improvement.

[3] Allocative efficiency maps the marginal cost set and the price set, technical efficiency maps the output and input sets, and economies of size efficiency maps

the marginal cost and output set. This study uses economies of size efficiency and technical efficiency because the literature and survey data reveal that prices may not serve as incentives to compensate small firms for their HACCP expenses.

[4]Taylor (1989) discussed other interesting pitfalls of the duality theory and possibilities to resolve them. The test for homotheticity and the model restrictions eliminate some specification errors.

[5]The homotheticity test is a very important statistical test to determine production structure before biased technical change and economies of size can be measured. It serves as a robust test of the functional form used, and determines the direction of technical change and the magnitude of size efficiency (Karagiannis and Furtan 1993).

[6]The fact that cost structures are non-homothetic imply biases in this study are due to both Hicksian and scale effects (change of expansion path or movements along the same expansion path).

## References

Aly, H.Y., Belbase, K., Grabowski, R. and Kraft, S. " The Technical Efficiency of Illinois Grain Farms: An Application of Ray-Homothetic Production Function." *Southern Journal of Agricultural Economics*, 19(1987) pp. 69-78.

Antle, M. John. "The Structure of U.S. Agricultural Technology, 1910-78." *Amer. Ag. Econ. Association.* 1984.

Binswanger, Hans P. "The Measurement of Technical Change Biases with Many factors of Production." *Amer. Econ. Rev.* 6H(5):964-976, 1974.

Bjerklie, Steve: "HACCP in your plant. What HACCP is, What it isn't, and how your operations will be affected." *The Business Journal of the Meat and Poultry Industry,* Feb. 1994, pp. 58-59.

Chambers, R. G. "Applied Production Analysis." New York: *Cambridge University Press*, 1988.

Crosby, Philip. *"Quality Is Free."* New York: *McGraw-Hill.* 1979.

Crutchfield, S. R. et al.. " An economic assessment of food safety regulations: The new Approach to Meat and Poultry Inspection. *ERS, USDA* (1997).

Deming, W. Edwards. "Out of Crisis." *Cambridge, MA:MIT Center for Advanced Engineering Study.* 1986.

Garcia, P., Sonka, S., and M. Yoo. " Farm Size, Tenure and Economic Efficiency in a Sample of Illinois Grain Farms." *American Journal of Agricultural Economics.* 64(1982): pp. 199-123.

Grabowski, R. S. Kraft, C. Pasurka and Aly. "A Ray Homothetic Frontier and Efficiency: Grain Farms in Southern Illinois." *Euro. R. Agr.Econ.* 17(1990): 435-448.

Hornbaker, H. R., Dixon, L. B., and Sonka, T. S. "Estimating Production Activity Costs for Multioutput Firms with Random Coefficient Regression Model." *Amer. Ag. Econ. Ass..* 1989.

Illinois Agricultural Statistics. "Annual Summary Supplement." *Illinois Department of Agriculture.*(1993).

Illinois Department of Agriculture." Firm sizes and volume." *Illinois State Bureau of Meat and Poultry Inspection, Springfield IL.* (1993).

Ishikawa, Karou. *"What is Total Quality Control? The Japanese Way." Englewood Cliffs, N.J.: Prentice-Hall.* 1985.

Juran, M. Joseph. "The Quality Control Handbook." 3rd ed. New York: McGraw-Hill. 1974.

Karagiannis, Giannis and Furtan, Hartley W. " Production Structure and Decomposition of Biased Technical Change: An Example from Canadian Agriculture." *Rev. of Ag. Econ,* Vol. 15, No.1, Jan. 1993.

Karr, Keeley J. et al. "Meat and Poultry Companies Assess USDA's Hazard Analysis and Critical Control Point System." Meat and Poultry Magazine. 1994.

Laffont, Jean-Jacques and Tirole, Jean. "Using Cost Observation to Regulate Firms," *Working paper.* DP 27 (1984) CERAS.

Lau, L.J. and P.A. Yotopoulos. "A Test for Relative Efficiency and an Application to Indian Agriculture." *American Economic Review.* 61(1994): 94-109.

MacDonald, M. James, Ollinger, E. Michael, Nelson, E. Kenneth, and Handy, R. Charles. "Structural Change in the Meat Industry: Implication for Food Safety Regulations." *Economic Research Service, USDA.* 1995.

Mazzocco, A. Michael. "HACCP as a Business Management Tool." *American Journal of Agricultural Economics,* 78(1996).

Mazzocco, A. Michael and Cloutier, Martin L. "Relative Efficiency in Food Processing Industry: An Approximation of Data Envelopment Analysis." Working Paper series, University of Illinois.

McClelland, J. , M. Wetzstein, and W. Musser. "Returns to Scale and Size in Agricultural Economics." *WJAE.* 11(1986):129-33.

National Restaurant Association. "Managing a Food Safety System." The Educational Foundation, 1994.

Neff, D. L., Garcia, P. and Hornbaker, R.H. "Efficiecy Measures Using Ray Homothetic Functions: A Multiperiod Analysis." *Southern Journal of Agricultural Economics,* 23(1991) 113-121.

Neff, D. L., Garcia, P. and Nelson, H.C. "Technical Efficiency: A Comparison of Production Frontier Methods." *Journal of Agricultural Economics,* Vol. 44, No. 3, September 1993.

Nganje, Mazzocco, and McKeith. "Simulating HACCP Cost Hurdle for Small Meat Plants" *FAM Working Paper Series, University of Illinois.* 1995.

Ray, S. C. "A Translog Cost Function Analysis of U.S. Agriculture, 1939-77." *Amer. J. Agr. Econ.* 64(1982): 490-98.

Santana, Miriam Blau R. "Improving Yield Forecast- Reducing Production Costs." *Microwave & RF.* V. 26, Jan. 1987, pp. 83-84.

Scott et al. "HACCP and Total Quality Management - Winning Concepts for the 90's: A Review." *Journal of Food Protection.* 55(6) 459-462. 1992.

Seitz, Wesley D., Nelson, Gerald C., and Halcrow Harold G. "Economics of Resources, Agriculture, and Food." *McGraw-Hill Inc.* 1994.

Taylor, Robert C. "Duality Optimization, and Microeconomic Theory: Pitfalls for the Applied Researcher." *W JAE, 14(2):200-212. 1989.*

Toda, Y. "Estimation of a Cost Function When the Cost is Not Minimum. The Case of Soviet Manufacturing Industries, 1958-1971." *Review of Economic statistics.* 58(1976):259-268.

Tribus, Myron. "Deming's Way (High Quality at Low Cost)." *Mechanical Engineering.* V. 100, Jan. 1988, pp.26-30.

USDA. "Meat and Poultry Inspection Directory" *National Agricultural Statistics Service* (1994a).

USDA. "Meat and Poultry Inspection Directory." *Food Safety and Inspection Service,* Inspection Operations. (1994b) pp. 596-599.

USDA. "Report on a margin of safety." *USDA Report* (1990).

U.S. Navy Procurement. "Review Identifies $500 Million Savings in Trident Program." *Aviation Week and Space Technology*, V. 129, pg. 23.

WHO/ICMSF (World Health Organization/International Commission on Microbiological Specifications for Foods). 1982. Report of the WHO/ICMSF Meeting on Hazard Analysis of Critical Control Points System in Food Hygiene. WHO, Geneva.

# PART II
International Perspectives on HACCP Costs and Benefits

Chapter 16

# HACCP, Vertical Coordination and Competitiveness in the Food Industry

Gerrit W. Ziggers[1]

## Introduction

The food industry in general is confronted with rapidly changing markets, new technologies, world wide competition, and now legislation on liability and food safety. As a consequence, markets have become more dynamic and more complex. It affects all levels of the supply chain. These developments may have an important impact on transaction costs and will increase as one moves along the food supply chain. The desire to shift transaction costs to others in the chain will increase and will be an impetus for institutional innovation. Quality assurance systems, such as HACCP, have evolved and can contribute to reduce transaction costs to the extent that these systems become standard business practice. As a consequence, they also can serve as effective trade barriers. However, these systems are only considered as facilitating factors. The actual competitiveness of the food industry will be determined by the cooperating ability of firms, which means the ability to develop partnerships. The objectives of this paper are:
1. To identify the importance of HACCP for the Food Industry;
2. to identify the relationship between HACCP and transaction costs; and
3. to identify the impact of HACCP on the structure (vertical coordination) of the food industry.

## The Importance of HACCP for the Food Industry

**The History of HACCP**

In the EU the food industry is obliged to have a Hazard Analysis Critical Control Point system in place. In 1987, cooperation between the International Commission on Microbiological Specifications for Foods (ICMSF) and the World Health Organization (WHO) was established. In 1988 the results of this cooperation were laid down in a book (Thatcher et al. 1988). Since then, the system and its application has also been described in documents of Codex Alimentarius, n.d. (1995a and 1995b). The European Union adopted the HACCP approach in Directive 93/43, which means that HACCP is now part of national legislation in all member states of the European Union (Anon 1993). The system itself is a management system, which is based on systematic identification and assessment of hazards and risks associated with the production of food. Although originally developed to control microbiological hazards, the approach can also be used to identify and assess hazards of chemical and or physical nature. When all hazards are known, Critical Control Points (CCPs) can be identified; points or steps in the production process at which control can be applied and a food safety hazard can be prevented, eliminated, or reduced to an acceptable level. Following the identification of CCPs, control measures must be put in place. Those CCPs and its control measures have to be monitored. Corrective action must be taken when monitored values are above established critical limits. As with other quality systems, verification on a regular basis is required.

Much experience with the development and implementation of HACCP system has now been accumulated. It appears, however, that there are still some major problems associated with implementation, especially in small companies, where there is often not enough (scientific) knowledge available. There is also still debate in scientific circles concerning the extent to which risk has to be assessed. Very often it is a matter of setting priorities, since not all identified hazards can be eliminated in one operation. Priorities are often set on the basis of budgetary considerations rather than on knowledge.

It is appropriate to highlight the chemical hazards. In contrast to physical or microbiological hazards, a chemical hazard does not often lead to customer complaints. The risk associated with chemical hazards are often not of an acute toxicological nature, but rather have to do with long term exposure of consumers to low levels of unwanted substance. If a company is recording complaints, these recordings can often be used to estimate the occurrence of physical or microbiological hazards, but mostly there are insufficient data on chemical hazards. Chemical hazards are often associated with raw materials, for instance in the case of the presence of contaminants in such commodities. Numerous countries have now introduced legislation setting maximum levels for a great number of contaminants. Contaminant levels often can not be influenced during the production of food and not much knowledge on the fate of contaminants during processing is available. This means that these hazards often must be controlled using specifications for the raw materials. Such specifications must be realistic. There is no point in specifying zero tolerance, when it is clear that such a demand can not be fulfilled by the supplier (Hoogland et al. 1998). Currently, HACCP is even expanded beyond food safety and is used as a tool to improve product quality. It is also expected that the development of supply chain-based HACCP systems will be an important topic in the near future.

**The Importance of HACCP**

There are a number of reasons why, especially in the food industry, implementation of quality assurance systems, such as HACCP, is an issue of importance:
- Agricultural products are often perishable and subject to rapid decay due to physiological processes and microbiological contamination.
- Most agricultural products are harvested seasonally.
- Products are often heterogeneous with respect to desired quality parameters such as content of important components (e.g., sugars), size and color. This kind variation is dependent on cultivar differences and seasonal variables that can not be controlled.
- Primary production of agricultural products is performed by a large number of farms operating on a small scale.

Despite the progress in medicine, food science, and the technology of food production, illness caused by foodborne pathogens continues to present a major problem in terms of both health and economic significance. In 1990 an average of 120 cases of foodborne illness per 100,000 population were reported from 11 European countries, and estimates based on a more recent study indicate that in some European countries there are at least 30,000 cases of acute gastro-enteritis per 100,000 population (Notermans and Van der Giessen 1993; Hoogland et al. 1998). Nevertheless, only a small proportion of cases of foodborne illness are brought to the attention of food inspection, control, and health agencies. Estimates have been made of the economic consequences of foodborne illness, where individuals who become ill incur costs, their employers, families, health care agencies, and the food businesses involved. For example, in England and Wales in 1991, some 23,000 cases of salmonellosis were estimated to have resulted in overall costs of £40-50 million (Sockett 1991).

Significant changes are taking place in animal husbandry, large-scale food production, and distribution methods. The increased use of a range of raw materials and products originating from a wider range of countries has increased the potential for a geographical spread of diseases associated with particular contaminants. Many new processing techniques have been introduced which alone or in combination offer distinct product quality advantages, e.g., milder thermal processing, microwave heating, ohm heating, and high processing techniques. All present new food safety challenges which must be fully evaluated.

There is increasing demand for convenience foods requiring minimal handling or preparation in user friendly packaging. In addition, consumers are seeking foods that are more "fresh" and with enhanced natural flavors that inevitably challenge the industry to use less harsh processing and production regimes. While there is little evidence that such trends have led to increased foodborne illness, it must be appreciated that these foods will require greater care in their production, distribution, storage, and preparation prior to final consumption.

There is a significant world wide trend towards increased consumption of food outside the home. A major increase in the frequency of international travel for business and vacation purposes also means that more people are in touch with new types of products. Population changes are taking place. The young and the aged are groups at risk with regard to foodborne illness, and in a number of countries the population is ageing resulting in a stronger focus on health related aspects.

**HACCP and Quality Management**

Quality assurance is of paramount importance to all companies and organizations involved in the production, sale, and handling of food. Modern trading conditions and legislation require food businesses to demonstrate their commitment to food quality and establish an appropriate product quality program. Such a program should take into account the role of business in the food chain, i.e., whether they are primary producers, manufacturers, retailers, or caterers. A product quality program contains four primary elements:

1. It meets the expectations of the consumer;
2. it fits within the strategy of the company;
3. it ensures that a company is clearly committed to the quality of its products; and
4. it aims for the highest quality achievable, in terms of effectiveness and efficiency.

Such a program should highlight where improvements are necessary and can usefully be applied to both organizational and technological issues. All company employees, from senior management to food handlers, should be aware of the significance of a quality program. The program should identify the key tools and their application to all stages of production, distribution, and sale (see Figure 1) (Hoogland et al. 1998)

The analysis and assessment process is likely to identify issues for adjustment, modification, and improvement. This is an on-going review process designed to ensure compliance with changing regulations, customer needs, and developing techniques, and will further enhance both processing and the assurance of food quality and safety. Above all this will contribute to competitiveness as shown in Figure 2.

# HACCP and Transaction Costs

**Food Safety and Transaction Costs**

Transaction costs are the costs associated with the process of exchange and include transaction specific investments such as the costs of an information search, negotiation, and monitoring and enforcement costs of undertaking an exchange or the costs related to asset specificity, transaction frequency, and uncertainty (Williamson 1979). These costs encompass all aspects of the contractual relationship between the customers and supplier. Transaction costs can be affected by food safety regulations in several important ways. Product liability law is important because it specifies manufacturers' legal responsibilities and specifies the burden of proof. A key example is the legal standard to meet the due diligence requirements of product liability law. The requirement that firms practice due diligence (in Anglo-Saxon law as practiced in the Britain and

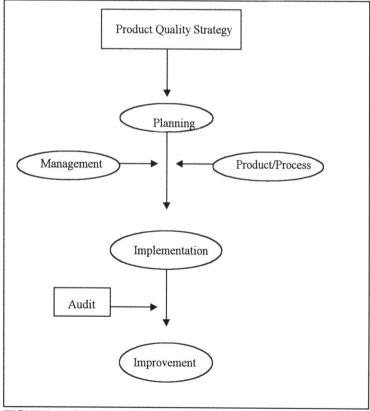

FIGURE 1. Quality improvement program.

in the United States) simply means that a firm must have taken all necessary steps to assure the safety of its products. (in other countries, such as the Netherlands and France, this is known as a "state-of-the-art" defence.) As Hobbs and Kerr (1992) conclude, "Whatever the form of (due diligence) procedure adopted, it (meant) an increase in costs." Due diligence, of course, comes to be defined by judicial decisions in previous legal actions. Due diligence may be satisfied by a guarantee from suppliers that delivered products are safe. Or, due diligence may require firms to be pro-active in assuring the safety of food in their possession, both the food they handle as well as the supplies they receive. Due diligence, or other aspects of food safety technical regulations, may require that firms be able to trace the product through the production process and to the potential sources of contamination or adulteration. One such example is the residual of prophylactic veterinary medicines.

As one moves up the food chain, transaction costs resulting from food safety technical requirements and quality assurance may increase. For example, slaughterhouses, as first processors, need to address not only the food safety practices within their production processes, but they must also verify the food safety practices of their raw material suppliers (beef cattle producers). To satisfy due diligence requirements, slaughterhouses may need to be well informed about the production practices used by farmers.

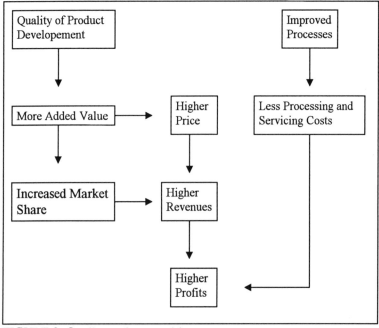

FIGURE 2. Quality and competitiveness.

(Do the cattle come from BSE-free herds? What type of medication have the cattle received? What type of feed have the cattle eaten?) Not only is the slaughterhouse concerned about collecting this type of information, but due diligence requires the slaughterhouse to monitor its suppliers' processes as well. As soon as the slaughterhouse accepts the beef cattle, it accepts legal responsibility for its safety. Moving up the food chain to the second stage processors, the meat processors in this case, the information and monitoring requirements expand. This is because the meat processors need to maintain their own food safety programs in addition to monitoring their slaughterhouse suppliers. The meat processors may also need to collect information about the raw material producers, the beef cattle producers. In short, tractability is of paramount importance. Consequently, as the product becomes further processed, firm information and monitoring responsibilities increase, as do tractability responsibilities, thereby increasing the transactions between the buyers and sellers. All of this contributes to an increase in transaction costs. In this way, food legislation requirements can increase transaction costs as one moves along the food chain (Bredahl and Holleran 1997).

**Transaction Costs and Transfer Pricing**

The impetus for institutional innovation and the desire to shift transaction costs and quality assurance costs to others in the food chain will increase as one moves up the food chain. Third stage processors and food retailers would face the greatest transaction costs and so have the greatest interest in institutional innovations that reduce transaction costs, or to shift them elsewhere in the food chain. Examples of specific transaction costs occurring during transactions are:

1. Supplier identification. Due diligence requires firms to ensure the safety of their supplies; therefore, firms must identify competent suppliers. This entails information search, which can take time, particularly if a firm needs either a variety of supplies or supplies that have inherent food safety hazards. Therefore, firms are expected to exercise caution in the search and selection of suppliers, thereby incurring informational search costs that are transaction costs.
2. Contract negotiation. During the contract negotiation stage, customers want to set product and process specifications, as well as to ensure that suppliers are able to meet them. Consequently, site visits and other forms of contract negotiation increase the level of communication and transactions between customers and suppliers, thereby increasing transaction costs.
3. Contract verification and enforcement. Trading partners desire to protect themselves from the hazards of exchange, such as incomplete contracts. A second important element of contract

verification and enforcement is the act of inspecting incoming products to ensure the safety and specifications of the incoming products. A third element of contract verification and enforcement is customer returns. Firms need to have systems in place to address these product rejects. Finally, one can not discuss exchanges and contract verification without discussing legal services. To set up exchanges and draw up contracts, verify them, or enforce them, firms may require legal services. Food safety regulations that require due diligence result in increased legal monitoring and legal assistance, particularly when a firm is responsible for the safety of the supplies it procures, as well as for the safety of the food it handles. Additionally, legal concerns arise as a result of drawing up product and process specifications.

A key issue therefore is actually the distribution of costs and benefits along the supply chain. It is desirable to gain insight into the way costs and benefits along the various stages of the chain are affected and distributed. These insights may help to develop effective transfer pricing instruments. Since prices are considered to be efficacious incentives in affecting economic decisions, they can be used to serve as appropriate signals to transmit customer preferences throughout all stages of the vertical system. In accordance with transfer pricing theory, effective payment systems should lead to economic decisions that positively affect chain performance, and give the separate participants the feeling that they are being fairly rewarded for the contribution they are making to the chain result.

**Institutional Innovation and Transaction Costs**

Increased buyer liability and traceability requirements stemming from food legislation and supplier quality assurance requirements increases communication and transactions between customers and suppliers. This, in turn, leads to increased transaction costs. Transaction cost analysis posits that one can reduce transaction costs by creating or adopting institutions. Institutions, therefore, emerge to facilitate exchange. It implies that in the food sector, institutions such as HACCP, or other (voluntary) quality assurance programs/standards can contribute to reduce transaction costs.

Quality assurance standards, such as HACCP and ISO, are nationally or internationally accepted procedures and guides initiated in order to maintain consistent quality. They are not substitutes for either product safety or other regulatory requirements. Rather, they specify the elements necessary for a quality system to consistently meet required specifications. They have a quasi-legal status because they can be used as a reference in product liability cases. The question of liability comes from many factors such as manufacturing, negligence, and regulatory violations that may lead to major or minor damage. One way to limit liability is to adopt a standard such as HACCP that can serve as a management tool for quality concerns (Bredahl and Holleran 1997).

## HACCP, Vertical Coordination, and Competitiveness of the Food Industry

**Quality Systems and Competitiveness**

There is no doubt that the food industry increasingly deals with competitive markets in which market orientation of products and services, and efficiency and reliability of their delivery, become decisive aspects for competitiveness. The food industry in general is confronted with rapidly changing markets, new technologies, and world wide competition. As a consequence, markets have become more dynamic and more complex. A retailer purchasing world wide perishable foodstuffs needs reliable partners. These trends are reinforced by legislation on liability and food safety. It implies that the development and introduction of new products can only be effectuated by joint investments and cooperation, which implies the need for vertical coordination in the food industry (Hughes 1994; Wierenga 1996; Zuurbier et al. 1996) Also with regard to these developments, firms are implementing quality assurance systems such as HACCP.

The specific characteristics of raw materials, products, and structure of this line of business as mentioned before advocates an integrated approach of quality assurance, or Integrated Quality Assurance (IQA). The surplus value of IQA can be related to the individual firm, the supply chain, and competing supply chains (Heyman et al. 1994). The surplus value for a firm lays in a more effective and efficient conduct of business. The second advantage for an individual firm to participate in an IQA program is assured sales. The surplus value for the entire chain of IQA concerns an increased profitability. This comes about by producing produce of high quality standards and by improving production throughout the entire supply chain, which also includes food safety. High quality products generated more value thereby optimizing production and reducing costs. This way IQA enables increased competitiveness with regard to competing supply chains.

The implementation of quality assurance systems is about management or can be considered a peculiarization of management. Management initiates, directs, and controls objective-directed activities (Kampfraath en Marcelis 1981; Ziggers 1993) aimed to improve products and production, to meet customer demands, and to optimize utilization of resources. It should reflect the policy of the firm, adjusted to the business environment in which the firm operates. Quality assurance at the firm level reflects this interrelation. Implementing IQA is more or less the same. Firms and groups of firms show only gradual differences along the dimensions that measure the degree of organization and therefore can be analyzed with the same concepts and theories as organizations (Godfroij 1981). With regard to IQA it means agreements about specifications, exchange of information, coordination, and control, or even redesigning the supply chain to realise superior customer value at minimal costs at the interorganizational level. This means to be successful, firms have to cooperate. The actual competitiveness of the food industry will not be determined by its (certificated) quality assurance systems or integrated quality assurance systems, which will only be a facilitating factor, but by the co-operating ability of firms. In other words, the ability to start partnerships. A partnership can be defined as "a set of interdependent firms that work closely together to manage the flow of produce and services along the supply chain, in order to realize superior customer value at minimal costs" (Wierenga 1996). In principle the partnership is temporarily and partial. Its organization and structure is a result of joint activities and the exchange of information, people, and resources (Zuurbier et al. 1996).

**Developing Successful Competitive Partnerships**

Partnerships can be put in a transactional and a co-operative action perspective. Within a transactional perspective, a partnership can be characterized as a regulatory system in order to establish an efficient and reliable flow of transactions. Actors are interdependent, but primarily directed towards their own objectives. Within a co-operative action perspective, a partnership can be characterized as a coordinating system in order to effectuate common objectives. Actors are interdependent in their common objectives as well as in their complementary contributions. This does not mean that transactions and co-operative action are mutually exclusive. Co-operative action might encompass contracting between actors and might therefore serve as a part and precondition of a transaction. In reverse, transactions often do not involve co-operative action. Bilateral and multilateral contracts can specify both procedural and actual aspects of transactions without referring to coordinated action (Godfroy 1993). Especially the development of partnerships in the perspective of co-operative action seems to be relevant for food and agribusiness as a result of the specific characteristics of this line of business. Nevertheless, the performance of partnerships will always be a compromising one. One has to bear in mind that objectives of partnerships depend to a large extent on the objectives and strategies of the participating actors and the nature of their interdependence. In cases where it has competitive elements the gains of one party are to some extent the losses of others.

Very often the decision to cooperate is inspired by the existence of mutual interdependencies. Several interdependencies are at issue, e.g., technical, knowledge, continuity, social, and capital interdependencies

(Kaman 1989). Therefore, the organizational structure and the derived transaction costs are not purely a matter of asset specificity, transaction frequency, and uncertainty, but are affected by forces which affect relations within partnerships. Hakansson (1985) mentions four forces:
- Functional interrelations: actors, activities, and resources are a system of interrelated supply and demand issues.
- Power: actors found their power on controlling the activities and/or resources.
- Knowledge: the development of activities as well as the use of resources are related to the knowledge possessed by partners.
- Time: partnerships are the result of contacts, knowledge, and experiences from the past. Changes within the partnership have to be accepted at least by a part of the participants otherwise changes will be marginal.

These forces proceed from the latent relations, which are related to the manifest relations. Activities of actors are linked within a partnership and are aimed to be continued in the scope of efficiency, but through that interdependencies will develop and/or will be strengthened. The structure and organization of a partnership will act as a control mechanism, which makes some changes easier and others more difficult (Ziggers 1996). This can be illustrated with the Sainsbury's program called Sainsbury's Product Management System. This system, which has been devised to be totally compatible with ISO 9002 (the international standard for quality management), details a framework for product quality systems and identifies the key areas that Sainsbury's suppliers will be expected to control to ensure the consistent production of high quality products for Sainsbury's. The goal of the program is to ensure that Sainsbury's deals with capable and competent suppliers. The Sainsbury's System provides a framework for the standards against which the suppliers will be audited to either become approved suppliers or to maintain supplier status. Note that the Sainsbury's System does not replace government regulations. By doing this, functional interrelations develop and asset specific investments are required. Suppliers have to meet the standard, but it also provides assured sales for them. On the other hand, Sainsbury has the power to introduce such a program but also need reliable suppliers to prove to come up to expectations of customers. Due to these functional interrelations, entry and exit barriers may develop.

Basically, the key factors for successful partnerships are clear benefits for all participants, trust, and openness, besides a good strategic fit of partners and (organizational) flexibility (Hughes 1994; Zuurbier et al. 1996). The firmness of the relationship will be determined by the extent to which partners come to an agreement about objectives, strategies, resources, coordination, facts, and remaining questions (Commandeur et al. 1989; Zuurbier et al. 1996). It implies also that a partnership can not be considered as a collective, but that one has to consider the individual position of firms as well. Therefore, the actual performance of partnerships might depend more on power and mutual interdependence of actors than on efficiency (Godfroij 1993; Ziggers 1994; Zuurbier et al. 1996). This can be illustrated by research of Kuiper and Meulenberg (1997) on the impact of modest price changes (10 percent) in a specific stage of the Dutch pork chain on prices of other companies in the supply chain. Results indicated that:

- Changes of retail prices bring about changes in prices of all companies in the channel: slaughterhouses, fatteners, and breeders. It took about ten months before all prices approached their new equilibrium by less than 1 percent.
- Changes of slaughterhouse prices influence pork pig and feeder prices substantially, but affected retail prices to a minor extent.
- Changes of feeder prices generated short-term changes in pork and pig prices (and the other way around), but did not bring about price changes of slaughterhouse and retail companies.

These results suggested that:

- Price coordination from retailer to feeder farm did not indicate structural bottlenecks. This result was considered as in line with the popular view that retailers have a strong power in the Dutch pork supply chain.
- Price coordination from feeder farm to retailer did indicate, in the short run, a bottleneck between the multiplier and finisher on one hand and the slaughterhouse and retailer at the other hand. This bottleneck, however, did not persist in the long run.

Partnerships will only last if they are able to meet the key factors of successful partnerships. Changes occur especially during economical recession in order to maintain continuity and control of gained positions, with functional, strategic, and political issues also involved (Kickert and Van Vught 1984). If common objectives are going to be different and coordinating actors are losing their influence, then the partnership might decline. Examples in Dutch food and agribusiness are the development of the "VTN" Food Horticulture Holland (food horticulture), the Greenhouse (ornamental flowers), and DUMECO (meat production, a merger of the meat processing firms Coveco, ENCB, and Gupa). These new organizations meant new and different positions for and relations between growers, auction, the central Board of Horticulture -- auctions and customers, respectively, and co-operative pig farmers and meat processors.

## Potential Impacts on Competitiveness of the Food Industry

The food industry in general is confronted with rapidly changing markets, new technologies, and world wide competition. Food safety, the structure of the food industry, and the evolution of private labels and/or brands increase transaction costs. On one hand, (voluntary) quality assurance standards, such as HACCP, and retailer quality assurance programs have evolved to reduce those costs. To the extent that these voluntary schemes become standard business practice, they could serve as effective barriers to imports from other countries. On the other hand, competitiveness in the food industry will be about the relative effectiveness of partnerships and the extent that it is able to produce to specification. The need to produce to specification will increase and the ability to measure product characteristics also will be enhanced. Thus, the cost of producing to meet customer demand likely will be lower in a more closely coordinated system.

Improved control over product quality, safety, and quantity in general, and the focus on product differentiation to supply to increasingly discriminating (niche) markets, may be considered as primary motives for vertical coordination modes in the food industry. (Hanf and Wright 1992; Barry et al. 1992; Den Ouden et al. 1996, Ziggers 1997). Product and market innovations are necessary to maintain continuity in terms of marketshare and turnover. Success depends to a large extent on the distribution of power, agreements about common objectives, and agreements about the sharing of costs and benefits. Developing partnerships in the perspective of co-operative action seems to be relevant for food supply chains. Motives for this are development of competitive power, need for quality, safety and sustainability of food produce, and flexibility to react fast to changing markets. Several forms of risk can be reduced this way: risk of fluctuating prices, risk of quantity/quality features (e.g., transport of pork, scheduling of pork finishing capacity with slaughterhouse, and meat products processing capacity), and risk of food safety and hygiene. Moreover, quality assurance systems may be considered as facilitating factors.

Partnerships are likely to extend across food supply chains from input supplier through producer to processor and distributor as is currently occurring in pork production and horticulture. It implies that market position and financial performance will depend increasingly upon management rather than on ownership of assets. Management not only in the form of operations and strategic management skills internal to the firm, but also in the form of successful negotiation of linkages with suppliers and distributors and having the proper external partners (Boehlje et al. 1995). In addition, the organizational structure of both firm and partnership and its ability to be responsive to changing customer needs and business environmental challenges are also important. These can be considered as critical skills, which are more difficult to develop, but might have a profound contribution to a sustainable competitive advantage (Boehlje et al. 1995). The implications will be a diversity of partnerships characterized by an enhanced asset specificity (technical, knowledge, s-kills) and higher exit and entry barriers. As a consequence those who are not motivated or capable to adapt to this new business environment might be effectively excluded from international markets that require efficient, reliable, and above all market-oriented business practice as a standard.

## Conclusion

In this paper it was argued that HACCP serves as a management tool to control both food safety and food quality. It also contributes to process improvement. Both will contribute to competitiveness. It also was argued that due diligence increases transaction costs along the supply chain and that institutional innovation emerge to facilitate exchange, which implies that in the food industry, HACCP or other quality assurance programs/standards can contribute to reduced transaction costs. However, they merely serve as a facilitating factor. The actual competitiveness of the food industry is determined by cooperation among firms. Finally, it was argued that improved control over product quality; safety, and quantity in general, and the focus on product differentiation to supply to increasingly differentiated markets are primary motives for vertical coordination modes. As a consequence, those who are not motivated or capable to adapt to this new business environment might be effectively excluded from international markets.

## Notes

[1]G.W. Ziggers is Head Department of Quality and Supply Chain Management, State Institute for Quality Control of Agricultural Products (RIKILT-DLO), P.O. Box 230, NL-6700 AE Wageningen, the Netherlands,
E-mail: g.w.ziggers@rikilt.dlo.nl

## References

Barry, P.J., Sonka, S.T. and Lajili, K., 1992. Vertical Coordination, Financial Structure and the Changing Theory of the Firm. In: *American Journal of Agricultural Economics:* Volume Number 74, 1219.

Bredahl, M.E., and E. Holleran. 1997. Food Safety, Transaction Costs and Institutional Innovation. In: *Quality Management and Process Improvement for Competitive Advantage in Agriculture and Food,* ed. G. Schiefer and R. Helbig, University of Bonn, Bonn.

Boehlje, M., Akridge, J. and Downey, D., 1995. Restructuring Agribusiness for the 21st Century. In: *Agribusiness: An International Journal:* Volume Number 11, nr 6, 493-500.

Godfroij, A.J.A., 1981. *Networks of Organization; Strategies, Games and Structures.* VUGA, Den Haag.

Godfroij, A.J.A., 1993. Interorganizational Network Analysis. In: P.Beije, J. Groenewegen en O. Nuys. *Networking in Dutch industries.* Garant, Apeldoorn.

Hakansson, H., 1985. *Industrial Technological Development. A Network Approach.* Croom Helm, London.

Hanf, C.H. and Wright, V., 1992. *The Quality of Fresh Food and the Agribusiness Structure.* Working paper. University of New England, Armidale N.S.

Heyman, E.A.C., Jansen, M.G.C., en Maijers, W., 1994. *Solutions for the Implementation of Integrated Chain Control.* NEHEM, 's-Hertogenbosch (in Dutch).

Hoogland, J.P. and A. Jellema and W.F.M. Jongen. 1998. Quality assurance. In: *Innovation of food production systems,* ed. W.M.F. Jongen and M.T.G. Meulenberg, Wageningen Pers, Wageningen.

Hughes, D., 1994. *Breaking with tradition: building partnerships and alliances in the European Food Industry,* Wye College Press, Wye.

Hobbs, J. and W. Kerr. 1992. Costs of monitoring food safety and vertical coordination in agribusiness: What can be learned from the British Food Safety Act 1990? In: *Agribusiness: An International Journal:* 8, 575-584.

Kamann, D.J.F., 1989. Actors in Networks. In: F.W.M Boekema en D.J.F. Kamann (eds). *Social-Economical Networks.* Wolters-Noordhoff, Groningen (in Dutch).

Kampfraath, A.A. and Marcelis, W.J., 1981. Management of Organizations. Kluwer, Deventer (in Dutch).

Kickert, J. en Vught F. van, 1984. Networks of Policy and Public Management. In: Bekker A. en Rosenthal U. (eds.), *Networks and the Public Administration.* Alphen a/d Rijn, Samson (in Dutch).

Kuiper, W.E. and M.T.G. Meulenberg. *A vector error-correction model of price time series to detect bottle-neck stages within a marketing channel.* Paper presented at the International Conference on "Vertical Relationships and Coordination in the Food System," June 12-13, Piacensa 1997.

Notermans, S. and A. van der Giessen. 1993. Foodborne diseases in the 1980's and 1990's – The Dutch experience. *Food Content* 4: 122-124.

Ouden, M. den, Dijkhuizen A.A., Huirne, R.B.M., Zuurbier, P.J.P., 1996. Vertical cooperation in agricultural production-marketing chains, with special reference to product differentiation in pork. In: *Agribusiness: An International Journal:* Volume Number 12, 3, 277-290.

Sockett, P.N. 1991. Food poisoning outbreaks associated with manufactured foods in England and Wales: 1980-89. *Communicable Diseases Report 1. Review* no 10, r105-r109.

Wierenga, B., 1996. *Competing for the future in the Agricultural and Food Channel,* Paper presented at the seminar Agricultural marketing and consumer behaviour in a changing world, Wageningen.

Williamson, O.E., 1979. Transaction-Cost Economics: The governance of Contractual Relations. In: *Journal of Law and Economics*: Volume Number 22, 233-261.

Ziggers, G.W., 1993. *Agricultural Entrepreneurship in a Management Perspective.* PhD. Dissertation Agricultural University, Department of Management Studies, Wageningen (in Dutch).

Ziggers, G.W., 1996. Agricultural production in transition. In: Kostadinos Mattas, Evaggelo Papnagiotou and Kostantinos Galanapoulos. *Agro-Food Small and Medium Enterprises in a Large Integrated Economy.* Proceedings of the 44th European Association of Agricultural Economists (EAAE), 11-14 October, 1995, Thessaloniki, Wissenschaftverlag VAUK, Kiel.

Ziggers, G.W., 1997. Integrated Quality Assurance in the Pork Supply Chain. . In: *Quality Management and Process Improvement for Competitive Advantage in Agriculture and Food,* ed. G. Schiefer and R. Helbig, University of Bonn, Bonn.

Zuurbier, P.J.P., Trienekens, J.H. and Ziggers, G.W., 1996. Vertical Cooperation: Methods to Start Partnerships in Food and Agribusiness. Kluwer, Deventer.

Chapter 17

# The Vertical Organization of Food Chains and Health and Safety Efforts

Frank H. J. Bunte[1]

The introduction of HACCP in Dutch law is preceded by EU regulation. Dutch law prescribes HACCP, unless equivalent health and safety mechanisms such as Good Manufacturing Practices (Hygiene Codes) are applied. Given the presence of a wide number of Hygiene Codes in the Netherlands, the introduction of HACCP may be seen as a quality signal given the efforts firms have to make in order to obtain HACCP certification. Firms differ in the quality signal they give to the outside world for a variety of reasons. One of these reasons is related to the vertical organization of the food chains in which they are active. The vertical organization of food chains influences the profit incentives of firms, among other things, because of vertical externalities, one of which is related to investments in product health and safety. This paper explores the effects of the vertical organization of food chains on safety and health efforts. Particular attention is paid to the influence of demand and supply characteristics on these efforts. The paper also presents some solutions for the inefficiencies related to the vertical externalities identified.

The paper is constructed as follows. The first section briefly reviews the literature on vertical coordination. The second section analyzes both cooperative and non-cooperative decisions with respect to product health and safety in a simple analytical model. The influence of demand and supply characteristics is taken into account. This section is concluded with three testable hypotheses. The third section describes Dutch health law with respect to HACCP briefly and provides preliminary results of tests of the hypotheses. The fourth section concludes.

## Review

Whether a firm invests in product safety and health depends on the investment's effects on the firm's costs and benefits. The firm's benefits equal the increase in gross profits, i.e., the rise in product demand times the profit margin on the last unit sold. Product demand grows when perceived quality rises due to safety and health measures.[2] In this paper, the firm's costs coincide with the investment costs.

A well-known characteristic of vertically related food chains is the fact that a processor does not sell more than its distributors do (Spengler 1950). Coca-Cola's output corresponds exactly with its distributors' output. This characteristic explains the occurrence of vertical externalities with respect to health and safety measures, whether they are directed to output enhancement or the prevention of calamities and the resulting output and liability losses. When Coca-Cola stimulates output by marketing healthier cola or safer bottles, its distributors' profits rise as well. However, Coca-Cola is not likely to take the latter effect into account. It considers the effects on its own profits only. As a result, the investment in health and safety are too low from the chain's point of view, since a positive externality is neglected. Thus, non-cooperative vertically related firms invest too little in product health and safety, since they are inclined to consider individual profits only.[3]

Firms may resolve the vertical externalities through both quality and price mechanisms. Both mechanisms will be discussed briefly.

**Quality Mechanisms**

Firms may resolve the externalities by contracting upon mutually beneficial quality arrangements. However, these arrangements do not resolve the externality problem if firms are able to shirk without being caught (Tirole 1988). Agreements should be enforceable before court. Otherwise there is nothing which withholds firms from shirking. Quality arrangements do not work under the following circumstances. When court is not able to check whether the stipulations agreed upon have been met, then the quality arrangement is not (completely) enforceable. The variables concerned may not be measurable. When the shirking firm is able to blame bad performance to other non-controllable and non-observable causes -- general bad quality of catches -- its contracting partner is not able to make a strong case before court (Mathewson and Winter 1985). In principle, private parties may agree upon quality arrangements supplemented by monitoring and penalty systems. However, these arrangements only suffice if they are based on controllable and observable variables. Enforceability requires the ability to verify and to inflict penalties on those not meeting the stipulations agreed upon. The latter is only possible when default can be proved before court.

**Price Mechanisms**

When firms are not able to solve the externality problem by private quality arrangements, they have to rely on other mechanisms. The price mechanism may offer a way out: multi-part tariffs may be used to solve the vertical externality (Tirole 1988). In a multi-part tariffs system, more than one tariff (price) is charged. For example, the upstream firm charges both a wholesale price and a franchise fee to the downstream firm. The vertical externality involves two problems: 1) prices should be set such that joint profits are maximized and 2) joint profits are to be distributed over the supplier and its customer in one way or the other. Since two non-correlated objectives are to be met, two non-correlated policy instruments (tariffs) are necessary to obtain these goals.

When the vertical externality occurs downstream, the solution is simple. A two part tariff suffices to arrive at the joint optimum. When the wholesale price equals marginal cost, the downstream firm maximizes joint profits, since its profit margin coincides with the chain profit margin [problem (1)]. The franchise fee may be used to redistribute downstream profits to the upstream firm [problem (2)]. When the vertical externality occurs upstream, the mark-up charged by the downstream firm has to be regulated. The upstream firm has to regulate both the wholesale and the final price in order to maximize both individual and chain profits at the same time. The mark-up charged downstream should coincide with downstream marginal costs. When the wholesale and the final price are set such that downstream marginal costs are just covered, upstream and chain profit margins coincide. The upstream firm's decisions thus maximize chain profits [problem (1)]. The franchise fee may again be used to redistribute profits [problem (2)].

**Conclusion**

When quality arrangements cannot be relied upon, multi-part tariffs may be used to give firms the right incentives. These incentives depend on the wholesale price. The wholesale price should be set such that upstream and downstream incentives are balanced out. This is because when the wholesale price rises, the incentives of the upstream firm increase, while those of the downstream firm decrease. The balance between upstream and downstream incentives can and should be made dependent on the reliability of quality arrangements with the downstream and the upstream firm. When a quality arrangement with the upstream firm can be relied upon, while an arrangement with the downstream firm cannot, the wholesale price should be low. In this case, the downstream firm has large profit incentives, while the upstream firm's behavior is regulated via quality arrangements. When quality arrangements with both firms cannot be relied upon, more complex price and incentive mechanisms are needed.[4]

# Vertical Externalities

This section analyzes and compares the price and investment decisions upstream and downstream firms make in two situations. First, we analyze the situation in which firms do not coordinate their decisions. Second, we analyze the situation in which they do. The investment decisions are related to the market structure characteristics demand elasticity, industry concentration, and returns to scale.

**Non-Cooperative Decisions**

This paper analyzes a simple two-links-chain model (Figure 1). In this chain, the product flows from the suppliers of raw materials via the upstream and the downstream firms to the consumers. Suppliers of raw materials charge $P_v^r$ per product, upstream firms $P_v^u$ and downstream firms $P_v^d$. Both the upstream and the downstream link are monopolistically competitive. In order to simplify the argument, each firm sells one product only. Because of product heterogeneity, each firm has a monopoly on the variety it sells.

We may now study the price and investment decisions upstream and downstream firms take when they do not coordinate their decisions. Both firms decide on prices and investments unilaterally and do not bargain over either prices or investments. Consider Figure 1 once again. By assumption, downstream firms decide on prices and investments when wholesale prices are known. This implies that upstream firms decide on price and quality efforts before the downstream firms do. However, the model is solved backwards by first deriving optimal downstream prices as

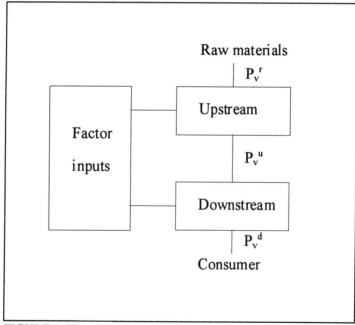

FIGURE 1. The chain.

a function of upstream prices. The reason for this order is the following. The downstream firms take all input prices and consumer demand as given. The upstream firms, however, know that they may influence downstream price decisions by their own price decisions. By analysing downstream equilibrium first, this influence is taken into account. In the second step, optimal upstream prices are derived, which are thus based, among other things, on the relation between upstream prices and downstream equilibrium prices.

Let us turn to the decisions on price and quality. Upstream and downstream profits are:[5]

(1)
$$\pi_v^u = P_v^u \cdot D_v(Q_v, P_v^d) - C_v^u(D_v(Q_v, P_v^d)) - S_v^u(X_v^u);$$

(2)
$$\pi_v^d = P_v^d \cdot D_v(Q_v, P_v^d) - C_v^d(D_v(Q_v, P_v^d)) - S_v^d(X_v^d)$$

where $\pi_v^u$ denotes upstream net profits on variety v and $\pi_v^d$ downstream net profits on variety v. $D_v$ indicates variety v's demand which depends on quality level $Q_v$ and consumer price $P_v^d$. $Q_v$ is a function of upstream and downstream quality efforts, $X_v^u$ and $X_v^d$ respectively: $Q_v = Q_v(X_v^u, X_v^d)$. Perceived quality does not depend on the quality efforts of rival producers: we do not analyze horizontal externalities in this paper. Production costs $C_v$ depend on demand and quality costs S are a function of quality efforts $X_v$. Superscripts u and d refer to upstream and downstream firms respectively. Note that $P_v^u$ is the price set by the upstream firm and paid by the downstream firm, while $P_v^d$ is the price set by the downstream firm and paid by the consumer.

The first order conditions for downstream profit maximization are:

(3)
$$\frac{\partial \pi_v^d}{\partial P_v^d} = D_v + \frac{\partial D_v}{\partial P_v^d} P_v^d - \frac{dC_v^d}{dD_v} \frac{\partial D_v}{\partial P_v^d} = 0;$$

(4)
$$\frac{\partial \pi_v^d}{\partial X_v^d} = \frac{\partial D_v}{\partial Q_v} \frac{\partial Q_v}{\partial X_v^d} P_v^d - \frac{dC_v^d}{dD_v} \frac{\partial D_v}{\partial Q_v} \frac{\partial Q_v}{\partial X_v^d} - \frac{dS_v^d}{dX_v^d} = 0.$$

The first first order condition may be rewritten to:

(5)
$$P_v^d\left[1 + \frac{1}{\epsilon_d^d}\right] = \frac{dC_v^d}{dD_v} = \epsilon_s^d.AC_v^d.$$

This expression relates the profit maximizing price level to marginal costs ($dC_v/dD_v$) and the price elasticity of consumer demand $\epsilon_d^d$ and is well known from the literature. Marginal costs equal the product of the price elasticity of downstream supply $\epsilon_s^d$ and downstream average costs $AC_v^d$.[6] One may use the first first order condition and the Lerner index in order to rewrite the second first order condition into:

(6)
$$-\frac{\partial D_v}{\partial X_v^d}\frac{P_v^d}{\epsilon_d^d} = \frac{\partial D_v}{\partial X_v^d}\left(P_v^d - \frac{dC_v^d}{dD_v}\right) = \frac{dS_v^d}{dX_v^d}$$

The quality effort pursued depends on the marginal cost of quality efforts and marginal benefits: the increase in demand times the profit margin on the last unit sold.

A similar analysis applies to upstream firms. The first order conditions now are:

(7)
$$\frac{\partial \pi_v^u}{\partial P_v^u} = D_v + \frac{\partial D_v}{\partial P_v^d}\frac{\partial P_v^d}{\partial P_v^u}P_v^u - \frac{dC_v^u}{dD_v}\frac{\partial D_v}{\partial P_v^d}\frac{\partial P_v^d}{\partial P_v^u} = 0;$$

(8)
$$\frac{\partial \pi_v^u}{\partial X_v^u} = \frac{\partial D_v}{\partial Q_v}\frac{\partial Q_v}{\partial X_v^u}P_v^u - \frac{dC_v^u}{dD_v}\frac{\partial D_v}{\partial Q_v}\frac{\partial Q_v}{\partial X_v^u} - \frac{dS_v^u}{dX_v^u} = 0$$

or

(9)
$$\frac{\partial \pi_v^u}{\partial P_v^u} = P_v^u \left[ 1 + \frac{1}{\epsilon_d^u} \right] = \epsilon_s^u \cdot AC_v^u;$$

(10)
$$-\frac{\partial D_v}{\partial X_v^u} \frac{P_v^u}{\epsilon_d^u} = \frac{\partial D_v}{\partial X_v^u} \left( P_v^u - \frac{dC_v^u}{dD_v} \right) = \frac{dS_v^u}{dX_v^u}$$

As before, the profit-maximizing price level is related to marginal costs and the price elasticity of derived demand; and the quality effort pursued to the marginal cost of quality efforts and marginal benefits: the increase in demand times the profit margin of the marginal product.

Both upstream and downstream quality efforts depend on three components: 1) the sensitivity of demand with respect to quality efforts: $\partial D_v/\partial X_v$; 2) the profit margin of the last unit sold: $P_v - \partial C_v/\partial D_v$; and 3) the costs of quality efforts: $\partial S_v/\partial X_v$. The first component is affected by the consumer-producer relation. When consumers face large uncertainty with respect to product quality due to moral hazard and adverse selection problems (Akerlof 1970), consumer demand will not be sensitive to quality efforts. The second component rises when:

1) The price elasticity of demand falls. The profit margin is negatively related to the price elasticity of demand, since the cost of a profit margin increase -- a loss in demand -- rises with the price elasticity of demand. This relation is reflected in the Lerner index: $L = (P_v - \partial C_v/\partial D_v)/P_v = -1/\epsilon_d$. The price elasticity of demand, of course, depends positively on the number of substitute products available.

2) The concentration in an industry increases. The profit margin is a negative function of the number of firms in the industry. Market power decreases with the number of rival firms. Cartelization is easy to achieve if the number of firms is small. This phenomenon is also related to the Lerner index.

3) The price elasticity of supply falls. Marginal costs may be rewritten as the product of the price elasticity of supply and average costs. When the price elasticity of supply is smaller than one, average production costs fall with output. It is easy to see that

increases in quality efforts are more profitable when there are returns to scale in production.

The above analysis again illustrates the trade-off between static and dynamic efficiency: quality efforts are positively related to profit margins. Appropriability is a necessary condition for quality enhancement. Appropriability is favored by a low price elasticity of consumer demand, industry concentration, and returns to scale in production.

**Cooperative Behavior**

In this sub-section, cooperative decision making is analyzed. We first show that health and safety efforts are higher when firms coordinate their price decisions. Thereupon, we discuss some of the coordination devices firms have available. Afterwards, we address the role of wholesale prices in coordinating investments in health and safety. Finally, we indicate the influence of market structure characteristics on the efforts pursued and the role of prices in restoring incentives to invest in health and safety.

*Cooperative Decision Making.* Non-cooperative quality efforts are given by equations (6) and (10). Cooperative quality efforts are given by the levels maximizing joint rather than individual profits. When the above analysis is repeated for cooperative decision making with respect to quality efforts, the following first order conditions may be derived after some substitution:

(11)
$$\frac{\partial D_v}{\partial X_v^d}\left[\left(P_v^d - \frac{dC_v^d}{dD_v}\right) + \left(P_v^u - \frac{dC_v^u}{dD_v}\right)\right] = \frac{dS_v^d}{dX_v^d}$$

(12)
$$\frac{\partial D_v}{\partial X_v^u}\left[\left(P_v^d - \frac{dC_v^d}{dD_v}\right) + \left(P_v^u - \frac{dC_v^u}{dD_v}\right)\right] = \frac{dS_v^u}{dX_v^u}$$

Equations (11) and (12) illustrate that cooperative effort levels depend on both upstream and downstream profit margins. Cooperative efforts are larger than non-cooperative efforts, since the positive effect of the demand increase on the other link is taken into account. Compare equations (11) and (12) with equations (6) and (10). The terms between brackets on the left hand side of the equations, i.e., the profit margin on the last unit sold, are larger in equations (13) and (14) than they are in equations (6) and (10). As a result, the right hand side of the former equations, i.e., marginal control costs, must be larger as well. And so, we have shown that cooperation increases quality efforts by internalizing vertical externalities.

*Coordination Devices.* Above, we argued that cooperative behavior may be induced by multi-part tariffs, e.g., franchise arrangements (Tirole 1988).[7] Consider the situation in which quality is only affected by downstream efforts. These efforts are maximum when downstream firms maximize individual and chain profits at the same time, i.e., when wholesale prices coincide with upstream marginal costs. In this case, the chain profit margin is reduced to the downstream profit margin. Upstream marginal costs are taken into account via the wholesale price. The upstream firms may benefit from downstream profits via the second tariff, the franchise fee. A similar tariff system may be used when quality is affected by upstream efforts only. In this case, the consumer price must also be controlled via resale price maintenance.

*Wholesale Prices.* However, when both firms affect product quality, two part tariffs are no longer sufficient. The wholesale price should be set such that both upstream and downstream firms maximize chain profits at the same time. This is a logical impossibility, since the profit margins of both firms should coincide with the chain profit margin, i.e., their sum. For this reason, the wholesale price actually set should balance out upstream and downstream incentives. The balance between upstream and downstream incentives can be made dependent on the reliability of quality arrangements. When, for example, a quality arrangement with the upstream firm can be relied upon (because of reliable monitoring activities), while an arrangement with the downstream firm cannot, the wholesale price should be low. In this case, the downstream firm invests optimally since it has large profit incentives, while the upstream firm's behavior is regulated via quality arrangements. The balance between upstream and downstream profit incentives, of course, also depends on the relative influence on product health and safety. When upstream firms have more influence on product health and safety, they should have larger profit incentives, i.e., the wholesale price should be high.

*Market Structure.* Above, we showed that quality efforts fall with the price elasticity of demand, industry concentration and diseconomies of scale. It is easy to see that this conclusion also holds when firms coordinate both their price and their health and safety decisions. The profit margins in equations (11) and (12) depend on the same variables as those in equations (6) and (10). They only enter equations (11) and (12) twice rather than once as they do in equations (6) and (10). When decisions are coordinated vertically, health and safety efforts depend on the price elasticity of final demand, industry concentration in both links and economies of scale in both links.

*Strategic Effects.* Above we argued that the wholesale price should be set such that upstream and downstream incentives to invest in product health and safety are balanced. If one of the demand and supply characteristics influences the incentives to invest in product health and safety negatively, wholesale prices may be set such that this influence is countervailed. If concentration is low downstream, downstream health and safety efforts may be too low from the chains's point of view. When upstream firms lower wholesale prices, downstream profit margins and health and safety measures increase. The latter raises upstream profits and compensates the loss in profits due to the decrease in the wholesale price. This argument corresponds with the discussion on strategic effects (Bulow et al. 1985). When there is a negative relation between the wholesale price and downstream quality efforts, equation (11) changes into:

(13)
$$\frac{\partial \pi_v^u}{\partial P_v^u} = P_v^u \left[ 1 + \frac{1}{\epsilon_d^u} + \frac{\partial D_v}{\partial Q_v} \frac{\partial Q_v}{\partial X_v^d} \frac{dX_v^d}{dP_v^u} \left( P_v^u - \frac{dC_v^u}{dD_v} \right) \right] = \frac{dC_v^u}{dD_v}.$$

Since $dX_v^d/dP_v^u$ is negative, the wholesale price corresponding with equation (13) is lower than the one corresponding with equation (9). The wholesale price affects final demand both via the price elasticity and the effect on downstream efforts. The latter effect is the strategic effect. Strategic effects influence profits via the changes induced in rival behavior.

**Hypotheses**

The theoretical solutions derived in the previous sections may be used to derive a number of hypotheses. In this section, we concentrate on three hypotheses on the relation between safety efforts and market structure. The next section provides a first test of the hypotheses. The first hypothesis is the following:

> H1: When industry concentration is high, safety efforts are high as well.

The private gains of safety and health measures also depend on the profit margin earned on the product concerned. When demand grows with perceived quality, private gains are larger the larger the profit margin is. The profit margin is high when market structure allows firms to charge high mark-ups. When industry concentration is high, competitive pressure is low and the mark-up charged will be high. Industry concentration affects the ability to appropriate profits. The larger this ability, the larger the incentive to take safety and health measures in order to boost consumer demand. The second hypothesis is as follows:

H2: When production is characterized by increasing returns to scale, safety efforts are high.

The effect of safety and health measures on firm profits also depend on the cost structure. Gross profits rise with product demand -- as a result of safety and health measures -- but more so when marginal costs fall rather than rise with output. In other words, when there are increasing returns to scale, safety and health measures are likely to be high. Note that measures for industry concentration and returns to scale are correlated. As a result, in future research, thorough empirical tests are needed to disentangle the importance of the two relations.

H3: Firms which coordinate their vertically related activities invest more in product health and safety than non-cooperating firms do.

Integration enables firms to change the incentive structure such that joint rather than private profits are maximized. This has a positive effect on safety and health measures, since aggregate profits normally exceed private profits. Vertical coordination may take place through multi-part tariffs, quantity forcing, or vertical integration.

## HACCP in the Netherlands

**Dutch Law**

The introduction of HACCP in Dutch Law followed the enactment of the EU Directive (93/42/EEC) on the Hygiene of Foodstuffs (Lugt 1997). The EU member states were obliged to implement this directive before the end of 1995. The Netherlands indeed implemented this directive at the last day allowed (December 15, 1995). The EU has two objectives with regard to the directive. First, the protection of human health is to be promoted. Second, the internal market is to be enhanced by promoting consumer trust in foreign food. The EU Directive obliges food companies to work according to the HACCP principles or to follow a relevant Guide to Good Hygiene Practice containing these principles. The introduction of HACCP seems to make business operators responsible for monitoring food safety. European food law may now be characterized as "enforced self-regulation" (Lugt 1997). However, the final responsibility remains with the national authorities.

Dutch health law is laid down in the Commodities Act (*Warenwet*). The act offers a framework for issuing decrees and ministerial orders containing concrete stipulations on food safety. Agricultural production is not regulated by the Commodities Act, but rather by the Agriculture Quality Act (*Landbouwkwaliteitswet*). In Dutch law, the possibility to follow approved Guides to Good Hygiene Practices (Hygiene Codes) rather than implementing HACCP oneself, is explicitly restated. The Hygiene Codes are drafted by representatives of food sectors, discussed with public and private interested parties, and approved by the Health Ministry. There are about 20 approved Hygiene Guides in the Netherlands. This number is fairly high compared to the United Kingdom and Germany where there is only one approved guide in each country (Lugt 1997). This difference is due to the corporatist tradition in the Netherlands.

### HACCP Implementation in the Netherlands

Since the Commodities Act has been revised in 1995, the Dutch Inspectorate for Health Protection considered 1996 as a transitional year in which the food industry was to be informed and its inspectors were to be trained. In 1997, the Inspectorate investigated the introduction of HACCP in four industries in which there are no Hygiene Codes. The results in Table 1 indicate that the implementation of HACCP is only beginning. Column 4 gives the share of firms implementing HACCP; column 5 gives the share of firms with operational HACCP systems. The Inspectorate of Health Protection investigated the performance of the operational HACCP systems by investigating one CCP. Column 6 presents the share of firms contolling this CCP at the moment of investigation. Note that the figures in column 6 refer to one CCP only. The gap between the operationality of HACCP systems and the control of the CCP merits two conclusions: 1)

**TABLE 1**
The Implementation of HACCP in the Netherlands
(Source: VMT (12-03-1998))

|  | Industry | | | |
|---|---|---|---|---|
|  | Oils and Fats | Crisps and Savouries | Vegetables and Fruits | Cocoa and Chocolate |
| **Investigated Companies** | 15 | 8 | 113 | 43 |
| **Share Large Companies**[a] | 64 | 52 | 29 | 26 |
| **Returns to Scale**[b] | 1.75 | 2.04 | 0.17 (0.92)[c] | 0.64 |
| **HACCP Implementation** | 93 | 75 | 66 | 56 |
| **HACCP Operational** | 73 | 62 | 37 | 42 |
| **Controlling CCP** | 40 | 37 | 16 | 23 |

[a]The share of large companies is measured as the number of companies with more than 50 employees divided by the total number of companies with more than four employees (Kamers van Koophandel 1995).

[b]Scale economies are measured as the minimum efficient scale (m.e.s.) In percentages of output for U.S. data (Sutton 1991).

[c]The m.e.s. for canned vegetables is 0.17; the m.e.s. for frozen vegetables 0.97. Only 5 percent of Dutch vegetables and fruits processing refers to frozen vegetables and fruits.

one should not rely on figures about operational HACCP plans as such and 2) there may be a gap between the HACCP plans intoduced and the health requirements of the Inspectorate of Health Protection.

The results in Table 1 indeed show that the introduction and implementation of HACCP is primarily observed in industries where the share of large companies and the minimum efficient scale are substantial. This observation accords with the hypotheses formulated above where we expected concentrated industries characterized by increasing returns to scale to perform more food and safety efforts than their counterparts. Recall that because of correlation problems thorough empirical testing is necessary to disentangle the importance of the two relations. Contrary to the hypotheses and the results in Table 1, we know from oral interviews[8] that Dutch multinationals refrain from applying HACCP up until now. Unfortunately, we do not have any data on the relation between vertical integration and food and safety efforts, as yet. From the interviews, we know that vertically related firms follow the firms first introducing HACCP in a particular food chain. Firms are aware of the fact that overall quality depends on the health and safety efforts performed in all links.

Table 2 presents the division of all 102 HACCP certificates granted January 1, 1998, over 13 industries. Table 2 gives only moderate support to hypotheses 1 and 2. On theoretical grounds, a positive correlation may be expected between column 3 and column 4. However, the numbers for beverages, vegetables and fruit, and meat products substantially diverge from their expected values. This divergence is due to differences in Hygiene Codes between the respective industries as well as other particular circumstances. The meat products industry, for example, is troubled by severe hygiene problems and, as a result, has a poor Hygiene Code. This implies that there is need and scope for improving product health and safety. The dairy industry, on the other hand, has one of the strictest Hygiene Codes. This implies that there is less reason to introduce HACCP in this industry.

## Discussion

Vertically related chains are characterized by the fact that upstream and downstream firms sell the same amount of output. Coca-Cola does not sell more than its distributors do. This characteristic causes vertical externalities. The health and safety measures performed in one link raise output and profits in other links as well. Since self-interested companies do not take these positive externalities into account, there is a tendency to underinvest in health and safety efforts from the chain's point of view (apart from the consumer's point of view).

The tendency to underinvest in health and safety measures may be overcome by quality arrangements between vertically related companies.

However, whenever companies are able to shirk, quality arrangements are not upheld. Shirking is possible when monitoring occurs infrequently due to monitoring costs or when bad performance may be blamed on other factors, such a the weather, the quality of catches, or the state of consumer demand. When bad performance may be successfully imputed to other factors, quality arrangements cannot be upheld before court, which reduces their impact to nil.

Vertical externalities may be overcome by (combinations of) coordination devices such as multi-part tariffs, resale price maintenance, quantity forcing, or even vertical integration. Given their positive impact

TABLE 2
Division of HACCP Certificates Over Industries in The Netherlands

| Industry | Share Certificates[a] | Share Companies[b] | Share Certificates Minus Share Companies | Share Large Companies[c] |
|---|---|---|---|---|
| Beverages | 3 | 6.1 | -3.1 | 65 |
| Oil and Fats | 1 | 1.5 | -0.5 | 64 |
| Dairy Products | 8 | 7.9 | 0.1 | 50 |
| Flour and Grain Products | 6 | 3.2 | 2.8 | 43 |
| Vegetables and Fruit | 12 | 3.8 | 8.2 | 29 |
| Bakery Products | 23 | 28.5 | -5.5 | 27 |
| Meat Products | 28 | 23.5 | 4.5 | 25 |
| Ready to Eat Meals | 2 | 3.9 | -1.9 | 20 |
| Fish Products | 1 | 5.6 | -4.6 | 11 |
| Additives | 2 | - | - | - |
| Egg Products | 3 | - | - | - |
| Hotel and Catering | 1 | - | - | - |
| Transport and Trade | 9 | - | - | - |
| Others | 2 | - | - | - |

[a]Source: Accreditation Council.

[b]Number of companies with more than four employees divided by total number of companies with more than four employees (Kamers van Koophandel 1995). This number is corrected for the fact that the corresponding shares in column 1 add to 84 percent.

[c]The share of large companies is measured as the number of companies with more than 50 employees divided by the total number of companies with more than four employees (Kamers van Koophandel 1995).

on health and safety measures, these devices should be evaluated benevolently from an antitrust perspective. Take a simple franchise arrangement between an upstream manufacturer and a downstream retailer in which a wholesale price and a franchise fee are specified. This arrangement is able to raise both upstream and downstream profits above the levels obtained when only the wholesale price is specified. A decrease in the wholesale price increases the incentives to invest in health and safety downstream which is related to the downstream profit margin. Chain (and downstream) profits rise with the increase in health and safety measures. The upstream manufacturer may capture part of the increase in chain profits through the franchise fee.

The incentives of both cooperative and non-cooperative companies depend on market structure characteristics. These characteristics determine the company's and the chain's profit margin. The profit margin determines the profitability of health and safety measures. The profit margin and health and safety efforts are high when the price elasticity of demand is low, industry concentration is high, and returns to scale are increasing. We expect product health and safety to be high in links and chains satisfying these conditions. Dutch experience gives moderate support for these predictions: HACCP tends to be introduced and implemented in concentrated industries characterized by economies of scale. However, current health standards have a major impact on the introduction of HACCP: Industries with severe health problems are among the first to introduce HACCP.

**Notes**

[1] F.H.J. Bunte is researcher at the Agricultural Economics Research Institute (LEI-DLO) in The Hague (Netherlands).

[2] Perceived quality only rises if the consumer is able to observe in one way or the other that product safety or health is improved. Perceived quality may not rise with safety and health measures when there are substantial informational asymmetries (Akerlof 1970).

[3] Horizontal externality may also lead to underinvestments in health and safety measures. When product demand depends on the overall quality level in the industry, product health and safety will tend to be too low due to the free rider problem (Telser 1960). Horizontal externalities have vertical side effects. When quality efforts are low in one link, chain demand will be low.

[4] Price mechanisms have other functions as well. The most important other function in vertical relations probably is the insurance function (Rey and Tirole 1986).

[5] In order to have a well defined profit maximization problem, we assume: (1) $\partial D_v/\partial Q_v > 0$ and $\partial^2 D_v/\partial Q_v^2 < 0$. Consumer demand rises with product quality, but increasingly less. (2) $\partial D_v/\partial P_v < 0$ and $\partial^2 D_v/\partial P_v^2 > 0$. Consumer demand falls with price and increasingly so. (3) $\partial Q_v/\partial X_v > 0$ and $\partial^2 Q_v/\partial X_v^2 < 0$. Product quality rises

with quality efforts, but increasingly less. (4) $\partial C_v/\partial D_v > 0$. Production costs rise with output and increasingly so. (5) $\partial S_v/\partial X_v > 0$ and $\partial^2 S_v/\partial X_v^2 < 0$. Quality control costs rise with the efforts undertaken and increasingly so.

[6] The price elasticity of supply is defined as: $\epsilon_s = (\partial C/\partial D)(D/C)$.

[7] Vertical integration and quantity forcing are alternatives for multi-part tariffs. Quantity forcing may take the form of shelf or minimum purchase requirements.

[8] The Director of the Inspectorate of Health Protection and the Director of the Accreditation Council have been interviewed.

**References**

Akerlof, George A. 1970. The market for "lemons": Quality uncertainty and the market mechanism. *Quarterly Journal of Economics* 84: 488-500.

Bulow, Jeremy I., John D. Geanakoplos and Paul D. Klemperer. 1985. Multimarket oligopoly: Strategic substitutes and complements. *Journal of Political Economy* 91: 488-511.

Kamers van Koophandel en Fabrieken (1995). *Adressen en bedrijfsinformatie catalogus '95/'96*. Woerden: Kamers van Koophandel en Fabrieken.

Lugt, Marieke (1997), *The shifting boundaries of public and private enforcement: Case study foodstuffs*. Utrecht: Utrecht University, mimeo.

Mathewson, G. Frank and Ralph A. Winter. 1985. The economics of franchise contracts. *Journal of Law and Economics* 28: 503-526.

Rey, Patrick and Jean Tirole. 1986. The logic of vertical constraints. *American Economic Review* 76: 921-939.

Spengler, Joseph J. 1950. Vertical integration and antitrust policy. *Journal of Political Economy* 58: 347-352.

Sutton, John. *Sunk costs and market structure*. Cambridge, Ma: MIT Press.

Telser, Lester G. 1960. Why should manufacturers want fair trade? *Journal of Law and Economics* 3, 86-105.

Tirole, Jean. 1988. *The theory of industrial organization*. Cambridge, Ma: MIT Press.

Voedingsmiddelentijdschrift (VMT), 12 March 1998.

Chapter 18

# Applying HACCP to Small Retailers and Caterers: A Cost Benefit Approach

Matthew P. Mortlock, Adrian C. Peters and C. J. Griffith[1]

## Introduction

The Food Safety Research Group at the University of Wales Institute, Cardiff (UWIC) is currently engaged in a project funded by the UK Ministry of Agriculture, Fisheries and Food entitled, *Evaluation of Barriers to the Usage of Food Hygiene Management Systems throughout the United Kingdom Food Industry*. This paper focuses upon the cost benefit approach adopted by the project in order to identify the economic impact of implementing HACCP based systems, particularly within food retail and catering businesses.[2]

The Hazard Analysis Critical Control Point (HACCP) system is the most widely discussed of all food hygiene management systems and is internationally recognized as representing the way forward for assuring food safety as we approach the millennium. The 1990s have seen the incorporation of the HACCP philosophy into food safety legislation across the European Union as a result of the horizontal directive on the hygiene of foodstuffs (DIR/93/43/EEC). In 1995, the United Kingdom introduced legislation to ensure its compliance with this directive. For the first time UK food business operators were obliged to use the basic principles of the HACCP system as their framework for controlling food hygiene (Food Safety [General Food Hygiene] Regulations 1995). In addition to such legislation, various government agencies and industry associations, most notably within Australia, Canada and the United States of America, have put in place a range of targeted food safety programs, all with HACCP at their core.[3]

The core principles of HACCP, to analyze the hazards inherent in a process, to identify the key points of control, and to establish the means of controlling those hazards, can be applied to any food related business, regardless of the products being handled (Jouve 1994). However, HACCP originally developed as a management tool within food manufacturing where the system has undoubtedly become an important international trading standard at both the mandatory and voluntary level (Caswell and

Hooker 1996). Yet despite this status, the HACCP approach has been less enthusiastically received among businesses in the UK retail and catering sectors.

This situation is reflected in the ongoing discussion surrounding the scope of the HACCP approach within these two sectors of the food industry, much of which has originated from sources within the United Kingdom (Sheppard et al. 1990; Meredith and Perkins 1995). These concerns, which will be expanded upon later in this paper, generally focus upon three broad areas: the potential financial implications of implementing HACCP-style systems, the wide range of food products handled which have to be included within the system, and the size/personnel base of such businesses.[4]

A number of strategies have been devised in the UK to try and help small catering businesses in particular to move towards a HACCP-based approach. Most notable of these is the *Assured Safe Catering* system (UK Department of Health 1993), developed by the UK government in conjunction with caterers to provide specific instruction on applying the principles of HACCP within this industry sector. Similarly, researchers at the University of the West of England, focusing initially upon catering practices within public houses, developed the *Small But Safe* approach, an informal HACCP-based system using a simple checklist approach (Meredith and Perkins 1995). This was designed to help small business owners comply with legislation while minimizing the input required from outside sources, for example Environmental Health Officers (EHOs) or food hygiene consultants.

Within the UK food retail sector, the debate about the use of HACCP has intensified as a result of the globally publicized outbreak of *Escherichia coli* O157, originating from a butchers shop in Central Scotland in November 1996, which claimed the lives of 20 victims. The report into the circumstances surrounding the outbreak suggested that the implementation of full HACCP systems should be accelerated within all butchers businesses with a licensing scheme being introduced in order to improve hygiene standards across the board and provide better consumer protection (Pennington Group 1997).[5] However, it is clear that if small retailers and caterers are to be successfully encouraged to adopt full HACCP-based systems in the future, the potential benefits and the costs of doing so must be clearly apparent to the business managers.

The cost benefit analysis method has been used extensively in the past as a decision making tool within environmental and transport policy. More recently it has been increasingly employed to help estimate the economic impact of food hygiene policy and practice relating to the implementation of HACCP-based systems. This trend has been most marked in the United States where data is available on the estimated costs of applying HACCP regulations across whole sectors of industry (Macdonald and Crutchfield

1996). While the cost benefit approach to food safety is less well developed in the UK, it has been used for some time by government to help estimate the impact of new regulatory proposals. For example, it has already been used to estimate the total non-recurring costs to industry, shown in Table 1, that would result from licensing and the subsequent introduction of HACCP within butchers businesses as proposed by the Pennington Group.

The range of costs presented in Table 1 depends upon the actual nature of the products handled by different butchers businesses. This reflects the common assumption that the more complex the control and monitoring procedures required, the greater the costs of implementing HACCP will be. The same source estimated the training requirements of industry at £6.5 million based on training all staff to a basic level and managers/supervisors to an intermediate level. With this added to possible licensing fees for businesses and the implementation costs listed in Table 1, it has been suggested that the maximum total cost to the butchers industry of licensing in line with the recommendations of the Pennington Group would be £38.5 million. This cost represents a relatively small

TABLE 1

Costs to UK Butchers of Complying With the Licensing Scheme[a]

| Category of Butchers Business | Total Number | Non-Recurring Costs Per Business |
|---|---|---|
| Manufacture & **wholesale** cooked meat products. | 673 | £1,000 - £5,000 |
| Manufacture a wide range of cooked meat products for **sale on premises.** | 1,309 | £1,000 - £5,000 |
| Manufacture a limited range of cooked meat products for **sale on premises.** | 2,646 | £1,000 - £5,000 |
| Buy in an extensive range of cooked meat products for resale. | 1,003 | £500 - £1,500 |
| Buy in a limited range of cooked meat products for resale. | 2,960 | Up to £1,000 |
| Other - do not sell cooked meat products. | 2,379 | Up to £500 |
| **TOTAL** | **10,800** | **£5 - £28 million** |

[a]The data presented in this table is taken directly from a "Draft Compliance Cost Assessment for Draft Regulations To Amend the Food Safety (General Food Hygiene) Regulations 1995 To Introduce An Annual Licensing Scheme For Retail Butchers," as produced by the UK Department of Health in February 1998.

fraction of the total annual turnover of the industry, estimated to be in excess of 2,440 million pounds.[6]

It was with this background in mind that we sought to further investigate the costs and benefits involved in applying HACCP-based systems to UK retailers and caterers. This project aimed to analyze both the direct and indirect costs involved in planning, implementing, and running HACCP-based systems, and the impact that having such systems in place had not only on the businesses finances, but also upon its working practices in general. The intention of this research was not to provide generic costings across whole sectors of the food industry, but rather to highlight the specific situations of various individual businesses who had tried to adopt a HACCP approach.

## Methods

An initial questionnaire was mailed to 1,650 representative UK food businesses across the retail, catering, and manufacturing sectors, the response from which helped to identify several of the barriers which exist to HACCP implementation. This survey was followed by face-to-face interviews with business managers in order to develop a wider appreciation of inherent difficulties of managing food hygiene and perceptions of HACCP-based systems. From these results, a selection of businesses that had implemented either HACCP or Assured Safe Catering were identified and invited to take part as specific case studies.

Businesses who agreed to take part in the study were visited, both to help them fully comprehend the purpose of the research and to provide us with a greater understanding about the mechanics of each business. Using information provided by the businesses about the areas covered within their HACCP plans, costing questionnaires were designed to elicit data about the costs involved in having the systems in place and any perceived benefits they bring. Managers were asked to complete the questionnaires providing cost data in terms of the direct expenditure and the value they placed upon the staff time spent in various activities. This method was designed to be as least burdensome as possible upon respondents' time, especially given that time constraints had been previously identified as a barrier to implementing HACCP in small businesses (Jouve 1994).

## Results and Discussion

The remainder of this paper has been divided into sections reviewing the specific issues that may face small retailers and caterers when attempting to implement HACCP systems. To support this discussion a variety of quotations taken directly from the transcripts of the face-to-face interviews with business managers will be included. Costing and attitudinal data from one case study will also be presented to provide further depth by detailing the costs of the practical application of the HACCP approach in a particular operation. Some general data about the case study business focused upon in this discussion can be seen in Table 2.

**Limited Financial Resources**

*If we were expected to comply with HACCP then we would be forced to close down because we simply haven't got the turnover to afford it* (catering business employing four staff).

It has been suggested that a pertinent problem for smaller businesses relates to the lack of financial resources at their disposal to support the implementation of HACCP (Jouve 1994). As can be seen in Table 2, the weekly turnover of the case study business was an average of £26,500 per week, this being relatively large by food retailing standards although nothing like on a par with the large supermarket chains. In order to provide a comparison with this turnover figure, the one off, non-recurring

**TABLE 2**
**Background Information About the Case Study Business**

| | |
|---|---|
| **Status of the Business** | Retail food hall within a larger department store. |
| **Staff Employed** | 12 full time and 19 part time food handling staff. |
| **Average Weekly Turnover** | £26,500 per week. |
| **Staff Involved in Design of the HACCP System** | Food hall manager and sales assistant *(Sales assistant holds a Diploma in Food Technology).* |
| **External Sources Used to Help Design HACCP** | Environmental Health Officers (EHOs), trade journals, HACCP textbooks. |
| **Reasons for Implementing HACCP** | To comply with requirements laid down by Environmental Health Officers (EHOs). No consideration was given to verification in the HACCP plan. |
| **Food Hygiene Practices in Place Before Implementing HACCP** | Cleaning schedules, pest control, stock rotation, temperature monitoring of food/equipment, formal food hygiene training, inspection of foodstuffs on delivery. |
| **Food Products Handled** | Fresh cream chocolates, dry goods, raw & cooked meats, bakery products, frozen cooked products, ice-cream. |

costs of planning and implementing the HACCP-based system are shown in Table 3.

As can be seen in Table 3, the total non-recurring cost to the business of implementing its new HACCP-based approach was £4,600. When compared with turnover at the weekly rate quoted in Table 2, this cost would equate to only 0.33 percent of annual turnover, below the scale of 0.5 to 1.5 percent quoted by the Department of Health costing of the licensing scheme in UK butchers. However, this figure should be treated with care because despite including the value to the business of the time spent by its staff in putting into place the new system, we can see that only £50 was spent in terms of additional wage bills beyond the norm. This suggests that the vast majority of the man hours spent developing and implementing the new system were carried out within the normal working day of the staff involved. As such, while the new system cost £4,600, if the value of the time spent is not included in the final analysis, then the tangible costs are nearer £2,500.

The manager of the case study agreed with the statement that, "the cost of implementing HACCP threatens the financial viability of the business." While this might seem a strange response given the relatively low proportion of the annual turnover which was used putting HACCP in

**TABLE 3**

**Costs of Planning and Implementing the HACCP-Based System**

| Activity | Man Hours Incorporated in the Cost | Cost |
|---|---|---|
| Formal staff meetings, preparation of background information, e.g. product specifications, flowcharts, HACCP plans, consulting external advice from literature and EHOs. | 50 hours | £1,000 |
| Additional staff training to familiarize staff with the new system. | 10 hours | £100 |
| Surplus wages to cover overtime claimed due to implementing the new system or to provide cover for staff involved. | 4 hours | £50 |
| Time and costs spent purchasing and installing any equipment required as a direct result of the new system.[a] | 18 hours | £750 |
| Cost of changes made to the fabric of shop, e.g. floors, walls, ceilings, storage units. | 80 hours | £1,500 |
| Auditing current suppliers and recruiting new ones. | 60 hours | £1,200 |
| **TOTAL** | **222 hours** | **£4,600** |

[a]Examples of this equipment include new boards and knives.

place, it is important to take into account the position of the business within the department store in which it is based. The food department manager reported that the higher management of the store was reluctant to see any change take place which affected the potential profitability of their food interests. If the costs of implementing and running HACCP in accordance with the wishes of EHOs had proven too high to bear, then the food department might well have been closed down in favor of goods that could deliver higher profit margins, for example fashion wear or cosmetics. As such, the manager did not believe that the system was cheap to implement although it had not been a waste of time and money.

**Lack of Time and Technical Resources**

*We started doing HACCP and we started analyzing critical control points in our spare time. When we had a bit of a lull in production we'd have a few hours on it* (food manufacturer employing nine staff).

Time is often cited as a constraint to the use of HACCP in small food businesses which are unable to devote personnel entirely to developing HACCP systems (Ehiri and Morris 1995). Business managers in retail and catering are often actively involved in the day-to-day functioning of the business and lack the time to spend on developing and running HACCP-style systems. As a result, the development of such systems, if it happens at all, becomes something that is done during any spare time, rather than being a key focus for the business. Unlike larger companies, smaller businesses lack the resources to employ technical staff with sole responsibility for food hygiene issues. This was clearly the case in the case study example where the majority of the time spent implementing the system took place within the normal working hours of those involved. In such situations, the notion of putting together a HACCP team made up of different areas of technical expertise (Mitchell 1992) has to be compromised in favor of simply making the best of the resources available within a given setting.

*It's very easy for large companies which have got margins on their products to get consultancy and things like that, but generally most consultancy is far too expensive* (seafood manufacturer employing 13 staff).

The case study manager agreed that the business would not have been able to implement the system without the support of external sources such as EHOs, yet also felt that hiring private consultants was too expensive. In this particular case, the business involved was perhaps fortunate that it had at its disposal a Sales Assistant who had obtained a Diploma in Food Technology and was equipped with the basic knowledge needed to help implement the system. Most small businesses would not be so blessed and would often have to rely on external consultancy to help implement such systems. Such consultancy would add a considerable extra burden onto the overall implementation costs depending upon the complexity of the individual situation. Furthermore, the business was also able to avoid the cost of training managers in the principles of HACCP to ensure that the system was understood not only by those designing it, but also by those with the responsibility of implementing it.

### Numbers/Range of Products Handled

HACCP-based systems are often said to be more difficult to apply to food retailing and catering due to the range of products often handled within such businesses (Bryan 1990).

> *HACCP is easy to apply because we have one product, or one type of product, in a very small space with very few stages*
> (ice-cream manufacturer employing three staff).

> *If you have a single type of product then HACCP is pretty easy*
> (caterer employing four staff).

In the case study example, the HACCP plan was able to incorporate the whole range of food products handled under five broad categories: dry goods, fresh chocolates, raw meats, cooked meats and other fresh cooked products, and frozen cooked products and ice-cream. This would certainly have helped reduce the costs of planning the system by ensuring that a range of different food products were covered by the same flow diagram. However, if we refer back to Table 3 and the total figure of £750 spent on purchasing and installing new equipment, breaking this down across the different categories of food products reveals that some £700 of this amount was spent on the raw meats and cooked meat/fresh cooked product areas alone.

### Does HACCP Reduce the Running Costs Associated With Food Hygiene?

The HACCP approach focuses the attention of food businesses upon specific points of control and is thus thought to have the potential to reduce the ongoing costs of managing food hygiene within businesses (Ehiri and Morris 1995). Table 4 details the ongoing running costs of the HACCP system within the case study.

The weekly running costs of £1,417 represent some 5.3 percent of the weekly turnover of the business. This seems a particularly large proportion although the majority of this cost is absorbed within the wage bill of the business as it accounts for 136 man hours in total. It is also worth considering that many business managers would not necessarily perceive the particular types of costs listed as something extra and unique to the HACCP system, but rather as the everyday practices which they are required to carry out simply to function as a business.

*I don't think that managing food hygiene costs me anything appreciable at all because we do it all in our general shop routine* (delicatessen employing four staff).

As with the implementation costs listed in Table 3, the running costs of the system differ across the various areas of the store. This time the raw meats and cooked meat/fresh cooked product areas accounted for just over half (£775) of the total running costs. The HACCP plan showed that these areas had the highest number of CCPs, reflected not only in the running costs, but also in the start up costs in these areas as mentioned previously. It is worth noting that Table 2 highlighted the fact that many of the activities now incorporated within the daily running of the HACCP system were already being carried out before it was put in place. Consequently, it is important to consider the change which has come about as a result of the system being in place.

The reported increase in the cost of managing food hygiene on a weekly basis was £440, inclusive of an increased manpower input of 39 hours. Once again it is important to be aware that the actual increase in the weekly wage bill was reported as an extra £100, suggesting that the majority of the extra time being spent on managing the system has been absorbed within the normal working hours of the staff. Finally, despite the notion that HACCP can reduce product wastage, the case study reported an increase in lost turnover due to product loss of some £400 per week. This figure seems particularly high, and represents a considerable

**TABLE 4**
**Ongoing Weekly Costs of Running the HACCP-Based System**

| Activity | Man Hours Involved | Cost |
|---|---|---|
| Cleaning and disinfection of the store (including the cost of cleaning materials). | 39 hours | £342 |
| Checking products and vehicles on delivery. | 39 hours | £480 |
| Monitoring & recording of "HACCP data" at CCPs and other control points. | 38 hours | £340 |
| Maintenance of refrigerated storage equipment. | 12 hours | £155 |
| Ongoing staff training and competency testing. | 8 hours | £100 |
| **TOTAL** | **136 hours** | **£1,417** |

reduction in weekly turnover of 1.5 percent. Given that the case study investigation was carried out within a month of the implementation of the HACCP plan, this would appear to be a short-term failure to control one or more of the CCPs. The manager has clearly identified the product loss and should be examining the control measures, monitoring procedures, critical limits, corrective action, and product disposition in order to remedy the situation. Control should be recovered very quickly unless there is a fundamental fault in the food operation, for example a chiller unit that cannot maintain food temperatures below the critical limit. This would again require further expenditure and re-evaluation of implementation and running costs.

### The Problem of Marketing HACCP

*It (HACCP) has cost far more to us but in many ways it was the only way this company could keep going...it is a basic necessity now* (seafood manufacturer employing 13 staff).

It was stated earlier in this paper that the HACCP approach had now gained international recognition and was now an important trading standard, particularly for companies operating on the world stage. Yet this is clearly of little relevance to small retail and catering businesses dealing within a limited catchment area and predominantly selling direct to the general public.

*After all, it's not as though our customers want to see evidence of HACCP* (caterer employing four staff).

Change within any industry is often motivated by consumer demands, with buyers increasingly demanding that manufacturers have HACCP systems in place. The costs of moving to a HACCP approach can be offset against either the securing of existing supply contracts or the winning of new contracts as a result. Nevertheless, it is hard to envisage a situation where HACCP becomes profitable by increasing demand for food products from retail and catering outlets (Beaumont 1991). While there is plenty of scope for such businesses to market their hygiene activity more positively, the extent to which this would prove cost effective and yield tangible results is doubtful (Leach 1996).

Various studies have suggested that consumers would be willing to pay more for guarantees of food safety. UK food consumers were found to be willing to pay a positive amount in order to secure a reduction in the risk of food poisoning, although the precise amount varied according to the food item in question and the demographic characteristics of the respondents (Henson 1996). It should be noted, however, that the contingent valuation (survey-based) approach, common to many "willingness to pay" studies including the one mentioned, has been subjected to considerable criticism because of its hypothetical nature and problems of information bias (Ajzen et al. 1996). Even if consumers can be proven willing to pay more for food products from establishments running HACCP-based systems, it would be a different matter to convince a small business owner that he could afford to charge a higher price than his competitor across the street.

*When we first opened there were only four other Indian restaurants, now everything is very competitive and we've got to be as competitive as possible* (Indian restaurant employing eight staff).

## Conclusions

While every business is unique and not all will have the good fortune to employ sales assistants with degrees in food technology, the results from the case study presented in this paper suggest that the implementation of HACCP-based systems can be achieved within the financial means of small retail/catering businesses. Non-recurring start up costs can be kept below 0.5 percent of annual turnover, and although the relative running costs of the system take out a substantial part of weekly turnover, many of these costs are not unique to the HACCP approach itself, representing instead the general management costs involved with running a food business which become incorporated within the system when it is put in place. The implementation of HACCP in the case study helped highlight the costs of maintaining good hygiene practices as a prerequisite to HACCP

The cost benefit approach at the micro level of individual businesses as described here is still relatively rare. Further applications of the method described in this paper will investigate the costs of HACCP implementation across a wider range of retail catering and manufacturing settings. Accompanying these ongoing case studies, willingness to pay experiments are being carried out to provide further data on the likelihood that the UK consumer, still relatively under-researched on this topic, will pay more for guarantees of safe food.

Businesses who were implicated in food poisoning outbreaks across Wales in 1997 and 1998 are being surveyed to provide data on the relative risks inherent in different businesses and the direct costs incurred by businesses who have had to deal with such an outbreak. Finally, the macro costs of adopting the HACCP approach across the Welsh butchery industry are also being investigated.

Worthwhile as detailed costings such as this are, it is doubtful whether cost benefit analyses alone can provide the necessary encouragement to other businesses to adopt HACCP-based systems. After all, the case study example was dominated by outgoing costs with no real mention of any tangible financial benefits. If such systems are to make a real and successful impact upon the UK retail and catering industries, then a whole culture change will need to take place. There is a still a general lack of understanding among food business managers of either the nature or purpose of the HACCP approach (Ehiri and Morris 1995). Many retail and catering managers in particular have often been doing their jobs for many years and are still able to look back to a time when controls were less stringent and yet food poisoning seemed rare.

> *Look at the people in the war, look how they lived, what they ate, what did they catch* (one man butchery business)?

> *I think we are overly hygienic, I was always brought up on the theory that a bit of dirt never hurt anybody* (delicatessen employing four people).

Opinions such as these reflect a lack of understanding about the real threats posed by the food we eat in the modern day and it is against this backdrop that industry leaders, trade associations, enforcement officers, and government alike must work. Yet the essence of the HACCP approach is about looking into the future to prevent problems from occurring and not to look back at how things once were. The adoption of the HACCP philosophy is driven first and foremost by the desire to improve food safety in a manner that goes beyond simple benefit-cost tests (Caswell and Hooker 1996). A new culture of moral responsibility is required to enforce the understanding that the actions of the individual hold the key to securing the maximum level of safety for us all.

### Notes

[1] Matthew Mortlock is a Research Assistant, Dr. Adrian Peters is an Honorary Reader in Food Microbiology, and Dr. Chris Griffith is Dean of Academic Affairs, all within the Faculty of Business, Leisure and Food at the University of Wales Institute, Cardiff (UWIC). The authors would like to thank MAFF for funding this research under strategic project FS1050, although the views expressed in the text are their own and do not necessarily reflect those of MAFF.

[2] A postal survey of 1,650 UK food businesses found that 69 percent of manufacturers, 13 percent of retailers and 15 percent of caterers claimed to be using full HACCP systems as defined by the seven principled approach adopted by CODEX.

³Such programs include the National Food Safety "Farm to Table" Initiative in the United States, the Food Safety Enhancement Program (FSEP) in Canada, and the Meat Industry Strategic Plan (MISP) by the Meat Research Corporation in Australia.

⁴The concerns listed about the financial impact of introducing HACCP-style systems and the problems of a limited personnel base could equally apply to small manufacturing businesses. However, the focus of this paper is upon small retailers and caterers in particular. The same postal survey of 1,650 businesses revealed that 75 percent of retailers and 80 percent of caterers employed fewer than eight food handlers.

⁵Full HACCP refers to the seven principle approach adopted by the joint FAO/WHO Codex Alimentarius Commission (WHO ALINORM 97/13A).

⁶This estimate is based on information provided by the UK Office for National Statistics for 1997.

**References**

Ajzen, I., et al. 1996. Information Bias In Contingent Valuation: Effects of Personal Relevance, Quality of Information and Motivational Orientation. *Journal of Environmental Economics and Management.* 30(1):43-57.

Beaumont, J.A. 1991. Industry Responsibility: Recognizing Priorities For the 1990s. *British Food Journal* 93(6):3-6.

Bryan, F.L. 1990. Hazard Analysis Critical Control Point (HACCP) Systems For Retail Food and Restaurant Operations. *Journal of Food Protection* 53(11):978-983.

Caswell, J.A. and Hooker, N.H. 1996. HACCP as an International Trade Standard. *American Journal of Agricultural Economics* 78(3):775-779.

Daniels, R.W. 1991. Applying HACCP to New Generation Refrigerated Foods at Retail and Beyond. *Food Technology* 45(6):122-124.

Ehiri, J.E. and Morris, G.P. 1995. *Implementation of HACCP Strategy: A Study of Food Businesses in Glasgow.* Department of Public Health: University of Glasgow.

*Food Safety (General Food Hygiene) Regulations 1995* (1995: S.I. No.1763) London: HMSO.

Henson, S. 1996. Consumer Willingness to Pay For Reductions in the Risk of Food Poisoning in the UK. *Journal of Agricultural Economics* 47(3):403-420.

Jouve, J.L. 1994. HACCP as Applied in the EEC. *Food Control* 5(3):181-186.

Leach, J.C. 1996. Raising Food Hygiene Standards - Could Customer Power and the New Laws Hold the Key? *Journal of the Royal Society of Health* 116(6):351-355.

Macdonald, J.M. and Crutchfield, S. (1996) Modeling the Costs of Food Safety Regulation. *American Journal of Agricultural Economics* 78(5):1285-1290

Meredith, L. and Perkins, H. 1995. Small But Safe: Hazard Analysis System For Small Food Businesses. *Environmental Health* 103(6):119-122.

Mitchell, B. 1992. How to HACCP. *British Food Journal* 94(1):16-20.

Pennington Group. 1997. *Report on the Circumstances Leading to the 1996 Outbreak of Infection With E.Coli O157 In Central Scotland, The Implications For Food Safety and the Lessons to be Learned.* Edinburgh: The Stationery Office.

Sheppard, et al. 1990. Hygiene and Hazard Analysis in Food Service. In *Progress In Tourism, Recreation and Hospitality Management*, ed. C. Cooper, 192-226. London: Belhaven Press.

Sperber, W.H. 1991. The Modern HACCP System. *Food Technology* 45(6):116-120.

UK Department of Health. 1993. *Assured Safe Catering: A Management System For Hazard Analysis*. London: HMSO.

WHO ALINORM. (1997/13A) Appendix 2: The HACCP System and Guidelines For Its Application. In the *Report of the 29th Session of the CODEX Committee on Food Hygiene.*, 30-37.

Chapter 19

# The 1996 *E. coli* O157 Outbreak and the Introduction of HACCP in Japan

Atsushi Maruyama, Shinichi Kurihara and Tomoyoshi Matsuda[1]

## Introduction

Japanese will remember the year of 1996 as the year they were shocked by a sudden surge of *E. coli* O157 infection among more than 10,000 people, with a death toll of 15. Since *E. coli* O157 was transmitted through foods served in school lunches in most cases, the majority of infected patients were children. Shortly after the outbreak, the Ministry of Health and Welfare (MHW) made an official announcement that radish sprouts (*kaiware*) was responsible for this massive food poisoning. In view of a drastic drop in *kaiware* demand following the announcement, the Japanese Greenhouse Horticulture Association (JGHA), in collaboration with the Ministry of Agriculture, Forestry, and Fisheries (MAFF) and the MHW, made a manual for sanitary management for *kaiware* production based on Hazard Analysis Critical Control Point (HACCP) in December 1996. This is the first case in Japan in which the HACCP system was applied in agricultural production.

The widespread infection by *E. coli* O157 highlights the need to re-examine the Japanese school lunch system. The MHW and the Ministry of Education, Science, Sports, and Culture (MESSC) immediately launched a nationwide assessment of the management of school lunch preparation in order to eradicate the causes of foodborne illness. The assessment shows that most of the schools have problems in their food preparation processes, that sanitary management levels are different among schools, and that there is a need for a manual that integrates sanitary management for improving food safety. An *ad hoc* committee of MESSC, which was formed for investigating this issue, concluded that the sanitary management in the school lunch system must be improved in order to ensure food safety. Manuals for safety lunch preparation based on HACCP were set up by the MHW and the MESSC in March and April 1997, respectively.

In this paper, we first outline the outbreak of foodborne illness by *E. coli* O157 in 1996 and its aftermath with special reference to the *kaiware* industry and the school lunch system. We then show the current state of the HACCP application in the *kaiware* industry, and present rough estimates of the cost of implementing HACCP measures in the industry. In the fourth section, we explain how the HACCP system is introduced in the school lunch system, analyze the "consumers' attitudes" toward the safety of school lunches, and conduct a simple cost-benefit analysis of the HACCP introduction into the school lunch system, using the data collected mainly in Chiba City.

## The Outbreak of *E. coli* O157:H7 Infection in 1996

Climatic conditions from the late spring to the summer in Japan, with high temperatures and high humidity, are best suited to the growth of various microbes, so that the incidences of foodborne illness by microbes were not rare even before the widespread outbreak of *E. coli* O157:H7 (*E. coli* O157 henceforth) in 1996. Table 1 shows the number of cases and patients therein of foodborne illness reported in Japan by type of causative bacteria for 1990-96. In terms of the number of patients, *Salmonella*,

### TABLE 1
Numbers of Cases of Foodborne Illness Reported in Japan by Type of Bacteria, 1990-96

| Type of Bacteria | 1990-95[a] No. of Cases | 1990-95[a] No. of Patients | 1996 No. of Cases | 1996 No. of Patients | No. of Patients Per Case 1990-95 | No. of Patients Per Case 1996 | % Share (Patients) 1990-95 | % Share (Patients) 1996 |
|---|---|---|---|---|---|---|---|---|
| *Salmonella* | 160 | 9,888 | 350 | 16,334 | 62 | 47 | 30 | 37 |
| *Staphyloccus aureus* | 79 | 2,037 | 44 | 698 | 26 | 16 | 6 | 2 |
| *Botulinum* | 3 | 6 | 1 | 1 | 2 | 1 | 0 | 0 |
| *Vibrio parahaemolyticus* | 214 | 5,757 | 292 | 5,241 | 27 | 18 | 18 | 12 |
| *E. coli* | 27 | 4,584 | 179 | 12,113 | 172 | 68 | 14 | 28 |
| *Welchii* | 18 | 2,177 | 27 | 2,144 | 122 | 79 | 7 | 5 |
| *Cereus* | 11 | 746 | 5 | 274 | 69 | 55 | 2 | 1 |
| *Campylobactoer jejuni* | 23 | 1,611 | 65 | 1,557 | 71 | 24 | 5 | 4 |
| Others | 3 | 96 | 6 | 46 | 30 | 8 | 0 | 0 |
| Sub-Total | 537 | 26,902 | 969 | 38,408 | 50 | 40 | 83 | 87 |
| Total Number[b] | 724 | 32,476 | 1,217 | 43,954 | 45 | 36 | 100 | 100 |
| Number Identified[c] | 619 | 27,253 | 1,046 | 38,683 | 44 | 37 | 84 | 88 |

[a] Average per year.
[b] Including non-bacterial and cause-unidentified incidents.
[c] Cases/patients for which/whom causes of foodborne illness are identified.
Source: MHW.

*Vibrio*, and *E. coli* O157 were the three largest sources of foodborne illness in Japan even during the early 1990s. The fact that the number of cases for most of bacteria types was larger in 1996 than in the previous years indicates the severity of food poisoning in general in 1996. In particular, the number of cases due to *E. coli* O157 increased in 1996 by more than six times as compared with the previous years. The number of patients due to *E. coli* O157 also recorded the largest increase in 1996, absolutely as well as relatively. As a result, the percentage share of patients due to *E. coli* O157 in the total number of patients who suffered from foodborne illness increased from 14 percent in 1990-95 to as much as 28 percent in 1996. This indicates that the outbreak of *E. coli* O157 in 1996 was extraordinary in scale.

The first mass infection by *E. coli* O157 occurred in Okayama Prefecture in May 1996 among school children who took school lunch (Table 2). The second incidence occurred in Gifu Prefecture in June, and the food that caused the infection was identified as vegetable salad served in school lunch. In July, the largest scale of mass infection occurred in Sakai City, Osaka Prefecture, involving nearly 6,000 school children. The radish sprouts (*kaiware*) in the vegetable salad served in school lunch were suspected as the causative food for *E. coli* O157 in this incidence. The scale and degree of the infection were large and serious enough to create a panic situation in the general public, particularly among the parents whose children went to schools and took school lunch.

The total number of persons infected by *E. coli* O157 amounted to 11,030 from May 1996 to February 1998, of whom 15 died. Nearly 80 percent of the patients were recorded during the three months after the outbreak in May 1996 (Figure 1). Within 1996, the death toll reached 12, significantly higher than the yearly average of 5.3 for 1990-95. The number of patients decreased sharply after August 1996, and became almost nil in the 1996-97 winter. However, it increased again in the spring and the summer of 1997. Although the number of patients was smaller than in the previous year, it was still serious enough that three casualties were recorded in 1997.

The major *E. coli* O157 incidences during this period involving more than 100 patients per case are listed in Table 3. It should be remarked that *E. coli* O157 infection occurred through school lunch in eight out of the ten major cases. It should also be noted that the causative foods have not been identified in five cases. From the cases for which contaminated foods were identified, some vegetables in the salads served, *kaiware* in particular, emerged as the source of contamination. In August 1996, the MHW made an official announcement in which they cast strong suspicions that *kaiware* was the food through which *E. coli* O157 was transmitted in the case of Sakai City, Osaka (Table 2). Later in April 1997, *E. coli* O157 was detected directly from *kaiware* which was served in Aichi and

## TABLE 2

Chronology of *E. coli* O157 Outbreak and Measures Taken, with Special Reference to *Kaiware* and School Lunch, May 1996 – March 1998

| Date | *Kaiware* | School Lunch |
|---|---|---|
| May 1996 | | The first outbreak of *E. coli* O157 in schools in Okayama prefecture (5/24). |
| June | | Salad identified first as contaminated food in Gifu outbreak. **MHW: Emergency instruction to schools and school lunch centers for better sanitary management.** |
| July | | The largest outbreak in Sakai City, Osaka, including 5,727 patients (7/12). |
| August | **MHW:** Interim report of Sakai outbreak (8/7): Pointed out *kaiware* as possible contaminated source. | **MESSC:** Emergency investigation of schools and school lunch centers. **MESSC's guidance to local govt.:** Implementation of biannual foodstuffs testing. |
| September | **MHW:** Final report on Sakai case (9/26): Reconfirmed *kaiware* as contaminated source. | |
| October | **MAFF:** HACCP-based manual for sanitary *kaiware* production. | |
| December | ***Kaiware* Producers Association:** Lawsuit against government for damage. | ***Ad hoc* committee of MESSC:** Common menu system has to be re-examined. **Management and Coordination Agency:** Investigation of schools and school lunch centers (to March 1997). |
| March 1997 | | **MHW:** HACCP-based manuals for meal preparation in larger facilities and homes. |
| April | Detection of *E. coli* O157: H7 from *kaiware* in Aichi and Kanagawa prefecture. **MHW:** Asked USA to examine *kaiware* seed from Oregon. [Reply] *E. coli* O157: H7 not detected (6/30). | **MHW:** Investigation of most of schools and school lunch centers (to May 1997). |
| June/October | **MAFF:** Development of effective disinfection method of *kaiware* seeds. | |
| March 1998 | **MHW's final conclusion:** *Kaiware* seeds contaminated by *E. coli* O157: H7. | |

Kanagawa Prefecture, and in March 1998, the MHW finally announced that the source of *E. coli* O157 was contaminated seeds which had been imported from Oregon, USA. It is not certain yet whether all the *E. coli* O157 infection cases were due to the same source, but, at least in a few cases, the same DNA pattern of *E. coli* O157 was commonly detected.

The outbreak of *E. coli* O157 has brought about chaotic situations in various spheres in Japan. Among them, the most seriously affected were certainly the *kaiware* industry and the school lunch system.

## The *Kaiware* Industry and HACCP

*Kaiware*, radish sprouts, is a kind of vegetable mostly used fresh as an ingredient in various salad dishes. It is a traditional vegetable crop in the western part of Japan. Incidentally, Sakai City, where the largest scale of *E. coli* O157 infection occurred, is one of the traditional *kaiware* growing areas; however, *kaiware* is a relatively recent addition to the dietary menu of Japanese in most other parts of Japan. Because of its handiness in preparation and other virtues, such as low calories, it quickly became an important food item not only in family kitchens, but also in restaurants and catering services. The most distinct feature of *kaiware*, however, lies in its

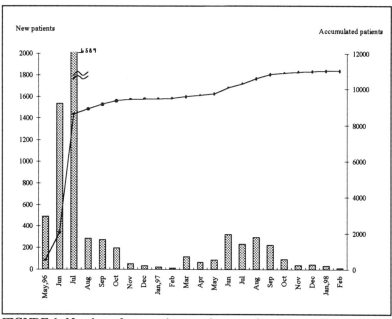

FIGURE 1. Number of new patients and accumulated due to E. coli O157 by month, Japan, May 1996 - February 1998. Source: http://www.mhw.go.jp

production side. Although it used to be grown on agricultural land, *kaiware* is produced now, without exception, in "factories" using hydroponic systems. Among the horticultural crops grown under hydroponic systems, *kaiware* production is characterized by its large scale of operation. A relatively small number of specialized farm-firms with large fixed assets produce *kaiware* for specialized marketing channels. Since its production is free from the constraints imposed by the small land size intrinsic in Japanese agriculture, *kaiware* has been considered a crop of hope that can be competitive in the international market.

### Demand Shrinkage and the Introduction of HACCP

Pin-pointed as the causative food for the *E. coli* O157 infection, the demand for *kaiware* dropped dramatically (Figure 2). The first decline in the demand, observed in August 1996 when the MHW mentioned *kaiware* as the source of contamination, was large enough that the amount going

TABLE 3

Large Scale (More Than 100 Patients Per Case) Cases of Foodborne Illness due to *E. coli* O157:H7, Japan, 1996-97

| Location | Date | No. of Patients | Contaminated Source (Poisoned Food) | Where Served?[a] | Where Prepared? |
|---|---|---|---|---|---|
| Aku, Okayama | 5/96 | 486 | School lunch (?)[b] | K, E | School lunch center |
| Gifu, Gifu | 6/96 | 395 | School lunch (Salad) | E | Own facility |
| Tojyou, Hiroshima | 6/96 | 185 | School lunch (?) | E | Own facility |
| Niimi, Okayama | 6/96 | 365 | School lunch (?) | E, J | School lunch center |
| Minato-ku, Tokyo | 6/96 | 191 | Box lunch (?) | C | Catering company |
| Sakai, Gunma | 6/96 | 138 | School lunch (?) | E | Own facility |
| Sakai, Osaka | 7/96 | 5,727 | School lunch (Radish sprouts) | E | Own facility |
| Morioka, Iwate | 9/96 | 124 | School lunch (Seafood sauce and salad) | E | Own facility |
| Obihiro, Hokkaido | 10/96 | 106 | School lunch (Salad) | K | Own facility |
| Okayama, Okayama | 6/97 | 171 | Meal lunch (Buckwheat noodles) | M | Own facility |

[a]E: elementary school; J: junior high school; K: kindergarten; C: company; M: medical institution.

[b]?: Cases in which it is not identified which food item in the "lunch" is poisoned.

Source : http://www.mhw.go.jp/

through the markets in Tokyo shrank to one-third within one month. After stagnating for several months, the demand started to decrease again in April 1997 when the MHW reported that the incidences of foodborne illness in Kanagawa and Aichi Prefectures were caused by *kaiware* seeds. Demand plunged eventually down to a level that was one-tenth of the level prior to the *E. coli* O157 outbreak. Such a shrink in *kaiware* demand occurred not only in Tokyo markets, but also nationwide.

In order to recover the demand, *kaiware* producers have made various efforts. They examined their own products, distributed leaflets explaining the safety of their products, and made sales promotions at the retail and wholesale levels. The Association of *Kaiware* Producers (AKP), covering 70 percent of all the *kaiware* producers in the country, made a petition to the Ministry of Agriculture, Forestry and Fisheries (MAFF) to make a sanitary management manual for *kaiware* production. In October 1996, an *ad hoc* committee formed in the MAFF, in collaboration with the MHW, to set up a HACCP manual. The Japanese Greenhouse Horticulture Association (JGHA) granted independent authentication to the producers who successfully implemented the HACCP measures in the manual, and allowed them to use a symbol mark of the authentication. Though not compulsory, more than 90 percent of *kaiware* producers implemented the HACCP system. Products with the symbol mark were ready to be marketed by the end of October 1996, and the JGHA advertised it in major newspapers.

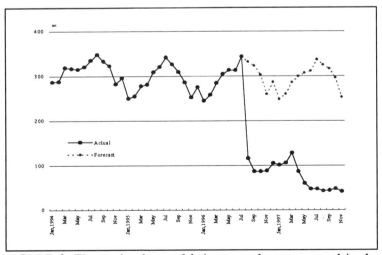

FIGURE 2. The total volume of *kaiware* products transacted in the markets in Tokyo, January 1994 - November 1997. Source: Tokyo Seika-butu Jyoho Center (Tokyo Information Center of Horticulture).

## Costs of HACCP Introduction for *Kaiware* Production

The production process of *kaiware* is relatively simple (Figure 3). Since it is produced under a controlled factory system with a production cycle as short as one week, it is relatively easy to implement the HACCP system by identifying its critical control points (CCPs) for possible hazards due to pathogenic germs. Important CCPs are the sterilizing of seeds and water used in the hydroponic system, and the prevention of the invasion of small animals, such as rats or moles, and insects. One remaining problem in this HACCP system is that there is no way to make perfect disinfection of the seeds fed from outside to the production system. Likewise, the sanitary control in the marketing process from *kaiware* firms through wholesalers and retailers to consumers is beyond the control of the producers. There is no HACCP manual specific to *kaiware* at the wholesale and retail levels yet. There is a HACCP manual for households, but it is not certain how well it is known to housewives.

What about the costs to implement the HACCP system for *kaiware* production? Rough estimates of the costs are presented in Table 4, based on the information obtained from two leading *kaiware* firms with authorized HACCP systems. These firms are of the largest scale in Japan producing 20 percent and 10 percent, respectively, of the total *kaiware* production in the country. Needless to say, the additional costs to implement HACCP systems vary depending on what facilities firms have

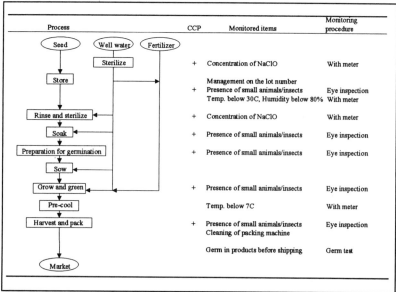

FIGURE 3. A typical process of *kaiware* production and critical control points for possible hazards.

before the introduction of the HACCP system. In a standard case, they need to install seed storage, since the HACCP manual stipulates that *kaiware* seeds must be stored in a separated zone. Various improvements are also necessary for the complete shielding of the production system against the intrusion of small animals and insects. The most important operational cost is for conducting microbial tests. The annualized costs for

**TABLE 4**
**Costs of Implementing HACCP for *Kaiware* Production**

|  | Investment Cost | | Annualized Cost[b] |
|---|---|---|---|
| **Cost Per Firm[a]:** | ...1,000 Yen... | | |
| **Equipment:** | | | |
| Seed storage | 3,000 | | 275 |
| Repair greenhouse/drainage systems for complete sealing against hazardous small animals/insects and disinfectors for hands and boots | 2,000 | | 183 |
| **Operation:** | | | |
| Feces test for employees/product test | | | 1,200 |
| Extermination of hazardous animals/insects | | | 300 |
| **Total:** | | | 1,958 |
| **Cost Per 100g Pack of *Kaiware*:** | | | Yen |
| Pre-O157 outbreak (production size = 3,000 mt/year/firm) | | (1) | 0.065 |
| Post-O157 outbreak (production size = 375 mt/year/firm) | | (2) | 0.52 |
| **Firm-Gate Price of 100g Pack of *Kaiware*:** | | | |
| Pre-O157 outbreak = Post-O157 outbreak | | (3) | 25 |
| **"Profit" Margin Per 100g Pack of *Kaiware*[c]:** | | (4) | 2.5 |
| **Percentage Share of HACCP Cost Relative to Price/"Profit":** | | | % |
| (1)/(3) | | | 0.26 |
| (1)/(4) | | | 2.6 |
| (2)/(3) | | | 2.1 |
| (2)/(4) | | | 21 |

[a]HACCP costs for a farm-firm with the production size of 3,000 mt/year.

[b]Investment costs are annualized by assuming the interest rate of 2.5% per year, a typical lending rate of commercial banks, and the usable life of 15 years for the facilities.

[c]"Profit" margin = returns to the management of farm-firm.

HACCP implementation amount to about two million yen per firm with the production size of 3,000 mt per year. It is remarked that more than 70 percent of the HACCP costs are recurrent costs, due partly to a low level of interest rate in Japan, which reduces capital investment costs.

*Kaiware* is marketed in a plastic packing case with the net weight of 100g. For a firm that produces 3,000 mt of *kaiware* per year, the HACCP cost per pack is estimated to be 0.065 yen. Since the price of *kaiware* received by the producers is 25 yen per pack, of which 10 percent is considered to be the returns to the management or profit margin, the unit HACCP cost is 0.26 percent of the unit price or 2.6 percent of the profit margin. In other words, a 2.6 percent increase in the sale of *kaiware* pays off the investments for HACCP implementation.

As shown in Figure 2, however, the *E. coli* O157 outbreak reduced the demand for *kaiware* to a level one-eighth of the pre-outbreak level. If the sale of the firm is reduced proportionally to the reduction in demand, the HACCP unit cost jumps to 0.52 yen/pack. There has been no change in the output price before and after the outbreak, so the percentage share of the HACCP unit cost with lowered sales in the output price is 2.1 percent. The fixity of the output price before and after the substantial shrink in demand may indicate that the supply curve of the *kaiware* industry is horizontal. If so, the profit margin would also remain the same. With this assumption, a 21 percent increase in sales from the post-outbreak level is required to pay off the HACCP cost. To the extent that this assumption is not satisfied, resulting in a lower level of profit margin, the cost recovery for HACCP would become more difficult.

The real cost burden of HACCP implementation for *kaiware* producers would lie between the upper and lower bounds explained above. In any case, it can be said that the HACCP cost is not so large. A small recovery in demand, say, to a level 100 percent higher than the depressed level (i.e., 20 percent of the pre-outbreak level), would make the cost recovery of the HACCP investments relatively easy. In the case of two firms from which we obtained the cost data on HACCP systems, produce with new antibacterial or sterilizing technology has been sold only since February/March 1998; therefore, how much the implementation of a HACCP system helps the *kaiware* demand recovery is yet to be seen. For demand recovery, it may be necessary to make additional investments in advertising the safety of *kaiware* with HACCP. The implementation of HACCP systems at the wholesale and retail levels, in conjunction with the production levels, may be required for more smooth demand recovery.

## The School Lunch System and HACCP

Forty-four years have passed since the School Lunch Act was enacted. In 1995, the school lunch system was adopted in 98 percent of elementary schools and 85 percent of junior high schools in Japan, and the number of students who take lunch under the system amounts to 13 million. The overwhelming majority of elementary and junior high schools in Japan are public schools run by local governments (the city-town-village level, below prefecture), and therefore the primary responsibility of managing the school lunch system falls under their jurisdiction. As in many other spheres in Japanese economy, however, the control by the central government of the primary and secondary education systems is strong. As such, the Ministry of Education, Science, Sports, and Culture (MESSC), through prefectural governments, oversees the educational system in which the school lunch system is a part. In addition, to the extent that it is related to the health of school children, "administrative guidance" as to the school lunch system also comes from the Ministry of Health and Welfare (MHW).

The school lunch system is classified into two sub-systems in terms of the way to prepare meals: 1) the own kitchen system in each school where lunch meals are prepared and served; and 2) the central kitchen system in the school lunch center where meals are prepared and distributed to schools. The school lunch system started with the own kitchen system, and later in 1964, the central kitchen system was introduced. The share of schools adopting the central kitchen system was 53 percent in 1995. Another point to be noted in the school lunch system is the way the school lunch menu is prepared. The common menu system, in which several schools or school lunch centers share the same meals, can be applied for several school lunch centers, as well as for groups of schools adopting the own kitchen system. The share of school children served under the common menu system was 44 percent in 1995 (Amemiya 1997). Because of the scale economy inherent in the central kitchen and common menu systems, the MESSC has been promoting these two systems under the School Lunch Rationalizing Policy.

### *E. coli* O157 Outbreak and Reactions

The outbreak of large scale *E. coli* O157 infection caused by school lunches in June 1996 shook the school lunch system seriously. Immediately after the outbreak, some schools suspended serving lunches. While conducting emergency inspections on sanitary conditions in schools and school lunch centers, the MESSC intended to resume from the second trimester starting in September. In the second trimester, however, several cases of foodborne illness occurred in schools not only caused by *E. coli* O157, but also by *Salmonella* and other bacteria. Especially noteworthy is the *E. coli* O157 incidence that occurred in an elementary school in Iwate Prefecture in September 1996, in which not only the contaminated food, but also the process and route of contamination in lunch preparation were identified (Table 3) (Shinagawa 1997).

In December 1996, the MESSC identified problems such as the insufficiency of the preventive measures against secondary infection and of the temperature management in cooking. The MHW, inspecting 83 percent of schools and school lunch centers in April/May 1997, found that 60 percent of the school lunch facilities preserved meals at normal temperatures and 38 percent had kitchens not clearly separated from the places handling pre-washed/pre-cleaned foodstuffs. The Management Coordination Agency investigated 73 schools and 12 school lunch centers between December 1996 and March 1997. All the investigations unanimously pointed out a general lack of sanitary management in the school lunch service. They also pointed out a need to reconsider the policy to promote the common menu systems.

In March 1997, the MHW sent out sanitary management manuals for the facilities dealing with a great number of meals, including those preparing school lunches. The manual, based on HACCP concepts, provides the model for food poisoning prevention by local governments. In April 1997, taking this MHW manual into account, the MESSC revised its circular on sanitary management of school lunches and made a management standard based on HACCP with checked items as to school lunch preparation. The CCPs in this MESSC standard and principal check points are shown in Figure 4. It is relatively easy to monitor the HACCP system in school lunch service, since school lunch centers and schools are under the MESSC's jurisdiction.

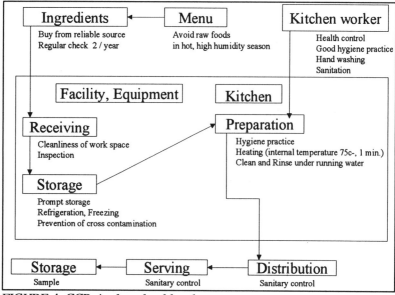

FIGURE 4. CCPs in the school lunch system.

According to the guidance and instructions from the ministries, local governments took measures to improve sanitary control/management in the school lunch system. Some of them distributed new equipment, such as sterilizers, refrigerators, and refrigerator cars to each facility preparing school lunches. Some governments made their own sanitary management manuals for school lunches and/or leaflets for parents for enhancing their consciousness of food hygiene at home. Additionally, several governments have plans to introduce the dry kitchen system, which is ideal for preventing secondary infection. For example, Morioka City proposed the plan to the City Assembly in December 1996. The MESSC encourages the introduction of the dry kitchen system by providing a special budget. The 1997 budget prescribed for this purpose was about three billion yen. However, the cost to install a new dry kitchen system is as expensive as 0.14 billion yen per school and 0.71 billion yen per school lunch center, so it would take a long time to install the new system in all the schools.

**Implementation of the MESSC Manual and its Costs**

To obtain information on the status of the implementation of HACCP in the school lunch system, we visited the School Lunch Office of the Board of Education which manages the school lunch system of Chiba City. There are 114 elementary schools, 54 junior high schools, and two schools for handicapped children in Chiba City, with a total enrollment of about 80 thousand taking school lunch. While all the elementary and the handicapped children schools adopted the own kitchen system, four school lunch centers served the junior high schools with the central kitchen system. On average, a school lunch center provided lunch to 13.5 schools and 6,900 students (Chiba City 1996).

Although *E. coli* O157 brought for Chiba City a damage of nine patients and one death (an infant) in 1996, there were no *E. coli* O157 patients due to school lunches. The school lunch system was therefore not suspended in Chiba City even during the peak of the *E. coli* O157 outbreak, but the City Government took several preventive measures following the instructions/recommendations by the MESSC. Examples of recommendations that were adopted are: keeping the temperature of the inner part of cooked food equal to or more than 80 degrees centigrade (higher by 5°C than the MESSC requirement); conducting microbial tests of ingredients at least twice a year for several randomly selected schools; and avoiding serving raw foods (e.g., serving boiled vegetables instead of fresh ones). The meal preparation facilities were examined by the public health center of the city. The City Government made leaflets for the parents of school children to call their attention to sanitation at home and to inform them of the safety measures for the school lunch system taken by the City Government.

Chiba City introduced the MESSC manual for the school lunch system and allocated funds for implementing it in the 1996 fiscal year (Case I in Table 5). It must be noted that the expenditures listed in this table are all

TABLE 5

Costs of Implementing HACCP in the School Lunch System

|  |  | Own Kitchen System | | Central Kitchen System | |
|---|---|---|---|---|---|
|  |  | Investment Cost | Annualized Cost[a] | Investment Cost | Annualized Cost[a] |
| Cost for Improving the Food Safety | | ...1,000 Yen... | | | |
| Case I. Emergency Measures Taken in Chiba City | | | | | |
| Equipment: | | | | | |
| New refrigerators | (1) | 339 | 31 | 1,847 | 168 |
| Operation: | | | | | |
| Additional supplies for sanitary control in meal preparation/ printing leaflets on O157 for parents | (2) | | 37 | | 638 |
| Feces test/the germ test of food materials | (3) | | 44 | | 679 |
| Total | (A)= (1)+(2)+(3) | | 113 | | 1,485 |
| Case II. Measures Recommended by the MESC | | | | | |
| Dry kitchen system | (4) | 140,000 | 9,333 | 712,000 | 47,467 |
| Total including the operational costs | (B)= (2)+(3)+(4) | | 9,415 | | 48,783 |
| Cost Per Meal[b]: | | ...Yen... | | | |
| Case I | (A) | | 1.6 | | 1.2 |
| Case II | (B) | | 133 | | 41 |
| WTP: | | | | | |
| Per month | | | 784 | | 799 |
| Per meal | (C) | | 36 | | 36 |
| Cost/WTP: | | | | | |
| Case I | (A)/(C) | | 0.04 | | 0.03 |
| Case II | (B)/(C) | | 3.7 | | 1.1 |

[a]Investment costs are annualized by assuming the interest rate of 2.5% per year and 15 years usable life for equipment. For the dry kitchen system, 50% of investment costs is assumed to be for buildings with usable life of 60 years and the rest for equipment.

[b]Annual cost divided by the number of meals served in a year. Assume that school lunch is served for 200 days per year, and that the number of meals per day is 353 per school in the own kitchen system and 6,000 per day in the central kitchen system.

Source: Cost data for Case I are from our hearing from the School Lunch Office of the Board of Education of Chiba City and for the own and central kitchen system in Case II are from Asahi Daily Newspaper (May 13, 1996) and Mainichi Daily Newspaper (April 21, 1998), respectively.

for equipment and operations, excluding the possibility to increase labor use. For instance, the checking and recording of the monitoring items as specified in the manual are supposed to be carried out by the cooks in the schools and the school lunch centers, in addition to their mandatory work of preparing about 200 meals per cook per day. The cost of operation for carrying out the system consisted mainly of expenses for microbial tests and supplies for sanitary management in meal preparation. As to equipment, new refrigerators were distributed to the school lunch centers (1,000 liter capacity) and the elementary schools (150 liter capacity). These refrigerators are used mainly to preserve meals already cooked for the microbial test, as stipulated by the MESSC manual, in order to keep them for a longer period to make it easier to identify, if necessary, a germ with a long incubation period such as *E. coli* O157. The annualized total cost is estimated to be 113 thousand yen per school under the own kitchen system and 1.5 million yen per school lunch center under the central kitchen system. Converting into cost per meal, it is 1.6 yen and 1.2 yen, respectively.

The expenditures above made by Chiba City are of emergency nature, and would therefore represent a minimum for implementing the MESSC manual. As mentioned before, it is necessary for more perfect safety controls in the school lunch system to replace the existing system with the modern dry kitchen system. The cost to introduce this ideal system is shown in Table 5 as Case II. It is assumed that operational costs are incurred as in Case I. Because of high investment costs, the estimated total annualized costs are much higher than for Case I; 9.4 million yen per school and 48.8 million yen per school lunch center. It is interesting to note that the unit cost per meal is substantially cheaper under the central kitchen system than under the own kitchen system, suggesting a strong scale economy for the former.

**Parents' Perceptions and their WTP for Improvements**

In order to collect data on the willingness-to-pay (WTP) of the parents for improving food safety in the school lunch system, we conducted a mail survey for those living in Chiba City in February 1998. Two sets of samples, one with elementary school children and the other with junior high school children, were randomly drawn, and 1,000 questionnaires were sent out in each set, of which 264 and 244 were returned, respectively. The total populations were 47,767 and 27,936 households, respectively.

Judging from the answers given by the parents who returned questionnaires, their concern about the safety of the school lunch system was high (Table 6). About 80 percent of them knew that the menus for school lunches were changed, such as removing fresh salads, after the *E. coli* O157 outbreak, and they thought that the preventive measures taken by the schools for improving the food safety in the school lunch system were appropriate. Among the possible measures, most of the parents with junior high school children thought it was better to change the central kitchen system (current kitchen system) to the own kitchen system, while the parents with elementary school children mostly requested to introduce new equipment for enhancing food safety. It should be noted, however, that very few had knowledge of HACCP. More than 70 percent of them also wanted more information about *E. coli* O157. Nanba and Odate (1997), who conducted a similar survey in Osaka in December 1996, report that more than 70 percent of the respondents expressed dissatisfaction with the state of knowledge and information available to them. Giving more information to the general public seems to be imperative.

The MHW also produced a manual for households in April 1997, referring to the "Food Safety in Kitchen" by the USDA. The CCPs in the manual are purchase, storage, pre-preparation, cooking, eating, and

**TABLE 6**

**Parents' Perceptions on the Safety in the School Lunch System[a]**

| Questions | Answer | Elementary School | Junior High School | Total |
|---|---|---|---|---|
| Do you know changes in lunch menu? | Yes | 87 | 66 | 77 |
| How do you evaluate the safety measures taken for school lunch? | Appropriate | 90 | 69 | 81 |
| What measures do you think important? | Kitchen system | 5 | 38 | 29 |
| | Better facilities | 41 | 16 | 23 |
| | Organic foods | 24 | 20 | 21 |
| | Sanitary system | 29 | 24 | 25 |
| | Suspend school lunch | 0 | 2 | 1 |
| Do you know HACCP? | Yes | 6 | 3 | 4 |
| Do you want to get more information on O157? | Yes | 68 | 70 | 69 |
| What points do you pay attention in preparing foods at home? | Material acquisition | 56 | 41 | 49 |
| | Storage | 44 | 32 | 39 |
| | Pre-preparation | 68 | 57 | 63 |
| | Cooking | 82 | 77 | 80 |
| | Eating | 65 | 52 | 59 |
| | Leftovers | 58 | 52 | 55 |

[a]Based on the mail survey. Percentages of respondents who gave the answer.

leftovers. Eighty percent of the sample parents paid attention to cooking. Behind such perceptions would lie the fact that the insufficiency of heating in meal preparation was often pointed out as a cause of infection after the *E. coli* O157 outbreak. While more than 50 percent of them paid attention in the meal preparation stages of pre-preparation, eating, and disposal of leftovers, relatively less care was given to food storage.

As to WTP, the question asked was, "how much are you willing to pay for a school lunch fee per month including the payment for improving the food safety in the school lunch system as compared with the level without it?" Such an open-ended method was adopted because no prior information was available about the distribution of the WTP. The answer of zero WTP could be of three types: (1) honest response, (2) negative attitudes, and (3) refusal to play the game (protest zero) (Mitchell and Carson 1989). Protest zero had to be discarded because it did not represent a fair economic valuation. Since it is difficult to distinguish protest zeros from others, however, we included all the zero answers in our analysis. The total number of samples for which WTP and major household characteristics was available is 358.

The WTP for improving the food safety in the school lunch system was estimated to be 784 yen per month for respondents with children enrolled in elementary schools and 799 yen per month for those with children enrolled in junior high schools, but the difference was not significant statistically. Factors affecting the level of WTP were examined by regressing it on some household characteristics (Table 7). Dependent variables are the rate of increase of WTP over the pre-improvement level of school lunch fee for models I, II and III, and the difference between the post- and the pre-improvement levels for model IV. Estimation was carried out with the ordinary least squares method. The distribution of WTP was skewed significantly because of zero value, so the fitting of the regression is poor. Nevertheless, menu and information dummies, and income (or income squared) gave significant coefficients with positive signs in all the models. The menu dummy caught the knowledge of parents on the change in the school lunch menu after the *E. coli* O157 outbreak. The information dummy represented the willingness of parents to get more information on *E. coli* O157. The results, therefore, indicate that the parents who have more concern about the food safety in school lunch were willing to pay more.

Lastly, let us compare the WTP with the cost of improving the food safety in the school lunch system (Table 5). If the cost of implementing the MESSC manual for the school lunch system is as cheap as in Case I, the WTP far exceeds the cost for both the own kitchen and the central kitchen systems. Enhancing food safety in the school lunch system can be attained with the additional contribution to the system by the parents. Should the cost be as high as in Case II, however, the WTP would not be sufficient to finance the cost. This is the case even for the central kitchen system with a strong scale economy. Need arises, then, for justifying the investments to take into account the benefits that the HACCP generates in the society by reducing the cost of illness (Roberts and Marks 1995). Since children are vulnerable to infectious disease, the medical costs to cure

TABLE 7

Factors Affecting the Willingness-to-Pay for Improving Food Safety in School Lunch Service[a]

| Variables[b] | Model | | | |
|---|---|---|---|---|
| | Dependent Variable = WTP (%) | | | = WTP (Yen) |
| | I | II | III | IV |
| Menu Dummy | 0.066** | 0.062** | 0.061** | 242.347* |
| | (2.225) | (2.236) | (2.195) | (1.922) |
| Information Dummy | 0.080*** | 0.081*** | 0.082*** | 350.13*** |
| | (3.085) | (3.160) | (3.192) | (2.996) |
| School Dummy | -0.013 | | | |
| | (-0.516) | | | |
| Housewife Dummy | -0.024 | | | |
| | (-0.994) | | | |
| Sex Dummy | 0.025 | | | |
| | (0.483) | | | |
| Educational Level Dummy | -0.001 | | | |
| | (-0.061) | | | |
| Age | -0.001 | | | |
| | (-0.423) | | | |
| Ratio of Children | 0.089 | | | |
| | (0.745) | | | |
| Income | 0.012*** | | 0.012*** | |
| | (3.511) | | (3.647) | |
| Income$^2$ | | 0.001*** | | 2.527*** |
| | | (3.684) | | (3.388) |
| Constant | -0.010 | 0.011 | -0.043 | 122.86 |
| | (-0.085) | (0.333) | (-1.020) | (0.814) |
| Adj-R2 | 0.0590 | 0.0689 | 0.0682 | 0.0575 |

[a]Figures in parenthesis are t-ratio. ***, **, and * represent the significance of 0.01, 0.05, and 0.10, respectively.

[b]Variable definition: Menu dummy: knowledge on the change in school lunch menu after the *E. coli* O157 incidence, known = 1, not known = 0; information dummy: willingness to get more information on *E. coli* O157, yes = 1, no = 0; school dummy: elementary school = 1, junior high school = 0; housewife dummy: full-time = 1, no = 0; sex dummy: female = 1, male = 0; educational level dummy: higher than junior college = 1, others = 0; ratio of children: the ratio of children to the household size; age: age of respondents (center value of age classes); income: annual household income (center value of income classes).

them would be substantial once hazards occur, as indicated by the *E. coli* O157 infection in 1996. Not all the medical costs due to the 1996 outbreak are available, but, for the case of Sakai City in Osaka, the direct medical expenses alone for curing the school children infected, not taking into account the secondary medical costs due to the infection, are tentatively estimated to be 140 million yen (Sakurai 1997).

## Concluding Remarks

We outlined in this paper the *E. coli* O157 outbreak in 1996 in Japan and how it induced reactions by the MHW, the MESSC, and the MAFF, which has resulted in the implementation of HACCP in the *kaiware* industry and the school lunch system. Preliminary cost-benefit analyses of introducing HACCP in these two fields were also attempted.

It should be remarked that the database for the cost-benefit analyses in this paper is weak, and the analyses are not far beyond anecdotal, waiting for refinements in the future. Even so, the following conclusions from the analyses would remain intact. The cost of implementing the HACCP system in the *kaiware* industry is relatively low, and a small recovery in the demand would easily pay off the cost. In order to recover the demand, however, it would be necessary to make the concept of HACCP popular and familiar to the general public through advertisement and propaganda. In the case of the school lunch system, the cost of HACCP introduction varies widely depending on the degree of safety you want to ensure. If the food safety could be attained at the lower-bound cost, it would be far below the willingness-to-pay for improving food safety by parents whose children take school lunch. If the higher-bound cost must be assumed, however, the parents' contributions alone would not be enough to finance the HACCP investments. It is suggested that, even if such a case is the reality, the social benefits of HACCP in the school lunch system would be higher than its cost once the cost of illness is taken into account.

It should be emphasized that the concept of HACCP is not well known to the public, partly because it has just been introduced in the production arena in agriculture as well as in the school lunch system. This low publicity of HACCP must be a major reason why the recovery of the *kaiware* demand has been so sluggish. For the school lunch system, many people want to have more information on the management of food safety and the HACCP system. The economic effects of the HACCP system are expected to come about more clearly as it is adopted more widely.

### Notes

[1] A. Maruyama and S. Kurihara are assistant professor and T. Matsuda is Associate Professor, at the Faculty of Horticulture, the Chiba University. The

authors are grateful to Masao Kikuchi for his helpful comments on an earlier version of this paper.

## References

Chiba City. 1996. *Shisei Gaiyo* (The outline of Annual Activities and Expenses).

Mitchell, R. Cameron and Richard T. Carson. 1989. *Using Surveys to Value Public Goods: The Contingent Valuation Method*: Resources for the Future.

Amemiya, Masako. 1997. *Gakko Kyusyoku wo Kangaeru: O157 Jiken wa Naze Okirunoka* (Japanese School Lunch System: Why do O157 outbreaks occur?) Aoki-Shoten.

MHW. 1996. *Shokutyudoku Tokei*(The Statistics of Food Poisoning).

Nanba, Atsuko and Odate Junko. 1997. *Katei de no O157 Taisaku ni tsuite* (Preventive Measures at home). A paper presented at the 44th conference of the Japan Dietetics Society.

Roberts, Tanya and Suzanne Marks. 1995. Valuation by the Cost of Illness Method: The Social Costs of Escherichia coil O157: H7 Foodborne Disease. In *Valuing Food Safety and Nutrition*, ed. Julie A. Caswell, 173-205. Westview Press.

Sakurai, Takuji. 1997. Meat Sanitary Regulation in Japan: How has Japan been responding to new food safety concerns? A paper presented at International Conference of Agricultural Economists held at the Hyatt Regency Sacramento for August 10-16, 1997.

Shinagawa, Kunio, et al.. 1997. *Iwate-ken Morioka-shi ni okeru Taiou to Kadai* (Measures Taken and Problems Remained, in O157 Outbreak in Morioka City, Iwate Prefecture). Bull. Natl. Inst. Public Health 46(2): 104-112.

Chapter 20

# Analysis of Implementation and Costs of HACCP System in Foodservices Industries in the County of Campinas, Brazil

Marcia R. D. Buchweitz and Elisabete Salay[1]

## Introduction

Research conducted in the main urban centers of Brazil between 1995 and 1996 showed that the population allocates an average of 25.5 percent of their food budgets for food to be consumed outside the home (IBGE 1998). Indeed, the food service industry[2] in general plays an important socioeconomic role in the country. While, for example, the general level of jobs in the metropolitan region of São Paulo from April 1997 to April 1998 decreased some 1.6 percent, the sector of foodservices increased 7.8 percent (SEADE and DIEESE 1998). A great spurt in growth in the sector occurred in 1976 when the Brazilian government instituted a program of incentives to encourage companies to furnish meals for their employees (the Program of Food for Workers, or PAT) (Proença 1996). This PAT uses tax incentives to subsidize companies so they will guarantee that meals are provided for their employees, whether through company restaurants (either self-operated or leased to outsiders), distribution of meal tickets, or the provision of "basic food baskets" (containing some 20 kg of basic food staples such as rice and beans).

According to the Brazilian Association of Collective Meal Suppliers (ABERC), 30 percent of the 30 million registered Brazilian workers receive some sort of food subsidy from the companies where they work. Half of these workers are found in the state of São Paulo. In 1996 the market for such foodservice industries was estimated at 8.9 million meals per day: 2.7 million supplied by catering companies, 1.0 million served by company-managed restaurants, and 4.8 million obtained via meal tickets, a total value of US$ 6.0 billion.[3]

In Brazil there is still no informatized system covering foodborne diseases, although it is assumed that such illnesses are responsible for hundreds of deaths and thousands of hospitalizations (Almeida 1994). Government estimates indicate that 12 percent of the hospital stays in Brazil are due to infectious intestinal illnesses, especially from food (Livera et al. 1996).

Indeed, research shows that food safety problems frequently originate with foodservices. Thus, according to the studies analyzed by Coelho and Vanetti (1994), surveys of outbreaks of food poisoning have revealed the sector of foodservices to be responsible for the accidents which occurred. Moreover, consumer complaints to the Secretary of Health arising from problems with prepared food are frequent (Salay 1998).

Research with 200 foodservice entities in Brazil has revealed the existence of food processing conditions which could place the health of the population at risk: carelessness in operational conditions, lack of basic health safety conditions in relation to personnel and utensils, and a lack of maintenance of equipment. The lack of instruments for monitoring was also observed, as well as technical ignorance on the part of employees, a lack of interest and motivation, and insufficient training furnished to make the application of more advanced techniques of quality control feasible (Silva 1993).

One of the attempts to overcome failures of the market to guarantee food safety for the Brazilian population has been the adoption of the HACCP system. The national Minister of Health passed a law (regulation number 1428/93) which made the use of the HACCP system obligatory for all establishments involved in the preparation and distribution of food, and stipulated a deadline of August 1995 for adaptation to these norms. An extension of this deadline was later granted to companies in the area of collective food distribution to facilitate the introduction of Good Manufacturing Practices (GMP) norms. According to information from the Center for Health Control of the São Paulo State Secretary of Health, a gradual adoption of the GMP regulation is under way, with criteria established for the progressive implementation of these norms during the period of 1996 to 1997. Once this is accomplished, the gradual implementation of the HACCP system will be necessary (Franco 1997).

Little information is available about the implementation of the HACCP system and of the GMP norms in Brazil, or about factors related to the adoption of these methodologies. It is supposed that only certain companies have adopted the system since the Minister of Health has not yet developed an effective program for the implementation of the measure. The Ministry of Agriculture and Supply, however, is developing a program of incentives to encourage the industries of fish, meat, and milk to adopt the HACCP system, with an initial concentration involving exporting establishments (Salay and Caswell).

The present paper was designed to estimate the proportion of foodservice industries which utilize GMP norms and the HACCP system in the county of Campinas[4] of the state of São Paulo. It was also designed to evaluate the causes for non-adoption of the HACCP system, as well as to estimate to the costs of its implementation and maintenance.

## Data Collection and Methodology

The county of Campinas encompasses 22 cities with 2.3 million inhabitants and this number is increasing at a rate of 2.37 percent per year, some 6.7 percent of the population of the state of São Paulo, living in an area of 5,290 km$^2$. The majority of these inhabitants live in the cities (95.5 percent). In economic terms, the region of Campinas furnishes 10.0 percent of the tax resources of the state and encompasses 7.5 and 7.1 percent of the industrial and commercial establishments, respectively (SEADE 1996).

No list of all the foodservice establishments in the Campinas region exists, but a list of the names and addresses of those registered with the ABERC[5], the Program of Food for Workers (PAT), and the Regional Council of Dietitians (CRN-2) was obtained. These establishments represent a large portion of the relevant entities. An attempt was made to contact all of 160 of them. Eighteen were eliminated from consideration due to bankruptcy, 18 because they were only involved in packaging the "basic food basket," 18 because they could not to be located. Fifty-three were not included for other reasons (i.e. refusal to grant an interview, or delay in scheduling an interview, etc.). Thus, a total of 56 companies have been interviewed, which accounts for 52.8 percent of the total number (106) known to be in operation.

A questionnaire was developed to survey the general characteristics of the company and to verify if the GPM and HACCP systems have been implemented, as well as investigating the main reason for any non-adoption. A second questionnaire assessing cost estimates associated with implementation and maintenance of the HACCP system, based on an instrument developed by Colatore and Caswell (1997), was developed which evaluates the major items influencing costs: planning, expenses with employees, training, equipment, materials, laboratory analysis, monitoring, record keeping, review of critical control points (CCP) records, and revision of HACCP. After pre-testing the two questionnaires, they were applied and the people responsible for quality control of the companies were interviewed. Data collection took place during the months of March, April, and May of 1998. A team of eight students of the Faculty of Food Engineering were trained to collect the data; a senior research worker supervised the work and conducted three interviews about costs which were granted.

## GMPs and HACCP Adoption

Table 1 shows that the majority of the foodservice establishments studied were commercial ones (84.0 percent), including food contractors (62.6 percent of the sample), while institutional foodservices represent

16.0 percent. The latter include industry restaurants (7.1 percent) and hospitals (7.1 percent), A large number of the foodservice establishments analyzed (57.2 percent) were responsible for the production of only a relatively small number of meals, i.e., from 0 to 1,000 meals per day. Only 7.1 percent produced over 100,000 meals per day (Table 2).

Table 3 shows that the majority of the foodservice establishments had implemented neither GMP norms nor the HACCP system (53.6 and 58.9 percent, respectively). Only 23.2 percent reported that they had adopted GMP's regulations and 17.9 percent had adopted the HACCP system. The main reason given for non-implementation of the HACCP system was a lack of information about it (54.5 percent); followed by economic factors (15.2 percent), a lack of demand by customers (12.1 percent), and a lack of interest on the part of the company (12.1 percent). According to those interviewed, the lack of government control was not particularly important in their failure to implement the programs. In the case of a lack of utilization of GMP norms, the most important factor was again a lack of information (66.8 percent), followed by economic factors (13.3 percent), and a lack of demand by the consumer (10.0 percent) (Table 4).

### TABLE 1
Type of Foodservice Industries Functioning in County of Campinas. São Paulo, Brazil, 1998

| Type of Foodservice | N | (%) |
| --- | --- | --- |
| Commercial | 47 | 84.0 |
| Food contracts | 35 | 62.5 |
| Eating places | 12 | 21.4 |
| Institutional | 9 | 16.0 |
| Industries | 4 | 7.1 |
| Schools | 1 | 1.8 |
| Hospitals | 4 | 7.1 |
| Total | 56 | 100 |

### TABLE 2
Classification of Foodservice Establishments by Total Number of Meals Per Day in the County of Campinas, São Paulo, Brazil, 1998

| Meals Per Day | N | (%) |
| --- | --- | --- |
| Up to 1,000 | 32 | 57.2 |
| 1,001-10,000 | 16 | 28.6 |
| 10,001-50,000 | 3 | 5.4 |
| 50,001-100,000 | 1 | 1.7 |
| Above 100,000 | 4 | 7.1 |
| Total | 56 | 100 |

Table 5 shows that for the sample studied, the majority of foodservice establishments (76.7 percent) where GMP norms had not been adopted produced fewer than 1,000 meals per day. All of those serving more than 50,000 meals had either adopted them or were in the process of doing so. The same is true for the HACCP system, with larger establishments having implemented it.

Table 6 furnishes data about the type of foodservices offered by the establishments in the sample and the implementation of the GMP norms and HACCP system. It can be seen that the majority of those foodservices which have adopted the GMP norms (84.6 percent) were food contractors. The same was true in relation to the implementation of the HACCP system. Of the four foodservice establishments in hospitals studied, only one used the GMP norms, and none have adopted the HACCP plan.

## Cost Estimate for HACCP

Table 7 presents the general characteristics of the three foodservice establishments which provided information about the costs of adoption and

### TABLE 3
Implementation of GMP Norms and HACCP System in Foodservice Establishments in the County of Campinas, São Paulo, Brazil, 1998

| Situation Relative to Implementation | GMP Norms N | (%) | HACCP System N | (%) |
|---|---|---|---|---|
| Implemented | 13 | 23.2 | 10 | 17.9 |
| In implementation | 13 | 23.2 | 13 | 23.2 |
| Not implemented | 30 | 53.6 | 33 | 58.9 |
| Total | 56 | 100 | 56 | 100 |

### TABLE 4
Reasons to Justify Lack of Implementation of GMP Norms and HACCP System in Foodservice Establishments in the County of Campinas, São Paulo, Brazil, 1998

| Reason | GMP Norms N | (%) | HACCP System N | (%) |
|---|---|---|---|---|
| Economic factors | 4 | 13.3 | 5 | 15.2 |
| Little government control | 1 | 3.3 | - | - |
| Little customer demand | 3 | 10.0 | 4 | 12.1 |
| Lack of information | 20 | 66.8 | 18 | 54.5 |
| Lack of interest by company | 1 | 3.3 | 4 | 12.1 |
| Other reasons | 1 | 3.3 | 2 | 6.1 |
| Total | 30 | 100 | 33 | 100 |

maintenance of the HACCP system. One was a self-managed industry-restaurant serving 7,500 meals per day and two were food contractors, one with a meal production of 55 thousand (in the country as a whole) and the other serving more than 100,000 meals in Brazil (25,000 in the states of São Paulo and Paraná). The time necessary for implementation of the HACCP system varied from 3 to 24 months. As was found by Colatore and Caswell (1997), costs associated with the implementation of HACCP system of the companies were not separated from general expenses, so the results here have been based on estimates.

Table 8 presents the estimates of the main costs involved in the implementation of the HACCP system. For the food contractors, the process of implementation implied the hiring of a consultant and the investment of part of the time of a manager. These costs totaled some $37,925.85 to $67,228.95 per unit administered for implementation. The cost of training also varied quite a bit, with values varying from $1,191.80 to $11,075.00 per unit administered for the two food contractors. The investment in materials to initiate the HACCP (thermometers, educational materials, etc.) was much more similar for the two contractors: $615.00 and $713.05 per unit administered.

The self-managed restaurant hired a consulting firm to implement the HACCP system. This consulting firm developed the plan and trained the employees, and also provided follow-up services for maintenance of the system. In this case, the costs were in the neighborhood of $3,540.00 for implementation. Moreover, the company spent approximately $944.00 for

TABLE 5

Establishment Size and Conditions in Relation to GMP Norms and HACCP System in the County of Campinas, São Paulo, Brazil, 1998

| Meals Per Day | GMP Norms | | | | | | HACCP System | | | | | |
|---|---|---|---|---|---|---|---|---|---|---|---|---|
| | Implemented | | In Implementation | | Not Implemented | | Implemented | | In Implementation | | Not Implemented | |
| | N | (%) | N | (%) | N | (%) | N | (%) | N | (%) | N | (%) |
| Up to 1,000 | 4 | 30.8 | 5 | 38.5 | 23 | 76.7 | - | - | 7 | 53.9 | 25 | 75.8 |
| 1,001–10,000 | 4 | 30.8 | 6 | 46.1 | 6 | 20.0 | 4 | 40.0 | 6 | 46.1 | 7 | 21.2 |
| 10,001–50,000 | 1 | 7.7 | 1 | 7.7 | 1 | 3.3 | 1 | 10.0 | - | - | 1 | 3.0 |
| 50,001–100,000 | 1 | 7.7 | - | - | - | - | 1 | 10.0 | - | - | - | - |
| Above 100,000 | 3 | 23.0 | 1 | 7.7 | - | - | 4 | 40.0 | - | - | - | - |
| Total | 13 | 100 | 13 | 100 | 30 | 100 | 10 | 100 | 13 | 100 | 33 | 100 |

materials. Of the three cases analyzed, the lowest cost in percent of annual sales reported was 0.09 percent.

The process of maintenance involved disparate investments from $445.05 to $4,306.20 per month (Table 9). While one company invested relatively little in training and personnel, it faced additional expenditures in relation to laboratory analyses and cleaning supplies, another had relatively high costs for personnel, training and monitoring, but did not invest in laboratory analyses. When a consulting firm was involved, $1,062.00 per month was spent for the maintenance of the system (laboratory analyses, registration analysis, review of the HACCP and training). The maintenance cost in percent of sales was noticeably greater for the smaller self-managed restaurant (0.60 percent). None of the establishments investigated here underwent additional costs for equipment, corrective actions, or other capital investments.

## Discussion and Conclusions

Foodborne diseases afflict a large number of Brazilians, especially through foodservice systems. Adoption of HACCP systems can help prevent these diseases, although the efficiency of the HACCP system as a regulatory tool to guarantee the provision of safe foods on the market has been questioned. As Unnevehr and Jensen (1996) point out, the HACCP can be interpreted as set of standards regulating both the process and the product, while economists argue that it would be more efficient to define product standards and let individual firms decide what process to use to achieve these standards. On the other hand, the HACCP system offers

### TABLE 6

Type of Foodservice Establishment Analyzed and Implementation Conditions in Relation to GMP Norms and HACCP System in the County of Campinas. São Paulo, Brazil, 1998

| Type | GMP Norms | | | | | | HACCP System | | | | | |
|---|---|---|---|---|---|---|---|---|---|---|---|---|
| | Implemented | | In Implementation | | Not Implemented | | Implemented | | In Implementation | | Not Implemented | |
| | N | (%) | N | (%) | N | (%) | N | (%) | N | (%) | N | (%) |
| Food Contracts | 11 | 84.6 | 9 | 69.2 | 15 | 50.0 | 9 | 90.0 | 10 | 79.6 | 16 | 48.5 |
| Eating Places | - | - | 3 | 23.1 | 9 | 30.0 | - | - | 2 | 15.4 | 10 | 30.3 |
| Industry | 2 | 15.4 | - | - | 2 | 6.7 | 1 | 10.0 | 1 | 7.7 | 2 | 6.1 |
| School | - | - | - | - | 1 | 3.3 | - | - | - | - | 1 | 3.0 |
| Hospital | - | - | 1 | 7.7 | 3 | 10.0 | - | - | - | - | 4 | 12.1 |
| Total | 13 | 100 | 13 | 100 | 30 | 100 | 10 | 100 | 13 | 100 | 33 | 100 |

certain advantages for the industry. Instead of testing the final product, it monitors potential problems by identifying critical points (CCPs) which are controlled through regular monitoring and it can be implemented in various forms so that the most efficient modality can be selected. Moreover, it furnishes companies with greater chances of competition on the international market. From the point of view of the government, one of the main advantages of HACCP is that inspection can be more efficient as the CCPs and the HACCP plan can be checked instead of making extensive tests of the products (Ehiri et al. 1995; Unnevehr and Jensen 1996). For better or for worse, Brazil has adopted GMP norms and the HACCP system as obligatory methodologies for all food establishments.

The present research project has shown, however, that few foodservice establishments have actually adopted the GMP norms (23.2 percent), which proves that the intended goal of universal implementation of GMPs in all food establishments in the state of São Paulo by the end of 1997 (Franco 1997) has not been met. Moreover, it has shown that the application of the HACCP system is also limited to an even smaller number of foodservice establishments (17.9 percent). Although these results do not represent the country as a whole, or all sectors of the food industry, foodservice establishments in the region of Campinas are typical of those in urban centers throughout the country.

The main reason cited for the lack of implementation of GMP norms and the HACCP system was a lack of information about them. This is inexcusable, since information about any such procedures which are the object of governmental regulation should be widely disseminated to the companies involved to facilitate success of the program. The results of this study confirm the fact that in Brazil the inspection of food establishments by public agencies is sporadic and infrequent (Salay 1998). In reality, the policies of food safety in Brazil are better organized in relation to products destined for exportation (Salay and Caswell). Not even the simple provision of information about laws and programs is effective for companies whose products are destined for

TABLE 7

Characteristics of Foodservice Industries Analyzed in the County of Campinas, São Paulo, Brazil, 1998

| Characteristics | A | B | C |
|---|---|---|---|
| Type of Foodservice | Food Contractor | Food Contractor | Institutional/ Industry |
| Number of Units Administered | 70 | 40 | 1 |
| Total Number of Meals | 25,000.00[a] | 55,000.00 | 7,500.00 |
| Time of Implementation of HACCP (Months) | 15 | 24 | 3 |

[a]This company produces more than 100,000 meals per day throughout the country, however, it supplied only average costs of the combined states of São Paulo and Paraná.

domestic consumption.

This problem of a lack of information can also be seen with the Brazilian Program for Quality and Productivity, which was designed especially to enhance the national efforts to achieve international competitiveness through the improvement of quality and productivity in relation to goods and services. Contracted services in Brazil, including some of the food contractors linked to the ABERC, were largely unaware of the program (49.7 percent). Moreover, the majority of those establishments reported that whatever the quality evaluations undertaken are made informally without use of the required program and methodology (MICT 1996). Even in more highly developed countries, food companies have a limited understanding of the HACCP system and its implementation. It is thus necessary for the government to develop educational programs to assure uniformity of application of its principles (Ehiri et al. 1995).

A large number of the foodservice establishments in Brazil which have not adopted the HACCP system and GMP norms produce relatively few meals. The larger companies are generally more effective in terms of quality and productivity (BNDES et al. 1996). These large foodservice companies are involved in more extensive quality control programs, which include the HACCP system. Moreover, these companies have adopted this system with the intention of competing in the markets of regional blocks like the MERCOSUR (Southern Cone Common Market). Many are also concerned with guaranteeing

TABLE 8

Costs of Implementation of HACCP System in Foodservice Establishments in the County of Campinas, São Paulo, Brazil, 1998

| Type of Implementation Costs | A<br>Costs/UA[a]<br>$ | B<br>Costs/UA[a]<br>$ | C<br>Costs<br>$ |
|---|---|---|---|
| Consulting Firm | - | - | 3,540.00 |
| Additional Employee | 15,168.90 | 41,472.00 | - |
| Consultant | 22,756.95 | 25,756.95 | - |
| Training | 1,191.80 | 11,075.00 | - |
| Additional Material | 713.31 | 615.00 | 944.00 |
| Additional Transportation | 236.00 | - | - |
| Total | 40,066.96 | 68,636.16 | 4,484.00 |
| Total Costs of Implementation of the HACCP (% of the Total Annual Sales)[b] | 0.24 | 0.19 | 0.09 |

[a] Unit administered by food contractor.
[b] Average sales value per meal among three firms is $2.00.

their present market since, as one of them suggested, the link of foodborne diseases to the name of a company can cost the loss of five years of business.

The second reason given for not adopting the HACCP system was cost. Businessmen are concerned with the high cost of laboratories, training, and the operation of the system (Karr et al. 1994). The present research estimated the costs of HACCP for three types of foodservice establishments. More accurate values could be obtained from a larger sample. The results presented here, however, show a number of tendencies. Certain differences were found, suggesting that any individual company should adopt a specific plan of operation and investment for the implementation of the HACCP system. One may need to invest heavily in training, or plan development, while others may find contracting of a consulting firm to be the best alternative. In all cases, however, additional expenses for the purchase of materials can be expected to be relatively low, and none of the foodservice industries investigated required any new equipment. The estimation of the cost of HACCP per meal in the cases analyzed showed greater maintenance costs for the smaller companies, although during implementation the costs were greater for the larger companies. This investigation suggests that the costs of implementation and maintenance of the HACCP system in Brazil for the foodservice

TABLE 9

Maintenance Costs of HACCP System in Foodservice Establishments in County of Campinas, São Paulo, Brasil, 1998

| Type of Maintenance Cost | A Costs/UA[a] $/Month | B Costs/UA[a] $/Month | C Costs $/Month |
|---|---|---|---|
| Consulting Firm | - | - | 1,062.00 |
| Additional Employee | 37.31 | 2,420.00 | - |
| Training | 59.00 | 587.00 | - |
| Monitoring and Additional Registration | 3.54 | 1,175.00 | 1,416.00 |
| Analysis of Registration and Review of System | 118.00 | 77.75 | - |
| Additional Laboratory Analysis | 70.80 | - | - |
| Additional Cleaning | 144.55 | - | - |
| Additional Material | 11.85 | 46.45 | 26.20 |
| Total | 445.05 | 4,306.20 | 2,504.20 |
| Cost of Maintenance of the HACCP (% of the Total Sales)[b] | 0.03 | 0.14 | 0.60 |

[a]Unit administered by food contractor.
[b]Average sales value per meal among three firms is $2.00.

industries as such are not especially high. Our results show that cost of implementation for the HACCP systems varied between 0.09 and 0.24 percent of the total annual sales, while the percent for maintenance is between 0.03 and 0.60 percent.

## Notes

[1]Marcia Buchweitz is a Professor at the Federal University of Pelotas working towards a doctorate in the Department of Food Planning and Nutrition of the Faculty of Food Engineering of the State University of Campinas. Elisabete Salay is an Assistant Professor in the Department of Food Planning and Nutrition of the State University of Campinas. The authors would like to thank Capes for furnishing the fellowship of the senior author and the GEPEA (Junior Company of the Faculty of Food Engineering of the State University of Campinas) for their collaboration in the field work.

[2]The National Restaurant Association (NRA) of United States divides the foodservice industry into three major groups: group one, commercial feeding, including eating and drinking places, food contractors, foodservice in lodging establishments, and other miscellaneous commercial foodservice retailers, group two, the institutional feeding group, including business, educational, government, or institutional organizations that operate their own foodservices, and group three, military feeding (Spears and Vaden 1985).

[3]http:www.quattro.com.br/nutrinews/apresent.htm 18/02/1998.

[4]The state of São Paulo has 42 administrative regions.

[5]ABERC is an association of companies with associates being food contractors. The CRN is a council of professionals in the area of nutrition which licenses professionals and foodservice establishments.

## References

Almeida, Rogéria C. C., Paulo F. Almeida and Arnaldo Y. Kuaye. 1994. Pontos Críticos em Serviços de Alimentação. *Higiene Alimentar* 8(30):17-20.

BNDES. Banco Nacional de Desenvolvimento Econômico e Social, CNI. Confederação Nacional da Indústria, SEBRAE. Serviço Brasileiro de Apoio as Micro e Pequenas Empresas. 1996. *Qualidade & Produtividade na Indústria Brasileira*. Rio de Janeiro.

BCB. Banco Central do Brasil. 1988. *Boletim do Banco Central do Brasil*. Brasília.

Citibank. 1990. Brasil se mantém como 8ª economia. *Folha de São Paulo*, 19 February.

Coelho, Ana I. M. and Maria C.C.D. Vanetti. 1994. Avaliação Microbiológica de Carnes Preparadas em Restaurante Industrial. *Higiene Alimentar* 8(32):27-32.

Colatore, Corinna and Julie A. Caswell. 1997. An Economic Analysis of HACCP in the Seafood Industry. Paper presented at NE-165: Private Strategies, Public Policies, and Food System Performance. Washington, D.C., June 2.

Ehiri, John E., George P. Morris and James McEwen. 1995. Implementation of HACCP in Food Businesses: The way ahead. *Food Control* 6(6):341-345.

Franco, Vera L. S. 1997. Atividades de Inspeção pelo HACCP no Estado de São Paulo (Portaria 1428/93). Paper presented at Simpósio Sistema HACCP, São Paulo. 13-14 October.

IBGE. Fundação Instituto Brasileiro de Geografia e Estatística. 1998. *POF 1995-1996*. Rio de Janeiro.

Livera, Alda V. S., Ana C.O. Santos, Enayde A. Melo, Josedira C. Rêgo and Nonete B. Guerra. Condições Higiênico-Sanitárias de Segmentos da Cadeia Alimentar do Estado de Pernambuco. *Higiene Alimentar* 10(42):28-31.

Karr, Kelley J., Audrey N. Maretzki and Stephen J. Knabel. 1994. Meat and Poultry Companies Assess USDA's Hazard Analysis and Critical Control Point System. *Food Technology* 48(2):117-121.

MICT. Ministério da Indústria Comércio e Turismo. 1996. *Pesquisa de Avaliação do Programa de Qualidade e Produtividade*. Brasília.

Ministério da Saúde. 1993. Portaria Nº. 1.428. *Diário Oficial da União*, 26 November.

Proença, Rossana R.C. 1996. *Aspectos Organizacionais e Inovação Tecnológica: Uma Abordagem Antropotecnológica no Setor de Alimentação*. Ph.D. Thesis. Florianópolis: Universidade Federal de Santa Catarina.

Salay, Elisabete. 1998. Qualidade de Alimentos e Cidadania: O Embate Esperado. *Suma Agrícola & Pecuária* 334:8-9.

Salay, Elisabete and Julie A. Caswell. Developments in Brazilian Food Safety Policy. *The International Food and Agribusiness Management Review* (forthcoming).

SEADE. Fundação Sistema Estadual de Análise de Dados.1998. *São Paulo em Dados*. São Paulo.

SEADE. Fundação Sistema Estadual de Análise de Dados and DIEESE. Departamento Intersindical de Estatística e Estudos Sócio-Econômicos. 1998. *Pesquisa e Desemprego na Região Metropolitana de São Paulo*. São Paulo.

Silva, Eneo A. 1993. Aplicação do Método de Análise de Riscos por Pontos Críticos de Controle em Cozinhas Industriais. *Higiene Alimentar* 7(25):15-22.

Spears, Marian C. and Allene G. Vaden. 1985. *Foodservice Organization*. New York: Macmillan Publishing Company.

Unnevehr, Laurian J. and Helen H. Jensen. 1996. HACCP as a Regulatory Innovation to Improve Food Safety in the Meat Industry. *American Journal Agricultural Economics* 78:764-769.

Chapter 21

# Costs and Benefits of Implementing HACCP in the UK Dairy Processing Sector

Spencer Henson, Georgina Holt and James Northen[1]

## Background

There is currently a great deal of concern about the safety of the UK food supply, reflecting an increase in the recorded incidence of food poisoning over time and a number of high profile outbreaks associated, in particular, with *E.coli* 0157. As a result there has been intense scrutiny of regulatory controls imposed on the food industry by government and the efficacy of the systems operated by food suppliers for the control of food-related hazards. In particular, there is interest in the extent to which the food supply chain is subject to food safety controls along the lines of Hazard Analysis Critical Control Point (HACCP), which is widely regarded as the base standard for good hygiene practice, and the extent to which regulatory controls require/encourage the adoption of such systems.

It is evident that the implementation of HACCP within the UK food processing sector has become increasingly widespread over recent years, particularly in larger processing facilities (Henson and Northen 1998; Agriculture Committee 1998). This can be attributed to changes in the nature of food safety regulation on the one hand, and private incentives for rigorous systems of food safety control on the other. In reality, these two factors have been inter-related, producing a complex system of public and private incentives for the introduction of HACCP in the UK food processing sector.

Food safety regulation in the UK has increasingly required food suppliers to be proactive in ensuring the safety of the products they manufacture/distribute. The Food Safety Act 1990 introduced a defense of "due diligence" against food safety offences, defined as:

> Such a measure of prudence, activity or assiduity, as is properly to be expected from, and ordinarily exercised by, a reasonable and prudent man under the particular circumstances; not measured by any absolute standard but depending on the relative facts of the special case.

Under this defense, suppliers of food must demonstrate that they have done all that is reasonably possible to ensure that the food they handle, and any food obtained from upstream suppliers, conforms with statutory food safety standards (Caswell and Henson 1997; Henson and Northen 1997; Hobbs and Kerr 1992). In practice, this requires that a firm has a quality assurance system which is adequate given the products being produced, the nature and range of potential problems, and the perceived and actual risks of failure, and that the firm ensures that this system is working properly (Fidler 1990). In most contexts this means HACCP.

Simultaneously, however, the European Union has attempted to harmonize hygiene regulations across member states which has, almost inevitably, introduced a series of new (or at least revised) product and process standards. In the case of the UK dairy processing sector, these are laid down under the Dairy Products (Hygiene) Regulations 1995, which implement EU Directive 92/46. One requirement under these regulations is for dairy processing plants to have in place a food safety control system which is widely recognized as conforming to a six-point HACCP plan. While there are no statutory requirements for dairy processing plants to verify that this system is working effectively (as would be required under a full seven-point HACCP plan), it is recognized that verification is required if the system is to provide an effective "due diligence" defense.

While regulatory incentives for the implementation of HACCP have clearly been important, simultaneously, a system of private incentives has operated. First, the major multiple food retailers have enforced increasingly strict systems food safety controls (Bredahl and Holleran 1997; Henson and Northen 1997; Henson and Northen 1998). In most cases such systems are based on the fundamental principles of HACCP. As a result, food processors that manufacture own-branded products on behalf of major food retailers are invariably required to have HACCP in place. Second, industry-level codes of good hygiene practice have fast become the minimum standard for participation in major food markets. Those processors that do not are generally only able to supply peripheral and lower priced markets.

While HACCP has been implemented in many food processing plants within the UK, there are concerns that incentives may be inadequate to stimulate the adoption of HACCP by particular categories of business (e.g., small and medium-sized enterprises) and in certain sectors (e.g., with low operating margins). On the one hand, there are concerns that the costs of implementing HACCP may be prohibitive and/or that economies of scale are significant. On the other, it is recognized that the benefits to food businesses themselves may be limited, or at least of a largely unrecognized and/or intangible nature.

There have been very few studies of the motivation to implement HACCP in the UK and the resultant costs and benefits to food businesses.

As a consequence, it is difficult to evaluate the extent to which the balance of costs and benefits to businesses acts as an incentive/disincentive for the further adoption of HACCP. Although some indication of the costs and benefits of implementing HACCP can be derived from published studies on ISO 9000 certification (see for example Small Business Research Trust 1994; Carlsson and Carlsson 1996; Institute of Quality Assurance 1991; Buttle 1996; Vanguard Consulting 1993; European Observatory for SMEs 1995), the qualitative differences between these two quality assurance systems masks many of the salient issues.

The aim of this paper is to explore the process of HACCP implementation in the UK food sector, using the dairy processing sector as a case study. The paper reports the initial results of a survey that addresses the following questions:
- What motivates companies to implement HACCP in their plants?
- What are the major costs of implementing and operating HACCP?
- What problems are encountered in implementing HACCP?
- What are the major benefits of HACCP?

The dairy processing sector has been chosen as the case study for two main reasons. First, it is one of the few sectors within the UK food processing industry that is subject to a legal requirement to implement a HACCP-based food safety control system. Second, because of the commodity-type nature of many dairy products, which makes it easier to discern the process by which HACCP is implemented. While the results of the survey are clearly specific to the dairy processing sector, they provide information which is more widely applicable to the food processing sector as a whole.

## Survey

A mail questionnaire was designed on the basis of the existing literature on the implementation of HACCP and ISO 9000 and discussions with trade organizations, providers of HACCP consultancy services, and companies that have actually implemented HACCP. The questionnaire was piloted on a random sample of 30 plants in the UK dairy processing sector. Respondents returning fully completed questionnaires were telephoned to obtain feedback on the issues covered by the survey and ambiguities in the wording of questions.

A total of 1,196 plants that were listed under the dairy or ice cream processing sectors in the UK Yellow Pages directory were mailed a questionnaire in March 1998.[2] A reminder was mailed to non-respondents at the start of May 1998. A total of 240 questionnaires was returned, of which 192 were fully completed. This represents an overall valid response rate of 16 percent.

A follow-up survey of 114 non-respondents was conducted at the end of May 1998. Of these non-respondents, 79.8 percent were non-processing establishments to which the questionnaire did not apply. A further 7.9 percent had ceased trading. This reflects the nature of the address list utilized for the survey, which included import agents, retail establishments, and storage facilities, as well as processing plants. The results of the follow-up survey suggest that the proportion of processing plants that returned a completed questionnaire was well in excess of 16 percent, although it is not possible to calculate the actual response rate. Nevertheless, the results of the survey should be viewed as the collective experiences of companies which had implemented HACCP in their plants rather than a representative assessment of the costs and benefits of implementing HACCP in the UK dairy sector as a whole.

The products most widely processed by respondents were milk and cream. The majority (51.7 percent) of plants processed only one product, while a further 33.6 percent processed two products. A relatively small number of plants processed non-dairy as well as dairy products, of which the most widespread were fruit/fruit products (21.3 percent), which were presumably used as ingredients in dairy products such as yogurt.

## HACCP Status of Survey Respondents

To facilitate a reliable assessment of information on the costs and benefits of HACCP provided by the survey, an attempt was made to identify the degree to which respondents had fully implemented HACCP. In total, 74 percent of respondents claimed that they had fully operational HACCP systems in place (Table 1). Of these, 97 percent claimed to have full documentation. Further, 93 percent claimed that their HACCP systems had been verified through external audits. Consequently, we can be reasonably confident that this group of respondents had fully operational HACCP systems.

**TABLE 1**

**HACCP Status of Plants Responding Survey**

| HACCP Status | Number (%) |
|---|---|
| Fully Operational | 142 (73.9%) |
| Being Implemented | 37 (19.3%) |
| Planned But Not Implemented | 2 (1.0%) |
| No Plans to Implement | 11 (5.7%) |

A further 19 percent was implementing HACCP, but their systems were not fully operational at the time of the survey. Of these, 53 percent anticipated that their HACCP systems would be fully operational within six months, while 94 percent anticipated that their HACCP systems would be fully operational within 12 months.

We can be confident that those respondents with fully operational HACCP systems had reasonable experience of the costs and benefits of implementing and operating such systems. Around 75 percent had operated HACCP systems for at least 12 months, while only 11 percent had operated HACCP systems for six months or less (Figure 1).

## Process of Implementing HACCP

Respondents were asked to estimate how long it had taken to implement their HACCP systems from the time they had first started planning until they were fully operational (Figure 2). Around 80 percent of respondents estimated that it had taken 12 months or less to implement HACCP in their plants. However, a small but not insignificant minority (around 12 percent), estimated that it had taken more than 18 months.

Respondents were presented with a list of issues which previous studies have suggested can motivate the decision to implement quality control/assurance systems such as HACCP or ISO 9000 (for example Small Business Research Trust 1994; Institute of Quality Assurance 1991; Buttle 1996; Vanguard Consulting 1993; Martin et al. 1995). They were then asked to indicate how important each of these issues had been in the

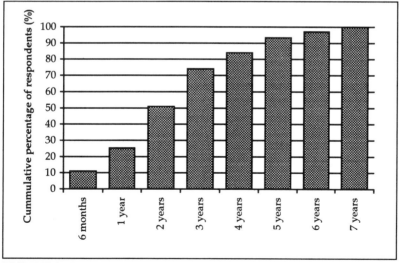

FIGURE 1. Amount of time since respondents had first implemented fully operational HACCP systems.

decision to implement HACCP in their own plants using a seven-point Likert scale ranging from "very important" (7) to "very unimportant" (1). Those respondents that had not implemented HACCP were presented with a similar exercise, although the results are not valid because of the small numbers involved.

The issues judged by respondents to have been most important in their own decisions to implement HACCP were the need to meet legal requirements and the need to meet the requirements of major customers (Table 2). This suggests that the implementation of HACCP by these companies was much dictated by external pressures. This is supported by the fact that 74 percent of respondents claimed that their major customers required them to have HACCP, which is maybe not surprising given that over half produced own-branded products for major multiple food retailers.

A variety of internal issues were also important in the decision to implement HACCP. These included improvements in control of the production process and improvements in product quality. In general, however, these appear to have played a secondary role in relation to legal requirements and the requirements of major customers.

In the UK, voluntary industry codes of good hygiene/manufacturing practice and third party certification have become widespread within the food processing sector (Henson and Northen 1997; Henson and Northen 1998; Bredahl and Holleran 1997). Many of these standards require HACCP in some form, although this may not extend to a fully documented

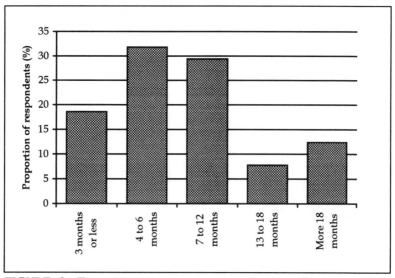

FIGURE 2. Time taken by survey respondents to implement fully operational HACCP systems.

and verified system. Around 79 percent of respondents complied with some form of voluntary industry standard or had been certified by a third party agency (for example, EFSIS). The most common standard, which almost 50 percent of respondents complied with, was that operated by the Dairy Industry Federation (DIF), the largest trade organization in the UK dairy processing sector. This provides a mechanism for meeting the requirements of the Dairy Products (Hygiene) Regulations 1995, and as such includes a requirement for a six-point HACCP system.

Approximately 26 percent of respondents were certified to ISO 9000. All of these plants considered the implementation of HACCP to be easier with ISO 9000 in place. Presumably, in most cases this will have been based on the perceptions of such respondents regarding the difficulties of implementing HACCP without ISO 9000 in place, rather than actual experience.

To make better sense of the issues judged by respondents to have been most important in their decision to implement HACCP, the importance scores were subject to factor analysis. A total of four factors had Eigenvalues exceeding one and collectively accounted for 69.6 percent of the variance in importance scores across the 14 issues presented to respondents. Loadings were derived for each of these factors using a varimax rotation (Table 3). On the basis of the factor loadings, these four factors can be interpreted as follows:

### TABLE 2

**Mean Importance Scores in Descending Order for Factors in Decision to Develop/Implement HACCP**

| Factor | Mean Score |
| --- | --- |
| Meet Legal Requirements | 6.11[a] |
| Meet the Needs of Major Customers | 6.02[a] |
| Generally Regarded as Good Practice | 5.86[b] |
| Improve Control of Production Process | 5.78[b] |
| Improve Product Quality | 5.70[b] |
| Attract New Customers for Products | 5.37 |
| Hold Onto Existing Customers for Products | 5.11[c] |
| Reduce Customer Complaints | 4.92[c] |
| Improve Efficiency/Profitability of Plant | 4.60 |
| Reduce Product Wastage | 4.12 |
| Needed for Plant to be Third Party Accredited | 3.70[d] |
| Reduce Need for Quality Audits by Customers | 3.53[d] |
| Recommended by Trade Organization | 3.24 |
| Access New Overseas Markets | 2.58 |

Note: Items denoted by same letter are not significantly different to each other at 5 percent level based on Wilcoxon sign rank test.

- **Factor 1:** The issues which loaded most heavily on this factor included "to improve efficiency/productivity of plant," "to improve product quality," "to reduce product wastage," "to improve control of production process," and "to reduce customer complaints." This suggests that this factor is associated with **internal efficiency** as a motive to implement HACCP.
- **Factor 2:** The issues which loaded most heavily on this factor were "reduced need for quality audits by customers" and "needed for plant to be third party accredited." The issues "hold onto existing customers for products" and "attract new customers for products" also had relatively heavy loadings. This suggests that this factor is associated with **commercial pressure** as a motive for the implementation of HACCP.
- **Factor 3:** The issues which loaded most heavily on this factor were

TABLE 3

Factor Loadings for Motivation to Implement HACCP Derived from Varimax Rotation

| Reason for Implementation of HACCP | Factor 1 | Factor 2 | Factor 3 | Factor 4 |
|---|---|---|---|---|
| Meet the Needs of Major Customers | 0.20309 | 0.10489 | 0.86958 | -0.1168 |
| Meet Legal Requirements | 0.25224 | -0.11613 | 0.67422 | 0.14781 |
| Improve Control of Production Process | 0.72702 | -0.15132 | 0.26039 | 0.32195 |
| Reduce Product Wastage | 0.73191 | 0.41077 | 0.04386 | -0.07699 |
| Reduce Customer Complaints | 0.71473 | 0.14502 | 0.24533 | -0.02252 |
| Improve Efficiency/Profitability of Plant | 0.79307 | 0.33742 | 0.09828 | 0.02411 |
| Improve Product Quality | 0.78448 | -0.01306 | 0.16029 | 0.11101 |
| Recommended by Trade Organization | 0.27873 | 0.19826 | -0.0921 | 0.62467 |
| Reduce Need for Quality Audits by Customers | 0.17572 | 0.78031 | -0.05797 | -0.10355 |
| Needed for Plant to be Third Party Accredited | 0.09244 | 0.73078 | 0.1311 | -0.04555 |
| Generally Regarded as Good Practice | 0.12129 | 0.06797 | 0.02397 | 0.90701 |
| Hold Onto Existing Customers for Products | 0.11724 | 0.54693 | 0.47038 | 0.31314 |
| Attract New Customers for Products | 0.16469 | 0.53327 | 0.31706 | 0.46984 |
| Access New Overseas Markets | -0.153 | 0.44803 | -0.08584 | 0.29393 |
| Proportion of Variation Explained | 34.3% | 15.6% | 10.3% | 9.4% |

"to meet the needs of major customers" and "to meet legal requirements," which suggests this factor is associated with **external pressure** as a motive for the implementation of HACCP.
- **Factor 4:** The issues "generally regarded as good practice" and "recommended by trade organization" loaded heavily on this factor, suggesting this factor is associated with **good practice** as a motive for the implementation of HACCP.

The above factors illustrate the dual role of internal and external forces in the implementation of HACCP in the UK dairy processing sector. Work is continuing to investigate the types of businesses which were most influenced by each of these factors.

## Cost of Implementing and Operating HACCP

Respondents were presented with a list of costs which previous studies have suggested can be incurred when implementing quality control/assurance systems such as HACCP or ISO 9000 (for example Small Business Research Trust 1994; Institute of Quality Assurance 1991; Vanguard Consulting 1993; Martin et al. 1995; Nganje 1997; MAFF 1995; Williams and Zorn 1994; FSIS 1996). They were then asked to rank each of these according to importance relative to the overall cost of implementing HACCP in their own plant. If a cost had not been incurred, respondents were instructed to allocate a rank of zero.

There was great variation in the costs of implementing HACCP between individual respondents (Table 4). For example, while 16.7 percent of respondents judged external consultants to be the most

### TABLE 4
### Rank Scores for Costs of Implementing and Operating HACCP

| Cost | Proportion of Respondents Giving Zero Rank | Proportion of Respondents Giving Rank of One |
|---|---|---|
| *Implementation Costs* | | |
| External Consultants | 55.1% | 16.7% |
| Investment in New Equipment | 54.5% | 7.7% |
| Staff Training | 9.0% | 14.1% |
| Managerial Changes | 61.3% | 6.1% |
| Structural Changes to Plant | 53.9% | 6.5% |
| Staff Time in Documenting System | 5.8% | 45.5% |
| *Operational Costs* | | |
| Record Keeping | 12.8% | 41.2% |
| Product Testing | 24.5% | 25.9% |
| Staff Training | 11.1% | 18.2% |
| Managerial/Supervisory Time | 16.9% | 17.6% |

important cost associated with the implementation of HACCP, 55.1 percent had not incurred this cost. This suggests that the circumstances of individual firms and the standards to which they operated prior to the implementation of HACCP have a large influence on the costs of implementation. Considering the results as a whole, however, staff time in documenting the system was the cost most frequently incurred (ranked zero by only 5.8 percent of respondents) and which was most frequently ranked as the greatest cost.

Only 13.1 percent of respondents had formally estimated the costs of implementing HACCP prior to developing and implementing the system in their own plant. This is not to suggest, however, that respondents did not have a "feel" for the costs involved. Indeed, the majority considered most of the costs they had incurred when implementing HACCP to be in accordance with their prior expectations (Figure 3). A notable exception was staff time spent documenting the system, which 54.9 percent of respondents considered to be greater than expected.

Respondents were asked to rank the individual costs of operating HACCP in their plants in a manner similar to that described above. As with the costs of implementing HACCP, there was great variation between respondents in the importance of individual cost categories (Table 4). Considering the results as a whole, however, record keeping was the cost most frequently incurred (ranked zero by only 12.8 percent of respondents) and which was most frequently ranked as the greatest cost.

Taking account of the individual costs of operating HACCP (as detailed in Table 4), respondents were asked whether their total costs of production had changed as a direct result of implementing HACCP. While 50 percent of respondents indicated that their total costs of production had not changed, 44.7 percent indicated that their total production costs had increased. Further, there is evidence of significant economies of scale in the operation of HACCP. While 53.5 percent of respondents with 50 or less employees indicated that their costs of production had increased as a direct result of implementing HACCP, only 32.1 percent of respondents with over 50 employees indicated that their production costs had increased.

Only 8.3 percent of respondents had formally estimated the costs of operating HACCP prior to developing and implementing the system in their own plant. In most cases, however, the majority of respondents indicated that the costs of operating HACCP were approximately in line with prior expectations (Figure 4). The one exception was the cost of record keeping, which 52.3 percent of respondents indicated had exceeded their expectations.

The survey results suggest that there is significant variation in the costs of implementing and operating HACCP between individual plants. It is evident, however, that the most widespread costs are associated with intangible elements such as staff time spent documenting the system and record keeping, rather than tangible elements such as investment in new equipment, changes to plant structure, etc. Further, these costs typically exceed the prior expectations of those involved in the implementation of HACCP.

The reliability of these results, however, depends on the ability of respondents to clearly identify the costs of implementing and operating HACCP. Given that respondents may have made other changes to their plants simultaneous to the implementation of HACCP, this might have been difficult. For example, respondents may have associated any

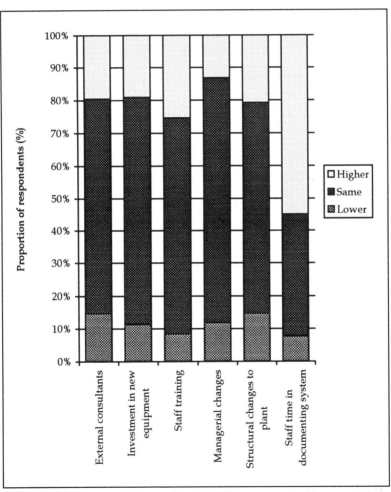

FIGURE 3. Costs of implementing HACCP compared with expectations prior to implementation.

structural changes that were made with other requirements of the Dairy Products (Hygiene) Regulations 1995, although the implementation of HACCP would have been difficult/impossible without them.

## Difficulties Implementing HACCP

Respondents were presented with a list of issues which previous studies have suggested can be problems in the implementation and/or operation of quality control/assurance systems such as ISO 9000 and HACCP (for example Small Business Research Trust 1994; Institute of Quality Assurance 1991; Vanguard Consulting 1993; Tompkin 1994). They were then asked to indicate how much of a problem each of these issues had been when implementing and/or operating HACCP in their own plants using a seven-point Likert scale ranging from "major problem" (7) to "minor problem" (1).

Mean problem scores were relatively low for all of the issues presented to respondents (Table 5), which suggests that no major difficulties were faced when implementing HACCP. However, given that all respondents had successfully implemented HACCP in their plants, this is not surprising. The issues judged to be most problematic were all associated with staffing. First, the need to retrain staff, in particular supervisory/managerial personnel. Second, the motivation of staff,

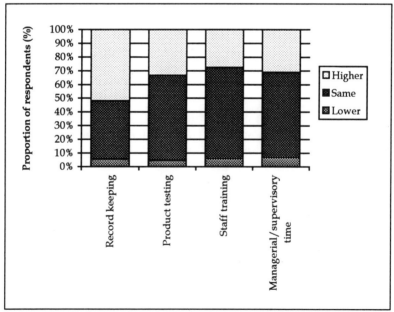

FIGURE 4. Costs of operating HACCP compared with expectations prior to implementation.

including not only those involved in production, but also supervisory/managerial personnel. The implementation/operation of HACCP was not perceived to reduce the flexibility of the production process, production staff, or the introduction of new products, or if it was, this was not perceived to be a major difficulty.

## Benefits of Implementing HACCP

Respondents were presented with a list of benefits which previous studies have suggested can result from the implementation and/or operation of quality control/assurance systems such as HACCP or ISO 9000 HACCP (for example Small Business Research Trust 1994; Institute of Quality Assurance 1991; Vanguard Consulting 1993). They were then asked to rank these benefits according to importance in terms of the overall benefit of implementing/operating HACCP in their own plants. If a benefit had not resulted, respondents were instructed to allocate a rank of zero.

Only a small minority of respondents (5.3 percent) indicated that they had not derived any benefits from the implementation/operation of HACCP in their plants. However, given that there was a relatively high proportion of zero ranks for all of the items presented to respondents, there were clearly significant differences in the benefits which had been derived (Table 6). The benefit which was cited by the highest proportion of respondents was the increased ability to retain existing customers (ranked zero by only 23.4 percent of respondents). This was also ranked most frequently as the benefit of most importance (ranked first by 34 percent of respondents). Increased ability to attract new customers and reduced product microbial counts were also frequently cited as benefits that had been derived from the implementation/operation of HACCP.

TABLE 5

Mean Importance Scores in Descending Order for Difficulties Faced When Implementing/Operating HACCP

| Factor | Mean Score |
| --- | --- |
| Need to Retrain Supervisory/Managerial Staff | 4.08[a] |
| Attitude/Motivation of Production Staff | 3.99[a] |
| Attitude/Motivation of Supervisory/Managerial Staff | 3.81 |
| Need to Retrain Production Staff | 3.72[b] |
| Reduced Staff Time Available for Other Tasks | 3.65[b] |
| Recouping Costs of Implementing HACCP | 3.12 |
| Reduced Flexibility of Production Process | 2.97[c] |
| Reduced Flexibility of Production Staff | 2.91[c] |
| Reduced Flexibility to Introduce New Products | 2.33 |

Note: Items denoted by same letter are not significantly different to each other at 5 percent level based on Wilcoxon signed rank test.

These results should, however, be interpreted with some caution. Given the intangible nature of many of the potential benefits of HACCP and the fact that a higher proportion of UK food processing businesses do not formally monitor the costs associated with quality assurance, many respondents may not have been aware of the impact that HACCP had on their operations. Further, it may have been difficult for respondents to isolate the impact of HACCP if other organizational changes had taken place at the same time. In practice, this means that respondents will have tended to emphasize the costs of implementing HACCP over the benefits.

The motivation for businesses to implement HACCP will reflect their prior expectations of the benefits which can be obtained; to the extent that the benefits associated with HACCP are underestimated, the level of motivation will be weaker. A relatively high proportion of respondents (around 30 percent) indicated that the benefits they had derived from implementing/operating HACCP were greater than expected (Figure 5). Within such businesses it might be expected that the motivation to implement HACCP will be weaker and/or external pressure will be more important as a factor influencing the implementation of HACCP.

**TABLE 6**

**Rank Scores for Benefits of Implementing/Operating HACCP**

| Benefit | Proportion of Respondents Giving Zero Rank | Proportion of Respondents Giving Rank of One |
|---|---|---|
| Reduced Product Wastage | 43.3% | 9.2% |
| Increased Product Shelf Life | 45.4% | 9.9% |
| Reduced Product Microbial Counts | 31.2% | 17.0% |
| Increased Product Prices | 61.4% | 0.7% |
| Increased Product Sales | 50.4% | 2.8% |
| Reduced Production Costs | 52.9% | 4.3% |
| Increased Motivation of Production Staff | 39.7% | 4.2% |
| Increased Motivation of Supervisory/Managerial Staff | 37.6% | 4.3% |
| Increased Ability to Retain Existing Customers | 23.4% | 34.0% |
| Increased Ability to Attract New Customers | 27.0% | 17.0% |
| Increased Ability to Access New Overseas Markets | 68.4% | 0.0% |

## Conclusions

The results of the survey raise a number of important issues with respect to the future development of HACCP in the UK dairy processing

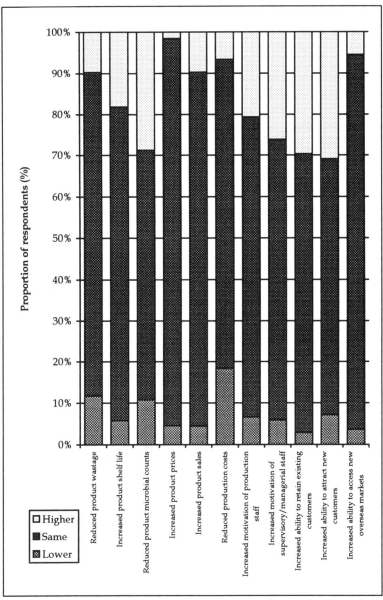

FIGURE 5. Benefits of operating HACCP compared with expectations prior to implementation.

sector, as well as the food processing sector as a whole, both in the UK and in other countries. It is anticipated that further analysis of the survey data will enrich the findings reported in this paper and provide more in-depth analysis of the costs and benefits associated with the implementation and operation of HACCP.

The motives for implementing HACCP differ between individual businesses, even within a sector which is subject to a legal requirement to implement HACCP-based food safety control systems. A number of the respondents had implemented HACCP prior to the promulgation of any legal requirement to do so. Further, it is evident that the requirements of major customers, pressure to conform with prevailing market standards of good hygiene/manufacturing practice, as well as the need to improve internal efficiency all played a part in motivating these businesses to implement HACCP.

The costs of implementing and operating HACCP vary between individual businesses according to their own particular circumstances and the prevailing standards to which they operate. However, the major costs associated with both the implementation and operation of HACCP seem to be associated with staff time rather than investment in new equipment or upgrading the structure of processing facilities. It is evident from the survey that in most cases businesses did not formally estimate the costs associated with HACCP prior to implementation, although they clearly had a "feel" for the cost involved. Consequently, many of the costs of implementing HACCP were approximately in line with prior expectations.

Respondents did not appear to have experienced major difficulties implementing HACCP. However, those difficulties that were experienced were mainly associated with the motivation and training of staff. This applied to both production and supervisory/managerial staff.

There was great variation in the benefits which respondents claimed to have derived as a direct result of implementing/operating HACCP. However, this probably reflects, at least in part, the intangible nature of many of the potential benefits of HACCP and the difficulty in separating out the effects of other changes which might have taken place simultaneously to the implementation of HACCP. The benefit that was most widely experienced, however, appears to be the increased ability to retain existing customers. This suggests that the main impact of HACCP is to enable businesses to meet the increasing demands of the marketplace rather than obtain any real competitive advantage over other suppliers.

The survey results raise a number of important issues with respect to the future development of HACCP in the UK dairy processing sector, as well as in the food processing sector as a whole, both in the UK and in other countries. It is anticipated that further analysis of the survey data will enrich the findings reported in this paper and provide a more in-depth analysis of the costs and benefits associated with the implementation and operation of HACCP.

**Notes**

[1] Spencer Henson is a Reader in Food Economics and Marketing, Georgina Holt a Research Fellow and James Northen a Research Assistant, Centre for Food Economics Research (CeFER), Department of Agricultural and Food Economics, The University of Reading, UK

[2] A copy of the questionnaire is available from the authors on request.

**References**

Agriculture Committee (1998). *Food Safety*. Fourth Report and Proceedings of the Agriculture Committee. Stationery Office, London.

Bredahl, M.E. and Holleran, E. (1997). Food Safety, Transaction Costs and Institutional Innovation. In: Schiefer, G. and Helbig, R. (eds) *Quality Management and Process Improvement for Competitive Advantage in Agriculture and Food*. University of Bonn.

Buttle, F.A. (1996). *Does ISO 9000 Work?* Manchester Business School, Manchester.

Carlsson, M. and Carlsson, D. (1996). Experiences of Implementing ISO 9000 in Swedish Industry. *International Journal of Quality and Reliability Management*, 13(7), 36-47.

Caswell, J. and Henson, S.J. (1997). *Interaction of Private and Public Food Quality Control Systems in Global Markets*. Paper presented at the conference "Globalisation of the Food Industry: Policy Implications," The University of Reading, September 1997.

European Observatory for SMEs (1995). *The European Observatory for SMEs Third Annual Report*. European Network for SME Research, The Hague.

Fidler, D. (1990). Due Diligence and Quality Assurance in the UK. *Food Control*, 7(2), 117-121.

FSIS (1996). Pathogen Reduction: Hazard Analysis and Critical Control Point (HACCP) Systems; Final Rule. *Federal Register*, 61(144), 38805-38989. Food Safety Inspection Service, United States Department of Agriculture, Washington.

Henson, S.J. and Northen, J. (1997). Public and Private Regulation of Food Safety: The Case of the UK Fresh Meat Sector. In: Schiefer, G. and Helbig, R. (eds) *Quality Management and Process Improvement for Competitive Advantage in Agriculture and Food*. University of Bonn.

Henson, S.J. and Northen, J. (1998). Economic Determinants of Food Safety Controls in the Supply of Retailer Own-Branded Products in the UK. *Agribusiness*, 14(2)

Hobbs, J.E. and Kerr, W.A. (1992). Costs of Monitoring Food Safety and Vertical Coordination in Agribusiness: What Can be Learned from the British Food Safety Act 1990? *Agribusiness*, 8(6), 575-584.

IQA (1991). *An IQA Survey on the Use and Implementation of BS 5750 Standards by Third Party Assessment Bodies as Seen by the End User*. Institute of Quality Assurance, London.

MAFF (1995). *Compliance Cost Assessment: Dairy Products (Hygiene) Regulations, 1995*. Ministry of Agriculture, Fisheries and Food, London.

Martin, S.A., Bowland, B.J., Calingaert, B. and Dean, N. (1993). *Economic Analysis of HACCP Procedures for the Seafood Industry.* Research Triangle Institute, North Carolina.

Nganje, W.E. (1997). *HACCP Costs of Small Meat Packing Plants: Summary of Survey Results.* Working Paper 970527, Food and Agribusiness Management Programme, University of Illinois.

Tompkin, R.B. (1994). HACCP in the Meat and Poultry Industry. *Food Control,* 5(3), 153-161.

Small Business Research Trust (1994) *Small Businesses and BS 5750.* Small Business Research Trust, Open University, Milton Keynes.

Vanguard Consulting (1993). *BS 5750 Implementation and Value Added.* Vanguard Consulting Ltd, Buckingham.

Williams, R.A. and Zorn, D.J. (1994). *Preliminary Regulatory Impact Analysis of the Proposed Rules to establish Procedures for the Safe Processing and Importing of Fish and Fishery Products.* Food and Drugs Administration, Washington.

Chapter 22

# HACCP and the Dairy Industry: An Overview of International and U.S. Experiences

Brian W. Gould, Marianne Smukowski and J.Russell Bishop[1]

## Introduction

In 1997, 155.2 billion pounds of raw milk were produced in the U.S. resulting in 1.1 billion pounds of butter, 7.3 billion pounds of cheese, 1.4 billion pounds of dry milk, and 1.3 billion gallons of ice cream manufactured (USDA 1997). With less than 5 percent of reported foodborne diseases originating from contaminated dairy products, the U.S. dairy industry has had an excellent safety record when considering the amount of dairy products produced (Johnson, Nelson, and Johnson 1990b). Before the advent of modern production techniques, milk and milk products were the source of significant foodborne pathogens. Diseases such as typhoid fever, scarlet fever, septic sore throat, diphtheria, tuberculosis, and brucellosis were commonly spread by milk (Marth 1985; Bryan 1983). A number of developments have occurred in the industry that have greatly improved the safety of dairy products, the most important being the widespread use of pasteurization. Other changes that have dramatically improved the production of safe dairy products include: improvements in on-farm and in-plant sanitation, virtual elimination of some dairy cattle diseases, adoption of the Pasteurized Milk Ordinance, adoption of standardized laboratory procedures, and increased availability of producer, processor, and consumer education programs (Marth 1985, 1981).

Although the U.S. dairy industry has not been the focus of recent federal HACCP initiatives, there are increasing efforts to make HACCP a standard practice with the anticipation that it will eventually be a mandated food safety system for the industry. So far, the industry's large processors have been the earliest adopters with smaller firms starting to embrace HACCP principles (Robinson 1997). This chapter provides an overview of the application of HACCP to the U.S. dairy industry with some brief comments about other country's experiences. With the importance of raw farm milk in determining product quality, there are a number of countries that are also proposing "cow-to-consumer" systems.

We provide a brief overview of the on-farm HACCP components of these programs.

## Potential Dairy Product Hazards

In-spite of the industry's excellent record, disease-causing bacteria have appeared in dairy-based finished products. Problems associated with the presence of Listeria monocytogenes, Salmonella enteritidis, Staphylococcus aureus, Escherichia coli and others have been documented.[2] The products affected have included cheese, ice cream, nonfat dry milk (NFDM), raw and pasteurized milk (Bryan 1983; Marth, 1985; Pearson and Marth 1990). Besides microbiological, potential physical hazards include metal, glass, insects, dirt, wood, plastic, and personal effects. Chemical hazards include natural toxins, metals, drug residues, food additives, and inadvertent chemicals (Smukowski and Brusk 1997). Table 1 contains a partial listing of potential microbiological, chemical, and physical hazards for selected dairy products.

### Examples of the Occurrence of Dairy-Based Diseases

A majority of dairy-based diseases have been the result of the use of raw or improperly pasteurized milk or milk products. This is evidenced by the epidemiological history of the largest single-source outbreak of salmonellosis in the U.S.. The outbreak resulted from the consumption of a national brand of ice cream (Schwan's) manufactured in Minnesota in 1994. Estimates are that over 224,000 persons nationwide became ill with S. enteritidis infections after consuming contaminated ice cream manufactured using "premix" from two suppliers. The premix was transported to the ice cream plant in tanker trailers, transferred to storage silos prior to use, and then moved into the production process. Although the ice cream premix was pasteurized by the original manufacturer, neither the premix nor other ingredients used in the final manufacturing process were pasteurized after delivery. No cause of the salmonella contamination could be found in the ice cream plant. The pathogen source was traced to tanker trailers used to transport the pasteurized premix from the original manufacturer and which had previously been used to transport unpasteurized liquid eggs containing S. enteritidis. The trucking company was found to have inadequate cleaning procedures between shipments, poor record keeping, and improper equipment maintenance (Hennessy et al. 1996). As a result of their analysis, Hennessy et al. (1996) recommended that food grade products be pasteurized after transport to the final production facility or that these ingredients be shipped in dedicated tanker trailers (e.g., no backhauls).

More than one-third of U.S. farm milk is used for cheese production. This percentage varies tremendously across states. In Wisconsin, for

TABLE 1
Potential Hazards Present in Selected Dairy Products

| Fluid Milk | | Cheese | | Ice Cream | Dried Milk Products | | Whey | Evaporated or Condensed Products | |
|---|---|---|---|---|---|---|---|---|---|
| Microbiological | Chemical | Microbiological | | Microbiological | Microbiological | | Microbiological | Microbiological | |
| Salmonella | Antibiotics | Salmonella | | Salmonella | Salmonella | | L. monocytogenes | Salmonella | |
| L. monocytogenes | Pesticides | L. monocytogenes | | Mold Spores | L. monocytogenes | | E. coli | L. monocytogenes | |
| S. aureus | Sulfonmides | S. aureus | | L. monocytogenes | S. aureus | | S. aureus | S. aureus | |
| S. enterotoxin | Physical | S. enterotoxin | | E. Coli | S. enterotoxin | | S. enterotoxin | S. enterotoxin | |
| C. perfringens | Insects | E. coli | | S. aureus | E. coli | | | C. perfringens | |
| E. coli | Soil | Campylobacter | | Chemical | C. botulinum | | | Yersinia | |
| Yersinia enterocolitica | Glass fragments | Shigella | | Non-Food | C. perfringens | | | Campylobacter | |
| Campylobacter | Wood slivers | Brucella | | Chemical Vapors | | | | | |
| B. cereus | Metal fragments | C. botulinum | | | Chemical | | | B. cereus | |
| Shigella | | Chemical | | | Sulfonamides | | | Shigella | |
| Brucella | | Nitrates, nitrites | | | Antibiotics | | | Brucella | |
| | | Aflatoxin | | | Pesticides | | | Chemical | |
| | | Pesticides | | | | | | Antibiotics | |
| | | | | | | | | Pesticides | |

Sources: International Dairy Foods Association 1996; Marth 1985.

example, more than 88 percent of farm milk is used to make cheese. Thus, there are significant resources devoted to assuring that safe cheeses are produced. In contrast to traditional make procedures, today most cheeses are produced using pasteurized milk. Some cheeses continue to be made from milk that has not been subject to any heat treatment or with the use of milk that is heat treated that is less severe than the full pasteurization process of 71.7° C for 15 seconds. The use of milk that has been subject to minimal heat treatment may destroy some micro-organisms, but bacteria such as L. monocytogenes, Salmonella and E.coli O157 may survive and contaminate the final product (Institute for Food Science and Technology [IFST] 1990). Table 2 contains a review of recent outbreaks of diseases resulting from the consumption of contaminated cheese. Unpasteurized milk was involved with the majority of the cases with the remaining the result of incorrect pasteurization procedures.

An example of how problems can arise in cheese production can be seen in the California L. monocytogenes problem of 1985. There was an outbreak of illness caused by the presence of this bacteria in California which was traced to a Mexican-style "semi-soft" cheese. The mortality rate was approximately 30 percent with 29 deaths. Consumption of cheese from a single plant was identified as the source of the listeriosis. The significance of the problem was shown when cheese samples with different expiration dates were found to contain L. monocytogenes. Some of the cheese was made with improperly pasteurized or raw milk. Evidence indicated that the cheese manufacturer may have mixed raw with pasteurized milk (Pearson and Marth 1990).[3]

In Canada there is an emerging problem with Salmonella infection originating from processed cheese products. S. enteritidis and S. typhimurium have been discovered in shredded mild cheddar cheese component of pre-packaged "Lunchmate" luncheon food and mild cheddar cheese blocks produced from the same vats as the Lunchmate shredded cheese. Between March 1 and April 20, 1998, there had been a reported 687 confirmed cases of Salmonella-based illnesses across Canada, 458 in Ontario, and by the end of March, three separate product recalls. There are a number of hypotheses as to the source of the pathogens. One possible source is that the cheese was produced using improperly pasteurized (thermized) milk. The second is the cross-contamination from the meat components of the prepackaged product. This outbreak represents the largest foodborne outbreak of Salmonella enteridis in Canada since 1984 when 2,000 cases of S. typhimurium occurred which were also linked to the consumption of contaminated cheese (Health Canada 1998).

A final example of problems arising from the consumption of contaminated dairy products can be found in the 1983 Massachusetts outbreak of listeriosis. Forty-nine patients were hospitalized with septicemia and meningitis caused by L. monocytogenes. Fourteen of the

TABLE 2
Reported Outbreaks of Foodborne Disease Due to Cheeses, 1983-98

| Year | Country | Pathogen | # of Cases | # of Deaths | Cheese | Description | Reference |
|---|---|---|---|---|---|---|---|
| 1983 | Netherlands, Den, Sweden, USA | enterotoxigenic E.coli | >3000 | NR | Brie | Use of raw milk | MacDonald et al.; Nooitgedagt and Hartog 1988 |
| 1983-87 | Switzerland | L.monocytogenes | >122 | 34 | Vacherin Mont d'Or | Made from thermalized milk | Bille 1990 |
| 1984 | Canada | Salmonella typhimurium | 2,700 | 1 | Cheddar | Salmonella survived up to 8 mos. in refrig. storage | D'Aoust et al.1985, D'Aoust 1994 |
| 1984-85 | Scotland | Staphylococcus aureus enterotoxin | >13 | 0 | Sheep Milk | Cause: mastitis and post-infection carriage by ewes | Bone et al. 1989 |
| 1985 | Switzerland | Salmonella typhimurium | >40 | 0 | Vacherin Mont d'Or | Hand-borne cross-contamination from pigsty | Sadik et al. 1986 |
| 1985 | USA | L.monocytogenes | >142 | 48 | Mexican Style | Contamination caused by addition of raw milk | Linnan et al. 1988 |
| 1988-89 | England | Unknown | 155 | 0 | Stilton | Decide to pasteurize milk for Stilton manufacture | Maguire et al. 1991 |
| 1989 | England | Salmonella dublin | 42 | 0 | Irish Soft | 4 cows were asymetric excreters, Salmonella detected in curd but not in raw milk nor factory | Maquire et al. 1992 |

## TABLE 2 (continued)
### Reported Outbreaks of Foodborne Disease Due to Cheese, 1983-98

| Year | Country | Pathogen | # of Cases | # of Deaths | Cheese | Description | Reference |
|---|---|---|---|---|---|---|---|
| 1989 | USA | *Salmonella javiana* and *S. oranienberg* | 164 | 0 | Mozzarella | Contamination: 0.36 cells/100g - 4.3 cells/100g | Hedberg et al. 1992 |
| 1992 | England | *Salmonella livinstone* | 10 | NR | Cheese | | Djuretic et al. 1997 |
| 1992-93 | France | Verocytotoxin forming *E. coli* | NR | 1 | Fromage Frais | | Anonymous 1994a |
| 1993 | France | *Salmonella paratyphi B* | 273 | 1 | Goats Milk | Micro. monitoring failed to detect for 2 months | Desenclos et al. 1996 |
| 1994 | Scotland | Verocytotoxin forming *E. coli* | >20 | 0 | Local farm produced | | Anonymous 1994b |
| 1995 | France | *L. monocytogenes* | 20 | 4 | Brie de Meaux | Disinfection and control measures reinforced | Goulet et al. 1995 |
| 1995 | Malta | *Brucella melitensis* | 135 | 1 | Soft cheese | | Anonymous 1995 |
| 1995 | Switz. and France | *Salmonella dublin* | 25 in Fr. | 5 in Fr. | Cheese from Doubs reg. | Problem eliminated with prod. control measures | Vaillant et al. 1996 |
| 1996 | Eng. and Scot | *Salmonella gold-coast* | >84 | 0 | Cheddar | | Anonymous 1997a |
| 1996 | Italy | *Clostridium botulinum* | 8 | 1 | Mascarpone | | Aureli et al. 1996 |
| 1997 | England | *E.coli O157* | 2 | 0 | Lancashire | | Anonymous 1997b |
| 1998 | Canada | *Salmonella typhimurium* | >650 | 0 | Cheddar, Swiss | Found in Lunchmates, cross contamination possible | Health Canada 1998 |

Source: IFST 1988. For a review of additional cases refer to Johnson, Nelson, and Johnson, 1990b.

patients died. Occurrence of the disease was correlated with the consumption of a particular brand of whole and 2 percent milk produced from cows diagnosed with listeriosis. It is unknown how the L. monocytogenes infected the pasteurized milk (Marth 1985).

## HACCP and the Dairy Industry

In spite of the occurrences described above, dairy products continue to be one of the safest classes of food. To maintain the quality of their products, the U.S. dairy industry has undertaken several initiatives to accelerate the adoption of HACCP principles by its manufacturers. The transformation of raw farm milk to finished dairy products implies that for most dairy processors the first critical control point (CCP) is the receipt of this milk. With the FDA placing primary responsibility for food safety on the food manufacturer or distributor, there is increasing pressure for these processors to work with their suppliers to minimize microbial contaminants and chemical residues (Cullor 1995a, 290).

### On-Farm HACCP

Compared to the European Union, the U.S. is playing catch-up with respect to the implementation of HACCP programs for the dairy sector. In the early 1990s, the Netherlands instituted a structured trace back/trace through producer funded program. Under this program all dairy stock are identified within 72 hours of birth and movement of these animals must be reported to the system. It is anticipated that the European Union will have a HACCP system for the dairy industry in place within 5-7 years given that the CODEX Food Hygiene Standards, which include HACCP, should be in place within 2-3 years.

The New South Wales Development Corporation (NSWDC) is promoting on-farm HACCP procedures as part of a "cow-to-consumer" system. Their on-farm HACCP system focuses on the following raw milk characteristics: somatic cell count (SCC), chemical residue, iodine contamination, sediment, high freezing points, standard plate counts (SPC), coliform counts, and antibiotic residue (NSWDC 1998).

A major participant in the world dairy trade is New Zealand. Recognizing the importance of meeting Codex sanitary standards, the New Zealand dairy industry has established an extensive food safety system for their industry. The Ministry of Agriculture administers the Dairy Industry Regulations 1990 (DIR) which defines the standards for food safety and truth in labeling. In terms of on-farm safety, there is an annual inspection of all dairy farms by HACCP inspection or quality management audit. This inspection is subject to the Farm Dairy Code of Practice which identifies how the DIR will be achieved (Beagley 1998; New Zealand Dairy Board 1998).

When compared to the adoption of HACCP by processing firms, the development of on-farm HACCP plans may be more difficult given the diverse types of dairy operations. An example of such diversity can be seen when comparing confinement versus rotational grazing systems. The types and degree of quality control problems will vary across production technology implying the need for technology specific HACCP systems. In addition, some have asserted the microbiological testing for verification will be expensive and record keeping may be very time consuming (Cullor 1995a).

As in any HACCP plan, there is a set of "prerequisites" that the producer should adopt which is defined as a control that must be in place before milk is harvested from the animal. This is in contrast to the farm operator's that must be implemented during milk harvesting. The prerequisites for a dairy farm operator may include the use of:
- accepted management practices with respect to building and equipment maintenance, manure removal, bedding and barn sanitation;
- adequate maternity pen and obstetric equipment sanitization;
- use of approved milking hygiene and management practices;
- adoption of accepted herd management (culling) procedures; and
- appropriate medical treatment and isolation procedures of cows with mastitis (Cullor 1995a,b).

Focusing on a specific problem, Cullor (1995b) provides the steps associated with the development of a generic on-farm HACCP program designed to control mastitis (Figure 1). Mastitis, a health problem faced by

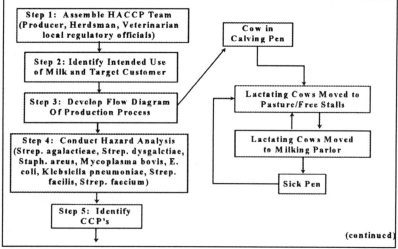

FIGURE 1. Steps for developing on-farm HACCP system to control mastitis. Source: Derived from Cullor 1995b.

every dairy farm, is an inflammatory disease of the cow's udder resulting from the presence of such bacteria as Streptococcus agalactiae, S. aureus, E. coli, L. monocytogenes, and Pseudomonas aerginosa. The food safety issue is that bacteria can be discharged in the cows milk. The generic program shown in Figure 1 is based on the seven principles of HACCP. A multi-person team will be formed to develop and implement the plan. We assume this milk will be used by a processor in the production of a consumer dairy product. Under the conventional milking system, cows are moved into the calving pen. The hazard analysis has identified mastitis as typically caused by the pathogens shown in Step 4 which may be present in the maternity pen, sick pen, free stalls or pasture, and milking parlor.

Under this model the only true CCPs would be temperature control of the milk and drug therapy administration. Other control points such as cleaning and sanitizing, housing and bedding, water and waste management, sick and calving pens, and treatment areas (environmental hygiene) are categorized as prerequisites. As in any HACCP plan, control points must be monitored, corrective action taken when required, and precise records kept, whether these points are classified as CCPs or prerequisites. Critical limits need to be established which determine whether remedial measures need to be adopted. These limits could include somatic cell count, standard plate count, coliform counts, and/or rate of herd infection. In order to determine whether these critical limits have been exceeded, evaluation procedures need to be adopted which include microbiological examination of milk and milk equipment. Corrective measures that could be undertaken if monitoring indicates a problem could

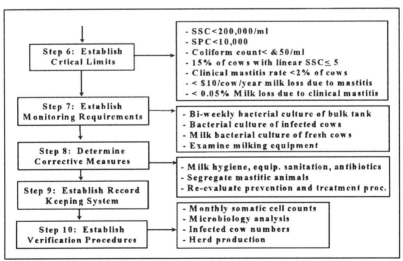

FIGURE 1 (continued). Steps for developing on-farm HACCP system to control mastitis. Source: Derived from Cullor 1995b.

include reviewing and revising procedures for milking, equipment sanitation, pen cleaning, antibiotic protocols, and segregation of infected animals. Verification of the efficacy of the established HACCP program can be achieved using a number of variables. Cullor (1995b) suggests variables such as monthly bulk-tank somatic cell counts, microbiology results, number of mastitis infected cows in the sick pen, and milk production records.

**HACCP and Dairy Processing**

The Food Safety Enhancement Program (FSEP) being implemented in Canada is a cooperative effort of the federal government and food industry to provide a uniform approach to implementing and inspecting HACCP systems. Although voluntary, there have been some indications at the federal level that it may be made mandatory sometime in the future. The FSEP applies to all federally registered processing establishments (including dairy plants) and egg grading stations. Given that it is HACCP-based, the FSEP moves away from the philosophy of control-based on testing (i.e., testing for a failure) to a preventative approach to food safety (International Dairy Foods Association [IDFA] 1996:2). Under the FSEP:

- plants are responsible for developing their own HACCP-based program;
- agriculture and Agri-food Canada assesses these HACCP-based programs and assists plants to meet FSEP requirements during the development and implementation of this program in their establishments;
- plant personnel are responsible for controlling, monitoring, and keeping accurate records for each CCP and ensuring proper procedures and controls have been followed; and
- agriculture and Agri-food Canada will review plant records, assess corrective actions and observed on-line processing at CCPs, take samples as appropriate, and verify that the overall HACCP plan is effective.

In spite of the voluntary nature of the program, red meat plants that ship to the U.S. have had to follow the same time frame for establishing a government-certified HACCP program as in the U.S.

The New South Wales Dairy Corporation has initiated a mandatory HACCP-based Quality Plus Quality Assurance System. Similar to the Canadian system, their Quality Plus system focuses on preventative measures rather than solely on final product analysis. Under this system, the role of the NSWDC is one of auditing the developed system instead of conducting the actual test of the quality of final product (Peta and Kailasapathy 1995; NSWDC 1998). The microbiological testing requirements are not limited to dairy firms in the corporation in that they encompass firms that utilize dairy products as raw materials. Problems faced with the implementation of this program have been most severe for smaller manufactures with problems such as: insufficient technical resources, concentration of functions, time, and lack of financing needed to implement their HACCP plans (Peta and Kailasapathy 1995).

The New Zealand Ministry of Agriculture is in the process of implementing a mandatory Codex-based HACCP system. All dairy factories are currently ISO 9001 or ISO 9002 certified. Under the proposed HACCP program, all dairy factories must be audited using a HACCP system operated by the dairy manufacturer, an external inspection that is based on HACCP principles by an approved organization, or inspection using good management principles based on HACCP principles and pre-approved by the Ministry of Agriculture. Under the first alternative, an annual HACCP audit is needed. Under the last two options, three external audits are required. With its implementation, the New Zealand Ministry of Agriculture will have the responsibility of approving audit procedures, recognizing organizations for validation of HACCP systems, inspection of dairy factories, and deciding whether additional inspection and/or monitoring will occur and the length of time of these audits when non-compliances are discovered (New Zealand Ministry of Agriculture 1998).

The U.S. dairy industry is responding to safety concerns by initiating the development of industry-wide HACCP procedures. The International Dairy Foods Association (IDFA), in conjunction with dairy processors and the Wisconsin Center for Dairy Research, developed a three stage "Dairy Product Safety System" which incorporates the following necessary stages: prerequisites (GMPs), HACCP, and employee training. Prerequisites, when applied to dairy processing, are controls that must be in place before milk/food/etc. is produced. In contrast, CCPs represent those controls that must be in place during processing and are "universal steps or procedures that control the operational conditions within a dairy plant, allowing for environmental conditions that are favorable to the production of safe dairy products" (IDFA 1996:15). The number of CCPs are determined, in part, by the quality of the GMPs implemented. The IDFA program has been widely accepted by the industry and regulatory agencies, and is the basis for proposed changes to the Pasteurized Milk Ordinance. Below we present two examples of generic HACCP programs, one for cheddar cheese and another for ice cream.

**Examples of Generic HACCP Programs**

Cheesemaking capitalizes on the coagulation of milk. First, milk is screened to ensure that there are no antibiotics or harmful agents present that could negatively impact the cheesemaking process. The raw milk may not have the desired characteristics (e.g., fat or protein) for the cheese to be manufactured. This raw milk can be standardized to the desired fat/protein profiles using a number of alternative techniques such as cream removal or addition of non-fat dry milk (NFDM). After standardization, the milk is heated and held at a given temperature for a short period to destroy unwanted bacteria (i.e., pasteurization). Starter cultures are then added to the heated milk which change a very small amount of the milk sugar into lactic acid. This acidifies the milk at a much faster rate and prepares it for the next stage. Rennet is then added to the milk which causes the milk to coagulate to a solid gel. This is cut with special cutting tools into small cubes of the desired size (e.g., curds) to facilitate expulsion of whey. Heat is applied to start a shrinking process which, with the steady production of lactic acid from the starter cultures, will change it into small rice-sized grains.

At a carefully chosen point the curd grains are allowed to fall to the bottom of the cheese vat; the left-over liquid, which consists of water, lactose, and albumen (now called whey) is drained off; and the curd grains are allowed to mat together to form large slabs of curd. When making cheddar cheese, after the whey is drained off, the curd is subjected to a special form of treatment -- the curd is turned upside down and stacked. This is referred to as cheddaring. When the whey acidity reaches a predetermined level, the blocks are milled into chips which are dry-salted before being placed in moulds. The cheese is pressed, either by its own weight or more commonly by applying pressure to the moulds. Treatment during curd making and pressing determines the characteristics of the cheese. Later, it is packed in various sized containers for maturing (Gösta 1996).

In Figure 2, we present a generic cheddar cheese process flow chart which identifies six possible CCPs: raw milk receiving, dairy ingredient storage, raw cream storage, pasteurization, culture media pasteurization, and metal detection.[4] Table 3 provides an overview of the results of the cheddar cheese plant hazard analysis which identifies these CCPs. As indicated above, the receipt of raw milk is the first CCP, and given the lack of progress of implementation of on-farm HACCP, this represents a major point by which pathogens can be detected and controlled. Any pathogens will be eliminated by the use of the appropriate pasteurization procedures. Temperature control is necessary to prevent Staph. toxin production. As noted in Table 2, a number of previous instances of foodborne illness have been due to incorrect pasteurization procedures. Should drug residues be detected in incoming milk, that milk will be rejected and should not be a concern for the remaining production process. At 45° F, milk can be stored for 72 hours.

Besides raw milk, there have been evidence of Salmonella contaminated NFDM. If cream is removed it is pasteurized given that it may be used for the manufacture of other products, sold, or used to standardize the cheese milk. Under the proposed program, the pasteurization of cheese milk occurs just before the milk enters the cheese vat and after any blending (standardizing) ingredients have been added. Inadequate temperature control may allow for the propagation of pathogens while the milk is stored. These pathogens will be destroyed by later pasteurization. In order to avoid contamination in the cheese vat, the starter culture media should be pasteurized. Given the mechanical nature of the milling process, milling equipment may break so there is the possibility of physical (metal) hazards. The last CCP is the metal detector

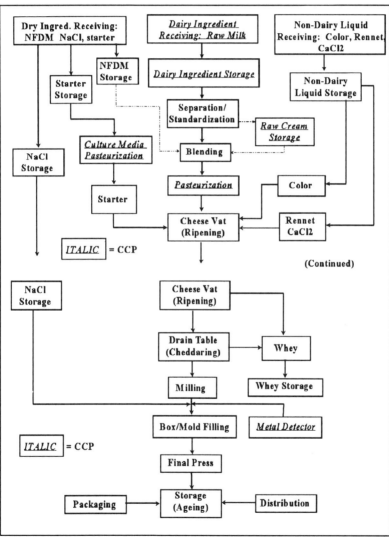

FIGURE 2. Process flow in cheddar cheese production and critical control points. Source: IDFA 1996.

TABLE 3
Cheddar Cheese CCP Descriptions

| CCP/Process Step | Potential Hazard | Control Criteria | Critical Limit | Monitor Frequency | Records (Location) | Responsibility | Corrective Action | Verification |
|---|---|---|---|---|---|---|---|---|
| Raw Milk Receiving | Microb. | Temp. | ≤ 45° F | Every Tanker | Load Ticket (QA/QC office) | Intake Operator | Hold and Evaluate Product | Therm. |
|  | Chemical-Drug Residues | 6-lactem Screening | No Positives | Every Tanker | Receiving Log (QA/QC office) | Intake Operator | Reject Milk | Calibrate Test Kit |
| Dairy Ingred. Storage | Microb. | Temp. Time | ≤ 45° F ≤ 72 hours | 3 times daily* | Recording Chart (QA/QC office) | QA Tech. | Hold Product, investigate cause and adjust | Recording, indicating therm. or calibration |
| Raw Cream Storage | Microb. | Temp. Time | ≤ 45° F ≤ 72 hours | 3 times daily* | Recording Chart (QA/QC office) | QA Tech. | Hold Product, investigate cause and adjust | Recording, indicating therm. or calibration |
| Pasteurization | Microb. | Temp. Time | ≥ 161° F ≥ 15 seconds | Continuous | Recording Chart (Production Office) | Pasteurizer Operator | Flow divert, recirculate and heat | Cut-in/cut-out checks, indicating therm. calib. |
| Culture Media Pasteurization | Microb. | Temp. Time | ≥ 161° F ≥ 15 seconds | Continuous | Recording Chart (Production office) | Pasteurizer Operator | Flow divert, recirculate and heat | Cut-in/cut-out checks, indicating therm. calib. |
| Milling | Microb. | Acid Development | pH ≤ 5.60 by 8 hr. after adding culture | Every Vat | Recoding Chart (QA/QC office) | QA Tech. | After pressing test for S. aureus if pH>5.60 | Calibrate pH meter before each use |
| Metal Detector | Physical-Metal Frag. | Metal Fragments | Limit of detection | Continuous | Detector Log (QA/QC office) | Filling Operator | Reject, Locate Cause | Detector calibration |

*Continuous recording thermometer. Sources: IDFA 1996; Bernard et al. 1997 which represents the last opportunity to detect such hazards. After the cheese is packaged, recontamination risk is low. Temperature abuse of cheddar cheese during shipping has not been associated with foodborne illness (Bernard et al. 1997:349).

The HACCP program described above assumes that the standard prerequisites have been implemented plant-wide.[5] These GMPs can be divided into nine general areas: pasteurization, post-pasteurization contamination, cross-connections, use of returned product and reclaiming operations, airborne contamination, plant environment, plant traffic, personal cleanliness, and sampling and testing. Under the cheddar cheese HACCP model proposed by Bernard et al. (1997) the following plant-wide programs should be routinely audited:

- plant layout, product flow and employee traffic patterns which minimize cross-contamination from raw material to post-pasteurization areas;
- potable water supply;
- pest control program;
- cleaning and sanitation standard operating procedures (SOPs);
- preventive maintenance program and SOPs for operating, maintaining, and calibrating equipment;
- recall procedure, including traceability of raw materials to supplier lots, coding of finished product, and traceability through distribution;
- SOPs for receiving and storing ingredients used in this product, including temperature control for bulk storage of milk;
- purchase specifications and letters of guarantee for compliance with regulatory requirements;
- antibiotic testing program for incoming raw milk for regulatory compliance;
- SOPs for shipping/distribution of product, including preventing cross-contamination due to backhauling; and
- training programs in general hygienic practice and HACCP implementation.

Pasteurization, as a CCP, obviously plays a critical role in the control of pathogens. A dairy plants should have an established pasteurization monitoring system. This system should be set so that it meets the "basic pasteurization principle" which states that every particle of milk or milk product be heated to at least a minimum temperature and held at that temperature for at least the specified time in properly designed and operated equipment. IDFA (1996) contains a listing of times and temperatures for milk to be used in alternative dairy products.

Staff in the Wisconsin Center for Dairy Research participated in a recent FDA pilot HACCP program conducted at a large cheese manufacturer located in central Wisconsin (FDA 1997). The participating plant formulated a HACCP program similar to that shown in Figure 2 for their American-style cheese production. In terms of costs, management noted that most were associated with plant personnel, given that implementation required considerable amounts of time as did daily verification procedures. Initial planning and setup of their HACCP program required slightly less than 800 man-hours with a cost of less than $20,000. Over the course of a year, management estimates that HACCP related validation and verification activities will require 600-720 hours with a cost of $14,000-$18,000. The primary benefits of implementing their HACCP program were identified as a general refocusing of the company on food safety; enhanced employee ownership and participation in the operations and attention to food safety; and, finally, enhanced employee ability to respond to process deviations in a timely manner (FDA 1997, 8).

As noted above, one of the indirect benefits of implementing a HACCP program is for the plant to operationalize an effective set of prerequisite programs. The participating cheese plant made a number of improvements to their prerequisites in order to reduce the number of CCPs. These included installing a computer system to monitor automatic cleaning systems, designating a specific individual to oversee rodent and pest control, improving their preventive maintenance program and documentation of such maintenance, improving their documentation system for manual cleaning and sanitizing operations, and developing an expanded operations manual covering required sanitation practices (FDA 1997, 15).

## Prospects for the Future of HACCP and the Dairy Industry

The dairy industry has learned a very valuable lesson from the seafood and meat industries and, as such, taken the initiative to develop their own HACCP/Dairy Product Safety system. The IDFA program has been officially endorsed by the FDA, and is widely used by the dairy manufacturing sector. If the dairy industry continues to implement HACCP principles at a fairly rapid pace, there should be little need for having HACCP regulations enacted by federal and state agencies. If rate of adoption of HACCP by the manufacturers/processors and producers does not continue to increase, HACCP may become a mandatory regulatory tool.

As with most food industry applications of HACCP, an important means of controlling potential hazards in dairy product manufacturing is the use of raw ingredients and materials that are hazard free. If these raw ingredients are hazard free, this implies that the only hazards that need to be controlled by a CCP are those that arise within the dairy operation itself (FDA 1997, 16). The importance of raw milk in the manufacturing process points to the necessity of having some type of HACCP-based system implemented at the farm level.

Besides the avoidance of a mandated program, what are the benefits to dairy farm producers and manufactures expected from HACCP adoption? For dairy farm operators there is a direct impact given that in the U.S., the price received for their product, raw milk, is determined in part by its quality, e.g., somatic cell count. The lower the somatic cell count, the greater the quality premium (or lower the somatic cell count deduction). Another obvious benefit for farm operators are reduced rejection rates of milk contaminated with drug residues. A third benefit is that with good herd health management, time required to care for sick animals is significantly reduced. For the dairy processor, HACCP adoption means the milk going into the cheese vat, dryer, ice cream premix, etc., is of a consistent quality enabling a better prediction of final product quality and lower rejection rates of final product. With less down time due to poor milk/product quality, plant efficiency should improve. Early HACCP adopters will have a competitive advantage over other manufacturers given a better ability to respond to increased consumer demand for quality assurance vis-a-vis nonadopters. As with other food processors, a fourth benefit from HACCP adoption is a decrease in the likelihood of being involved in legal disputes concerning foodborne diseases. Finally, U.S. dairy manufactures are increasingly recognizing that the major growth area for manufactured dairy products is the export market. Currently the U.S. plays a minor role in world dairy trade. Given that many countries have strict quality requirements for imported foods, having an industry-wide adoption of HACCP principles, makes it much easier to enter these foreign markets.

One of the biggest uncertainties concerning the adoption of HACCP principles by the dairy industry are the costs of such a program at the farm and industry level. Given the relative newness of the industry's HACCP program, little information has been collected concerning program costs. At the Wisconsin Center for Dairy Research, we have an evolving program of research and outreach activities in the area of dairy food safety and quality. As this program continues to grow, we will be undertaking a number of research projects which hopefully answers the question as to the cost of implementing a "cow-to-consumer" HACCP system for the Wisconsin as well as U.S. dairy industry.

### Notes

[1]B.W. Gould is Senior Scientist, Wisconsin Center for Dairy Research and Department of Agricultural and Applied Economics. M. Smukowski is Safety Outreach Specialist with the Wisconsin Center for Dairy Research, and J. Russell Bishop is Director, Wisconsin Center for Dairy Research and Professor in the Department of Food Science. All are located at the University of Wisconsin-

Madison. Financial support of the Wisconsin Milk Marketing Board and the College of Agricultural and Life Sciences, University of Wisconsin-Madison is gratefully acknowledged.

[2]More detail concerning the microbiological safety of cheese and other dairy products can be found in Johnson, Nelson, and Johnson (1990 a,b,c).

[3]In March, 1998 the presence of L. monocytogenes resulted in major recalls of several types of Hispanic style cheese in Illinois and California. In the Illinois case, a Queso Fresco cheese that was distributed nationally was recalled. The California recall mainly affected the Los Angeles area. No cases of illness have been reported (Cheese Reporter 1998).

[4]Much of this discussion is based on IDFA (1996), pp.104-113.

[5]Much of the following discussion is obtained from IDFA 1996:15-43. These GMPs are based on the sanitation principles laid out in the Pasteurized Milk Ordinance, the Code of Federal Regulations (Title 21) and the IDFA's Dairy Product Safety Manual.

## References

Anonymous. 1994a. Two Clusters of Haemolytic Uraemic Syndrome in France. Communicable Diseases Report, 4(7), 29.

Anonymous. 1994b. E.coli O157 Phage Type 28 Infections in Grampian. Communicable Diseases and Environmental Health: Scotland, Weekly Report# 28 (No. 94/46):1.

Anonymous. 1995. Brucellosis Associated with Unpasteurized Milk Products Abroad. Communicable Disease Report 5(32):151.

Anonymous. 1997a. Salmonella Gold-Coast and Cheddar Cheese: Update. Communicable Diseases Report, 7 (11):93-96.

Anonymous. 1997c. Verocytotoxin Producing Escherichia coli O157. Communicable Disease Report, 7(46):409-412.

Aureli, P., G. Franciosa, and M. Poursham. 1996. Foodborne Botulism in Italy. Lancet 348:1594.

Beagley, J. 1998. Personal Communication, New Zealand Dairy Board.

Bernard, D.T., W.R. Cole, D.E. Gmbas, M. Pierson, R. Savage, R.B.Tompkin and R.P. Wooden. 1997. Developing HACCP Plans: Overview of Examples for Teaching, Dairy, Food and Environmental Sanitation, 17(6):338-351.

Bille, J. 1990. Epidemiology of Human Listeriosis in Europe with Special Reference to the Swiss Outbreak. in Foodborne Listeriosis. eds A.J. Miller, J.L. Smith, and G.A. Somkuti, 71-74. Amsterdam: Elsevier.

Bone, F.J., Bogie, D. and S.C. Morgan-Jones. 1989. Staphylococcal Food Poisoning from Sheep Milk Cheese. Epidemiology and Infection, 103:449-458.

Bryan, F.L. 1983. Epidemiology of Milk-Borne Diseases. Journal of Food Protection 46:637-649. The Cheese Reporter. 1998. Listeria Prompts Recalls of Mexican-Style Cheeses by Two Companies, March 27.

Cullor, J.S. 1995a. Common Pathogens that Cause Foodborne Disease: Can They be Controlled on the Dairy. Veterinary Medicine, 90(2):185-194.

Cullor, J.S. 1995b. Implementing the HACCP Program on Your Clients' Dairies, Veterinary Medicine, 90(3):290-295.

D'Aoust, J-Y. (1994) Salmonella and International Trade. International Journal of Food Microbiology 24:11-31.

D'Aoust, J-Y., D.W. Warburton, and A.M. Sewell. 1985. Salmonella Typhimurium phage Rype 10 from Cheddar Cheese Implicated in a Major Canadian Foodborne Outbreak. Journal of Food Protection 48:1062-1066.

Desenclos, J.C. et al. 1996. Large Outbreak of Salmonella Enterica Serotype Paratyphi B Infection Caused by Goats' Milk Cheese, France: A Case Finding and Epidemiological Study. British Medical Journal 312:91-94.

Djuretic, T, P.G. Wall and G. Nichols. 1997. Large Outbreak of Salmonella Enterica Serotype Paratyphi B Infection Caused by Goats' Milk Cheese, France; A Case Finding and Epidemiological Study. Communicable Diseases Report 7(3):41-45.

Food and Drug Administration. 1997. Hazard Analysis and Critical Control Point (HACCP) Pilot Program for Selected Food Manufacturers: Second Interim Report of Observations and Comments, Center for Food Safety and Applied Nutrition, Washington D.C., October 31.

Gösta, B. 1996. Dairy Processing Handbook. Tetra-Pak Processing Systems, Lund Sweden.

Goulet, V. et al. 1995. Listeriosis From Consumption of Raw-Milk Cheese. Lancet 345:1581-1582.

Health Canada. 1998. Infectious Diseases News Brief. Division of Disease Surveillance, Health Canada, Ottawa, April 24.

Hedberg, C.W. et al. 1992. A Multistate Outbreak of Salmonella Javiana and Salmonella Oranienberg Infections Due to Contaminated Cheese. Journal of the American Medical Association 268:3203-3207.

Hennessy, T.W., et al. 1996. A National Outbreak of Salmonella Enteritidis Infections from Ice Cream, The New England Journal of Medicine 334(20):1281-1286.

Institute of Food Science and Technology. 1998. Current Hot Topics: Food Safety and Cheese, unpublished manuscript April.

International Dairy Foods Association. 1996. Dairy Product Safety System, Washington D.C.

Johnson, E., J.H. Nelson, and M.E. Johnson. 1990a. Microbiological Safety of Cheese Made from Heat Treated Milk, Part I: Executive Summary, Introduction and History. Journal of Food Protection, 50(5):441-452.

Johnson, E., J.H. Nelson, and M.E. Johnson. 1990b. Microbiological Safety of Cheese Made from Heat Treated Milk, Part II: Microbiology. Journal of Food Protection, 50(6):519-540.

Johnson, E., J.H. Nelson, and M.E. Johnson. 1990c. Microbiological Safety of Cheese Made from Heat Treated Milk, Part III: Technology, Discussion, Recommendations, Bibliography. Journal of Food Protection, 50(5):441-452.

Linnan et al. 1988. Epidemic Listeriosis Associated With Mexican-Style Cheese. New England Journal of Medicine 319: 823-828.

MacDonald, K.L. et al. 1985. A Multistate Outbreak of Gastrointestinal Illness Caused by Enterotoxigenic Escherichia coli in Imported Semisoft Cheese. Journal of Infectious Diseases 151:716-720.

Maguire, H.C.F. et al. 1991. A Large Outbreak of Food Poisoning of Unknown Aetiology Associated With Stilton Cheese. Epidemiology and Infection 106:497-505.

Maguire et al. 1992. An Outbreak of Salmonella Dublin Infection in England and Wales Associated With a Soft, Unpasteurized Cow's Milk Cheese. Epidemiology and Infection 109:389-396.

Marth, E. 1981. Assuring the Quality of Milk. Journal of Dairy Science. 64:1017-1022.

Marth, E. 1985. Pathogens in Milk and Milk Products. in Standard Methods for the Examination of Dairy Products, 15th edition. G.H. Richardson (eds)., American Public Health Association, Washington D.C.

New South Wales Dairy Corporation, 1998. Quality Plus Assurance System, unpublished document.

New Zealand Dairy Board, 1998. The New Zealand Dairy Industry, unpublished document, Wellington.

New Zealand Ministry of Agriculture, 1998. Standard MRD-Stan 14, Rev. 1, Wellington.

Nooitgedagt, A.J. and B.J. Hartog. 1988. A Survey of the Microbiological Quality of Brie and Camembert Cheese. Netherlands Milk and Dairy Journal 42, 57-72.

Pearson, L.J. and E.H. Marth. 1990. Listeria monocytogenes-Threat to a Safe Food Supply: A Review, Journal of Dairy Science 73:912-928.

Peta, C. and K. Kailasapathy. 1995. HACCP—Its Role in Dairy Factories and the Tangible Benefits Gained Through its Implementation. Australian Journal of Dairy Technology 50(11):74-78.

Robinson, R. 1997. HACCP Partners with Microbiological Testing. Dairy Field :33-35, June.

Sadik, C. et al. 1986. An Epidemiological Investigation Following an Infection by Salmonella typhimurium Due to the Ingestion of Cheese Made from Raw Milk. In Proceedings of the 2nd World Congress on Foodborne Infections and Intoxications 1:280-282, Berlin.

Smukowski, M. and N. Brusk. 1997. Dairy Products Industry. in Encyclopedia of Occupation Health and Safety, Vol 3., 4th ed., International Labor Office, Geneva.

United States Department of Agriculture. 1998. Dairy Products: Annual Report, April. Washington, D.C.

Vaillant, V., S. Haeghebaert, J.C. Desenclos, et al. 1996. Outbreak of Salmonella Dublin Infection in France, November-December 1995. Eurosurveillance 1(2):9-10.

Chapter 23

# Costs to Upgrade the Bangladesh Frozen Shrimp Processing Sector to Adequate Technical and Sanitary Standards and to Maintain a HACCP Program[1]

James C. Cato[2] and Carlos A. Lima dos Santos[3]

## Introduction

Bangladesh had a per capita Gross National Product (GNP) of US$230 million and a population of 124 million in 1996. The GNP increased 5.7 percent in Fiscal Year 1997, compared to 5.3 percent the previous year. Agriculture (including livestock, forestry, and fisheries) grew at 5.3 percent in 1997 after years of stagnation or negative growth. Bangladesh export growth was 14 percent in fiscal year 1997, with frozen shrimp and fish the fourth leading export item at 7.3 percent of the total. Fisheries contribute 9.8 percent of the portion of the GNP that is contributed by agriculture. Fisheries play a major role in nutrition, employment, and foreign exchange earning. About 60 percent of animal protein is supplied by fish and 1.2 million people are directly employed, with an additional 11 million people indirectly employed, in fisheries (Konuma 1998).

Bangladesh must continue to develop its fisheries and aquaculture resources and use shrimp and fish not only as a domestic food source, but as a contribution to export growth development. This mandates a focus on developing sustainable fisheries and aquaculture practices, and on producing safe export products which are competitive in the world's seafood markets. Safety and quality problems during recent years with frozen shrimp exported from Bangladesh have affected shrimp export markets. This paper provides estimates of the current and anticipated expenditures by the Bangladesh frozen shrimp sector to: 1) implement improved seafood safety procedures by upgrading processing plant and government facilities and personnel training and 2) maintain a seafood Hazard Analysis Critical Control Point (HACCP) program in Bangladesh for both industry and government.

# Frozen Shrimp Trade

**World-Wide**

Frozen shrimp and prawns are an important commodity in international commerce. Between 1987 and 1994, the volume of frozen shrimp and prawns imported on a world-wide basis increased 46 percent from 728,931 metric tons to 1,067,787 metric tons. The amount imported in 1995 and 1996 was slightly lower, at 1,002,198 and 1,019,741 metric tons, respectively. Total value of frozen shrimp imports peaked in 1995 at US$9.081 billion, or a 66 percent increase over 1987. 1996 imports were worth US$8.462 billion. Japan, the European Union, and United States frozen shrimp imports represent from 82 to 86 percent of the world total between 1987 and 1996. Japan and the European Union import approximately the same amount (about 30 percent), and the United States about 25 percent. Japan was the leader by a small amount from 1987 to 1990, but switched places as leader with the European Union from 1991 through 1996.

These three regions imported 83 percent of the world volume in 1996 as follows: European Union (32 percent); Japan (28 percent), and the United States (23 percent). These three markets represent about 90 percent of the world value in imported shrimp. Japan is the leader, with its purchases ranging from 43 percent of the world total in 1987 to 37 percent in 1996. The trend has been a slight decline in Japanese imported value. The United States value of imports in 1996 was 26 percent of the world total followed by the European Union at 25 percent. Value per kilogram imported is highest for Japan, followed by the United States and the European Union. Since 1987, the highest value per kilogram for all three regions occurred in 1995, with declines in value in 1996. Values per kilogram (1995; 1996) for each region are: Japan (US$12.07; US$10.83); the United States (US$9.85; US$9.57); European Union (US$7.16; US$6.63).[4]

Comparable data for 1997 across all three regions are not yet available, but preliminary data indicate that 1997 was a record year for shrimp prices due to limited supplies from aquaculture and a very strong United States market due to the strength of the United States economy and the dollar. United States imports expanded by 10 percent in 1997, over-taking Japan for the first time as the world's major shrimp market. Japanese imports fell by 7 percent in 1997 to 267,000 metric tons, the lowest figure in nine years, partly due to the weakened value of the yen (FAO/GLOBEFISH 1998).

**Bangladesh**

Asia[5] is the major frozen shrimp and prawns exporting region of the world. This region accounted for 55 to 62 percent of the volume of frozen shrimp exports world-wide from 1987 to 1995 and 55 to 64 percent of the value of frozen shrimp exports for the same period. For 1995, Thailand was by far the leading frozen shrimp exporter followed by Indonesia, India, China Mainland, Viet Nam, and the Philippines. Bangladesh frozen shrimp and prawn exports represent 3.8 to 4.5 percent of the Asia volume and 3.5 to 4.8 percent of Asia value from 1987 to 1995. On a world-wide basis, Bangladesh frozen shrimp and prawns exports represent 2.2 to 3.0 percent of the volume and 2.1 to 2.7 percent of the value. Average annual value per kilogram for Bangladesh exports has been below the average for the Asia region in seven of the nine years from 1987 to 1995. The only two years when Bangladesh value per kilogram was higher were 1988 and 1989. Since 1990, Bangladesh value per kilogram received has been from 7 to 16 percent lower than the Asia region.

Bangladesh's low percentage of the market, lower-valued product, and negative safety and quality reputation make it a likely price-taker instead of being a price-setter. One way Bangladesh can improve its own export position is to improve the safety and quality of its shrimp exports since the Bangladesh industry and government controls and can directly improve that particular element of its processing industry.

Bangladesh depends mainly on its inland fishery resources including aquaculture for domestic consumption and exports. Cultured shrimp is estimated to be about four times that of captured shrimp in export quantity. Shrimp represent about 90 percent of the value of Bangladesh marine product exports. Shrimp and prawns exported from Bangladesh are almost entirely in block frozen form. Shell-on packing is two kilogram or four pound cartons depending on the market. Freshwater prawns range in counts from under 5 to 51/60; cultured black tiger shrimp from under 5 to 61/70; and white shrimp from under 5 to 131/150. Other products are peeled and deveined, including mixed species which range in count size usually up to 131/150. Peeled and undeveined shrimp are packed in cartons with mixed species usually up to 300/500 count. Master cartons are 50 pound, 20 kilogram, or 18 kilogram. Private processing plants target principally the export market while government plants pack for both domestic consumption and export.

Bangladesh has increasingly relied on three major markets for frozen shrimp exports. The European Union has been the leading importer of Bangladesh shrimp from 1989 to 1991, and from 1994 through 1996, the most recent year for which comparable data are available. The European Union accounted for 34 percent, 40 percent, and 50 percent of total Bangladesh frozen shrimp exports these three years. The United States was slightly ahead in 1992 and 1993, second to the European Union all other years, and imported 38 percent, 32 percent, and 23 percent of all Bangladesh shrimp from 1994 to 1996. Japan has gradually increased its share, with 18 percent, 15 percent, and 26 percent from 1994 to 1996. Total value of Bangladesh frozen shrimp imports into these three markets reached a maximum of US$287.6 million in 1996 with the United States accounting for US$109.6 million, the European Union US$108.8 million, and Japan US$69.2 million. From 1989 to 1993, the value per kilogram of imports into the three markets was about the same. Some divergence occurred in 1994, with import values per kilogram for the United States and Japan increasing at a faster rate than the European Union. Japan prices fell to European Union levels in 1996, and even further in 1997, to 32 percent below that of the United States (Table 1).

## Safety and Quality Problems

Shrimp processed for the world market must be produced to meet minimum international standards. Standards followed should be consistent with those specified by the Codex Alimentarius, Commission (Codex 1995; Codex 1978 and subsequent revisions). The product must also meet buyer specifications and be produced to comply with regulatory requirements of the importing country. Consistently meeting minimum standards and buyer and importing country regulations also creates a "good" reputation for products from the exporting country.

Bangladesh has a reputation for producing seafood that sometimes does not meet the required standards of safety and quality. As in other developing countries, major constraints in Bangladesh include a lack of sufficient funds to invest in expensive mechanical equipment, fishing boats, pond grow-out facilities, buildings, and trained personnel. Insufficient and irregular supplies of electricity, high quality water and ice, and poor transportation facilities also hinder the use of modem sanitary practices.

Major safety problems begin mainly in "pre-processing operations," including the handling of raw shrimp (sorting by size and color, removal of heads, or peeling) in small plants, sheds, houses, or available open

### TABLE 1

**Average Value of Frozen Shrimp Imported into the United States, Japan, and the European Union from Bangladesh, 1989-1996**

| Importing Region | Years | | | | | | | | |
|---|---|---|---|---|---|---|---|---|---|
| | 1989 | 1990 | 1991 | 1992 | 1993 | 1994 | 1995 | 1996 | 1997 |
| | US$ Per Kilogram | | | | | | | | |
| United States | 7.69 | 8.45 | 7.48 | 8.08 | 8.32 | 11.32 | 13.27 | 11.87 | 13.56 |
| Japan | 7.53 | 8.62 | 7.70 | 8.22 | 8.24 | 10.42 | 13.75 | 10.82 | 9.21 |
| European Union | 6.90 | 6.70 | 7.97 | 7.84 | 8.34 | 9.93 | 10.95 | 10.56 | n/a |

spaces, often under conditions and in facilities unsuitable for food handling. "Pre-processed" raw shrimp is the raw material for industrial processing plants. Additional problems incurred during the actual processing at the plant level also often contribute to the final safety and quality of processed shrimp traded in world markets.

By the end of the 1970s, the Bangladesh seafood processing industry began to expand rapidly, but the use of new technology, sanitary facilities and processes, and trained manpower at the worker and management level did not match the rapid growth. As a result, the export of frozen seafood suffered considerable loss by rejection from 1975 to 1978, and seafood imports from Bangladesh were placed under automatic detention by the United States Food and Drug Administration (USFDA) (Ahmed 1998). International Aid Agencies, including the Food and Agriculture Organization of the United Nations (FAO) have recognized the potential for the expansion of the Bangladesh seafood industry for a number of years. FAO has also recognized the need for improvements in Bangladesh seafood safety and quality, with missions there in the early 1980s to assist Bangladesh in the development of national fishery product standards and regulations and fish inspection schemes. In 1983 the Bangladesh Government created a Fish and Fish Product Ordinance (Inspection and Quality Control), and in 1985 the quality control laboratory was updated and additional personnel hired (Ahmed 1998). Additional investments and improvements were made in 1997 and 1998 and are discussed later in this chapter.

More recently, FAO has initiated a project in Bangladesh to assist the development of the seafood industry. The project began in April 1996, and focuses directly on the preparation of a HACCP-based fish safety and quality[6] assurance program for shrimp and fish plants in Bangladesh. This project is designed to train key persons in the private and public sectors to design and implement HACCP-based safety assurance programs, to assist the government and shrimp industry in developing the HACCP concept, and to inform Bangladesh industry and government personnel about the new sanitary and quality requirements of major importing countries, particularly those that relate to the application of HACCP principles (FAO 1995). The project covers a wide range of aspects, from regulatory review (aiming at achieving the necessary "equivalence" with the rules-in-force in the chief importing markets, i.e., United States, European Union), up to the physical structure, finishing, and lay-out of seafood processing establishments. It gives a strong emphasis to the introduction of a HACCP-based preventive approach at industry and government levels

A parallel project carried out by INFOFISH with Common Fund for Commodities (CFC) funding focuses on the export promotion of value-added products and their sustainable development. This project began in mid-1996 and includes such activities as market feasibility studies to identify potential exporters, cost analyses, visits to target markets, pilot production, assessing consumer acceptability, exporting value-added products and monitoring the export activity, providing production manuals and industry training, and organizing investment seminars. A final objective includes improving the quality and sustainability of the processed products (INFOFISH 1996).

## Shrimp Safety and Quality in Major Markets

These assistance programs to Bangladesh are extremely timely. Bangladesh frozen shrimp exporters continue to have both real and perceived problems with buyers in the United States, the European Union, and Japan, concerning the safety and quality of its products. Bangladesh also continues to experience major problems in the reputation of its products with government import inspection programs in the United States and the European Union.

**Industry**

From the perspective of industry buyers, previous problems have been experienced with supply irregularity including poor delivery times and inability to execute orders in time, under-sizing, inclusion of mixed sizes, poor quality such as black spot, and under-weight shipments. Some buyers no longer purchase Bangladesh frozen shrimp imports except from time-- tested reliable suppliers. Buyers in the United States indicated that some shipments from Bangladesh were filthy and decomposed, product sourcing was inconsistent, Bangladesh products brought lower retail prices than those from some other countries, and insurance costs for the product were higher due to the refusal of some shrimp processors to take back product in quality disputes. Buyers in Japan also reported quality problems with previously imported shrimp from Bangladesh (CFC/INFOFISH 1997).

As part of the value-added INFOFISH project, visits by Bangladesh shrimp processors to Japan, the European Union, and the United States were sponsored in February and March of 1997. Potential buyers in Germany, France, and Belgium were interested in Bangladesh shrimp products as long as the processors could assure a constant supply and satisfy quality requirements. Successful initial exports of added-value shrimp products resulted by the end of 1997, particularly to Japan.

**Government Import Inspection Programs**

Bangladesh seafood exporters also experienced major problems during 1997 with seafood safety assurance and inspection programs in the European Union and the United States for Bangladesh frozen shrimp. These are summarized below.

**European Union.** On July 30, 1997, the European Commission banned imports of fishery products from Bangladesh into the European Union

(EEC 1997) as the result of European Commission inspections of seafood processing plants in Bangladesh. The concern resulted from serious deficiencies in the infrastructure and hygiene in processing establishments, and because there were not enough guarantees of the efficiency of the controls carried out by the competent authorities (Bangladesh government inspectors). The European Commission determined that this resulted in a potentially high risk for public health with regard to the importing and consumption of fishery products processed in Bangladesh.

Subsequent inspections and decisions recognized the Bangladesh Department of Fisheries, Fish Inspection and Quality Control, Ministry of Fisheries and Livestock as the competent authority in Bangladesh, indicated that Bangladesh quality assurance legislation was equivalent to that of the European Union and subject to certain provisions, and lifted the ban on seafood product imports from Bangladesh for six approved establishments for products prepared and processed after December 31, 1997 (EEC 1998). Other processing plants are now being inspected for approval.

**United States.** The United States Food and Drug Administration (USFDA) enforces United States laws which mandate that imported goods meet the same standards as domestic goods. Imported goods must be pure, wholesome, safe to eat, and produced under sanitary conditions. Products may be detained when imported into the United States without physical examination based on past history and/or other information indicating the product may be violative. Once a product is placed on automatic detention, normal entry may not resume until the shipper or importer proves that the product meets USFDA standards. During 1997, a total of 143 shipments of frozen shrimp from Bangladesh into the United States were automatically detained without examination. Eighty-seven percent of the shipments were detained between September and December, when most Bangladesh shrimp are imported. The reason for the 143 detentions were as follows: filthy, *Salmonella* (53); soaked, wet, filthy, *Salmonella* (28); *Salmonella*, decomposed (21); *Salmonella* (14); filthy, insanitary manufacturing, processing or packing (8); filthy (8); soaked, wet, *Salmonella* (5); filthy, *Listeria*, insanitary manufacturing, processing or packing (3); soaked, wet, filthy (1); filthy, label not bearing nutrition information (1); soaked, wet (1).[7] Automatic detention does not mean that the product is rejected. It means that each shipment is inspected and then allowed into the United States if the product is safe and meets minimum safety standards.

**Japan.** In contrast to the European Union and United States, government authorities in Japan have not recorded recent problems at time of import entry with shrimp from Bangladesh, although industry representatives have expressed concern as noted above. For the three-year period 1995 to 1997, no violations of Japan's food sanitation laws were recorded for shrimp imported from Bangladesh (Toyofuku 1998).

## Bangladesh Shrimp Sector Survey

HACCP cost data for this paper were derived from primary data collected through a three-page questionnaire faxed to Bangladesh frozen shrimp processing plants and the Bangladesh Department of Fisheries. Based on FAO records from training programs and personal visits to Bangladesh, and from the INFOFISH data base, a list of 86 seafood firms was compiled for Bangladesh. Fifty-one firms were identified as known or probable processors of frozen shrimp. A three-page questionnaire was faxed to these 51 firms located principally in Khulna and Chittagong, with a few to other locations, from FAO headquarters in Rome, Italy, during the first week of April 1998. Of this total, 36 faxes actually connected to a receiving fax machine, with 15 failing due to the fax number no longer existing. Of the 36 firms, 19 returned by fax a completed survey form to FAO by 22 April 1998. These 19 firms represent 30 percent of the volume and 41 percent of the value of Bangladesh frozen shrimp exports as determined by comparing 1997 reported volume and value of the 19[8] plants with average total frozen shrimp export secondary data from Bangladesh for 1993 to 1995. The Bangladesh Department of Fisheries, Fish Inspection and Quality Control, also returned by fax a separate questionnaire designed to determine the cost to the Bangladesh government of upgrading government laboratories, facilities and personnel to adequate technical and sanitary standards, and to implement HACCP as the principal monitoring program for processed shrimp safety.

Distribution by size of the 19 firms was representative across all segments of the industry. The 18 firms reporting volume (in metric tons) of frozen shrimp exports in 1997 was as follows: 0-99 (1); 100-199 (5); 200-299 (3); 300-399 (2); 400-499 (1); 500-599 (0); 600-699 (3); 700-799 (2); 800-899 (0); 900-999 (1). These 18 plants were principally shrimp processors, with frozen shrimp representing 90 percent of all seafood volume processed. Average volume and value per plant of frozen shrimp processed in 1997 was 388,135 kilograms valued at US$4,123,366.

## Status and Costs of HACCP Programs

### The "Cost" of HACCP

HACCP is based on seven principles: 1) conduct hazard analysis and identify preventative measures; 2) identify critical control points (CCPs); 3) establish critical limits; 4) monitor each CCP; 5) establish corrective action to be undertaken when a critical limit deviation occurs; 6) establish a record keeping system; and 7) establish verification procedures. When determining the "cost" of implementing HACCP in a seafood processing plant, it is assumed that the plant has in place the minimum technical and sanitary standards, follows standard sanitary operating principles, and that the plant can comply with buyer specifications and importing country safety and sanitary regulations. The plant's HACCP plan is then designed to reduce the probability of producing unsafe seafood.

In the case of Bangladesh, shrimp processing plants and government laboratories and inspection programs have not met minimum technical and sanitary standards until recently. This has made it necessary for both processing plant owners and the government to make substantial investments in facilities and training to upgrade to minimum standards. Only then can a HACCP plan be useful and effective. From the viewpoint of the industry and government in Bangladesh, these "upgrading" costs are viewed as a part of the costs of developing a HACCP program, since the investments must be made "to implement HACCP," as required by importing country regulations. In the case of processing plants in a country where plants already meet minimum standards, no costs would be incurred to upgrade, and thus these costs would not be considered a "cost" of HACCP. In the later case, and from the processing plant owner perspective, where adequate technical and sanitary standards already exist, the cost of HACCP would include only such items as the cost of training, HACCP plan refinement, sanitation audits, cost of implementing CCPs, equipment cleaning, record review, eliminating pests, administration costs, and rejected product, etc. From a government perspective, costs would include such items as training, addition of employees, and additional administration.

For shrimp processors and the government in Bangladesh, it has been necessary to make major investments to upgrade facilities and personnel to minimum technical and sanitary standards. These are referred to as "costs of upgrading." The cost to maintain a HACCP plan after upgrading most closely approximates the true "cost of HACCP" to the Bangladesh shrimp processing and government sectors. These costs are referred to in this paper as the cost of HACCP plan maintenance.

**Shrimp Processing Plants**

Seafood processing plants in Bangladesh have made serious attempts to implement HACCP plans. Out of 120 plants (not all process shrimp), the Bangladesh government indicates that 48 percent have been remodeled, spending from US$100,000 to US$220,000 each, 34 percent have prepared HACCP-based quality assurance manuals, and 5 percent have implemented HACCP principles since mid-1997. Fifty-eight plants have been primarily approved by the Bangladesh Department of Fisheries as meeting minimum quality standards, and six have been listed by the

European Commission to export products to the European Union (Ahmed 1998; EEC 1998).

Results of industry efforts are reflected through rapid changes which have been observed in the physical conditions of shrimp processing plants since July 1997.[9] Today most of the problems reported by European Commission inspectors in terms of plant and personnel hygiene, processing plant facilities, lay-out, fabrics, and finishing are solved. At the same time, industry is trying to solve problems linked to raw materials and all operations up to the factory deck. New strategies to eliminate unnecessary steps and to upgrade raw material quality are being attempted. The main change being implemented is the incorporation of de-heading and peeling operations in the exporter processing line to avoid pre-processing under unsatisfactory hygienic conditions.

Survey data from the 19 plants provided excellent cost information on the status of upgrading to meet adequate standards and of HACCP plan implementation. Eighteen of the 19 plants have HACCP plans in place, with initiation of the plans ranging from August 1997 to March 1998. The other plant is 50 percent complete with the plan, with a goal of August 1998 for completion. As of April 1998, the average plant had spent US$239,630 to upgrade the plant to adequate technical and sanitary standards. An additional US$37,525 in expenditures is anticipated per plant to complete the upgrade, for a total of US$277,155 per plant to be fully in compliance with minimum (basic) technical and sanitary standards. This compares reasonably with the US$100,000 to US$220,000 per plant referenced above from the report of the Government of Bangladesh. Of the total expenditures, plant repair and modifications represented 70 percent, lab installation was 13 percent, and added equipment cleaning was 10 percent. The average plant expects to spend US$34,875 each year to maintain the HACCP plan after the plant is upgraded and the HACCP plan is in effect (Table 2).

The total cost of industry upgrading, including some HACCP implementation for the 19 plants is US$5,265,945. These plants represent 30 percent of the industry volume. Assuming an identical distribution of plants in size and characteristics across the remaining 70 percent of the industry, the total one-time cost to the Bangladesh frozen shrimp processing industry to upgrade is US$17,553,150. The annual cost to maintain the HACCP plan per plant is US$34,875, or US$662,625 for the 19 plants, resulting in an industry-wide total cost of US$2,208,750.

For the 18 plants reporting volume and value data, the average value received in 1997 for exported shrimp was US$10.62 per kilogram. On a per kilogram (per pound) basis, the total one-time cost, mostly during 1997 and early 1998, for upgrading the average plant was US$0.7141 per kilogram (US$0.3239 per pound). This represents 6.72 percent of average price received in 1997. The cost to maintain the HACCP plan in future years represents US$0.0899 per kilogram (US$0.0408 per pound). This represents 0.85 percent of the average price received in 1997. Since the 19 plants represent 30 percent of industry volume and 41 percent of industry value, the responding plants are those that received higher value than the industry average. Thus, remaining and more marginal plants would have higher costs per kilogram of processed volume to implement and maintain HACCP plans.

This should also be considered a maximum estimate for Bangladesh shrimp plants, since cost maintenance estimates for the HACCP plan include plant repair/modification which may be necessary even if a HACCP plan is not in effect. Without plant repair/modification costs included, the lower estimate for annual HACCP plan maintenance is US$0.0327 per kilogram (US$0.0148 per pound), or 0.31 percent of 1997 price. Only one prior study on the cost of HACCP in shrimp plants is available. This study covered plants which processed breaded, cooked, and raw shrimp in the United States, and estimated the cost to maintain HACCP plans to be US$0.0009 per pound (National Fisheries Education

TABLE 2

Cost Per Frozen Shrimp Processing Plant in Bangladesh to Upgrade the Plant to Adequate Technical and Sanitary Standards during 1997 and Early 1998 and Expected Annual Costs in the Future to Maintain the HACCP Plan in U.S. Dollars[a]

| Cost Category | Cost to Upgrade Plant to Adequate Technical Sanitary Standards[b] | | | Expected Annual Cost to Maintain HACCP Plan |
|---|---|---|---|---|
| | Cost to Date | Additional Cost Anticipated | Total | |
| Consultants | 3,615 | 236 | 3,851 | 345 |
| Employee Training | 1,820 | 157 | 1,977 | 335 |
| Sanitation Audits | 6,266 | 779 | 7,045 | 557 |
| Plant Repair/Modifications | 165,920 | 26,966 | 192,886 | 22,175 |
| Added Equipment Cleaning | 25,522 | 3,462 | 28,984 | 2,177 |
| Rejected Product[c] | 2,546 | 322 | 2,868 | 7,256 |
| Lab Installation | 30,854 | 5,425 | 36,279 | 2,030 |
| All Other | 3,087 | 178 | 3,265 | 0 |
| Total | 239,630 | 37,525 | 277,155 | 34,875[d] |

[a]Plant Costs provided in Taka and converted to U.S. Dollars using 1997 average exchange rate of 43.78 Taka per Dollar.

[b]Average Per Plant based on 16 of 19 firms responding to survey.

[c]Due to following the HACCP plan in-plant.

[d]US$12,700 without plant repair/modification costs included.

and Research Foundation Undated). This cost per pound estimate is for plants with high technical and sanitary standards in place and which produce a more valuable per pound product. The cost per pound for the Bangladesh frozen shrimp processing industry of US$0.0408 per pound is much higher as expected due to the processing conditions and lower value product per pound.

A number of other questions asked each plant also provided insight regarding the attitudes of Bangladesh shrimp processors regarding HACCP plans. Comments across the plants surveyed are: 1) 100 percent indicated HACCP will cause shrimp buyers to be more confident in their frozen shrimp products; 2) 35 percent indicated HACCP will lower production costs per kilogram; 3) 67 percent indicated HACCP will lower marketing and sales cost per kilogram of shrimp; 4) 100 percent indicated HACCP will reduce product loss and detention at the importing country; and 5) 94 percent indicated that HACCP will improve the quality of shrimp produced in their plants. Firms were also asked if shrimp buyers from Japan, the United States, and the European Union asked if HACCP plans were in place when initiating the purchase of shrimp. For buyers from Japan, 22 percent always asked if a HACCP plan was in place, 17 percent asked over 50 percent of the time, and 61 percent asked less than 50 percent of the time. For the United States, 84 percent always asked if a HACCP place was in place, 11 percent asked over 50 percent of the time, and 5 percent asked less than 50 percent of the time. For the European Union, 83 percent always asked if a HACCP plan was in place and 17 percent asked more than 50 percent of the time.

**Bangladesh Fisheries Department**

The Department of Fisheries of the People's Republic of Bangladesh has adopted HACCP as the principal program used to monitor the seafood processing industry for safety and quality. HACCP was implemented nation-wide in December 1997. According to the Department of Fisheries, the quality of seafood products has improved and systematic hygienic operational systems are being followed by the seafood processing plants (Ali 1998). The Department of Fisheries has spent US$201,483 through December 1997 to upgrade its laboratories, hire additional employees, and train employees in order to achieve acceptable levels of technical and sanitary capabilities in its monitoring program. The principal costs are for new equipment and laboratories. An estimated US$180,676 will be spent in 1998, principally for new employees and laboratory chemicals and glassware. An estimated US$225,039 will be spent annually each year after, principally for new employee salaries. This annual cost closely approximates the added costs to use HACCP as the principal monitoring program after facilities and personnel are upgraded. Finally, the Fisheries Department proposes that a major investment program of US$14.9 million is needed to fully implement the highest levels of seafood HACCP programs within the government sector in Bangladesh (Table 3).

**Training Costs**

During 1996 and 1997, the FAO project in Bangladesh conducted four workshops on the design and implementation of HACCP-based quality assurance programs, training a total of 100 persons. Subsequently, the Bangladesh Department of Fisheries trained trainers (to qualify them to teach HACCP in-country), and organized five additional workshops

TABLE 3

Cost to Upgrade Government of Bangladesh Seafood Inspection Facilities to Adequate Technical and Sanitary Standards and Hire and Train Personnel, April 1996 to December 1997, Estimated Costs for 1998, and Estimated Annual Costs to Maintain HACCP Monitoring Program in U.S. Dollars[a]

|  | Cost to Upgrade Government Facilities and Personnel to Adequate Technical and Sanitary Standards | | Expected Annual Cost to Maintain HACCP Monitoring Program |
|---|---|---|---|
|  | April 1996 to December 1997 Actual | 1998 Estimated | 1998 Estimated |
| New Equipment | 104,612 | 34,262 | 11,461 |
| New Laboratories | 86,112 | 11,421 | 9,137 |
| Employees |  |  |  |
| New | 7,995[b] | 97,305[c] | 178,174[d] |
| Proposed[e] |  |  | 365,464 |
| Training[f] |  |  |  |
| New Employees | 0 | 3,426 | 6,852 |
| Existing | 1,827 | 2,284 | 2,284 |
| Chemicals and |  |  |  |
| Glassware | 937 | 31,978 | 17,131 |
| Total | 201,483 | 180,676 | 225,039[e] |

[a]Costs provided in Taka and converted to U.S. Dollars using 1997 average exchange rate of 43.78 Taka per Dollar.

[b]Seven employees.

[c]Seven employees + 71 additional employees for six months.

[d]Seventy-eight employees.

[e]The Bangladesh Fisheries Department proposes that a development project is necessary for the successful and sustainable implementation of HACCP in seafood processing. The estimated cost is $14,916,486 of which $8,382,823 would be paid by a donor agency. No donor has been identified to date. Total does not include proposed amount.

[f]Training costs only for in-country training.

training 120 persons from the industry. Fish inspectors were also sent to training activities outside the country.

The FAO project prepared a special training manual on HACCP and training materials on basic HACCP concepts and principles. Guidelines for the design and implementation of HACCP-based programs were developed by the Department of Fisheries in Bengali and distributed to the industry. Some processing plants developed HACCP training programs for their own workers including a training program organized by the industry in Khulna for shrimp producers, brokers, and transporters. The goal was to improve the quality and safety of shrimp raw material. Training materials on HACCP published in the United Kingdom were purchased and distributed to the industry. The FAO project has spent approximately US$72,000 for HACCP training programs in Bangladesh.

**Total Costs**

The total amount spent to upgrade facilities and equipment and to train personnel to achieve acceptable technical and sanitary standards in Bangladesh by industry, government, and external training partners amounts to US$18.0 million to date. The annual cost to maintain seafood HACCP each year after is US$2.4 million (Table 4). Based on the average annual value of industry-wide exports for 1993 to 1995 of US$190.0 million, the investment to date represents 9.4 percent of export sales for one year. To maintain the seafood HACCP program represents 1.26

**TABLE 4**

**Estimated Total Costs to Upgrade Bangladesh Frozen Shrimp Processing Plants and Government Inspection Facilities and Employees to Adequate Technical and Sanitary Standards, 1997, and to Maintain a Seafood HACCP Program in Future Years in U.S. Dollars**

|  | Upgrade Costs Through 1997[a] | Annual Cost to Maintain HACCP Program |
|---|---|---|
| Nation-Wide Industry | 17,553,150 | 2,208,750 |
| Government | 382,159[b] | 225,039[c] |
| External Training Programs | 72,000 | d |
| Total | 18,007,309 | 2,433,789 |

[a]A small amount of costs in early 1998 included.

[b]Includes $201,483 through 1997 and an additional $180,676 for 1998 estimated as "start-up" costs.

[c]These costs are a minimum estimate. Bangladesh is seeking a development program donor for funds that would allow HACCP program maintenance at a higher level per year (US$365,464) for a few years. After that, annual costs would more closely match the amount shown.

[d]Industry and government costs include in-country training programs. No cost estimate available for external training programs.

percent of export sales for one year. This estimate is a maximum estimate for HACCP costs since it includes plant repair/maintenance costs.

## Summary

Bangladesh frozen shrimp and prawns exports represent 2.2 to 3.0 percent of the volume and 2.1 to 2.7 percent of the value of world-wide shrimp exports. Shrimp represent about 90 percent of the value of Bangladesh marine product exports, and are a very important source of foreign currency to the Bangladesh economy. Major markets for Bangladesh shrimp are the European Union, the United States, and Japan, with 1996 sales to these three markets totaling US$288 million. Bangladesh shrimp exports have experienced safety and quality problems including detentions for inspection in the United States and a 1997 ban on imports into the European Union. The Bangladesh shrimp processing industry and the Bangladesh government have made major efforts to improve the quality of frozen shrimp exported from Bangladesh, although there is still room for improvement in some areas. Facilities are being upgraded and personnel trained in order to achieve at least minimum technical and sanitary safety and quality standards. Seafood HACCP programs are being implemented in the shrimp processing plants, and the government has implemented HACCP as the seafood safety program for the nation. Both industry and government have made major investments in more modern plants and laboratories and in personnel trained in HACCP procedures. Through 1997, a total investment of US$18.0 million has been made across all sectors for upgrading, and an additional expenditure of US$2.5 million per year is estimated to be spent in the future to maintain seafood HACCP programs. These annual maintenance costs represent about 1.26 percent of annual Bangladesh export sales value.

**Notes**

[1]Prepared for use by the People's Republic of Bangladesh, the Bangladesh frozen shrimp processing industry and HACCP training programs in Bangladesh, and the importers and consumers of Bangladesh shrimp. Presented during poster session at the "Economics of HACCP" Conference, Washington, D.C., USA, June 15-16, 1998. This paper was written during participation in the Food and Agriculture Organization of the United Nations Partnership Program with Academic Institutions and while affiliated with Florida Sea Grant project M/PM-12. No U.S. federal funds were used in the project. The Florida Sea Grant College Program is supported by an award from the Office of Sea Grant, National Oceanic and Atmospheric Administration, U.S. Department of Commerce, Grant Number NA 76RG-0120, under provisions of the National Sea Grant Programs Act of 1966.

[2]James C. Cato is Professor of Food and Resource Economics and Director of the Florida Sea Grant College Program, University of Florida, Gainesville, Florida, USA.

[3]Carlos A. Lima dos Santos is Senior Fishery Industry Officer, Fish Utilization and Marketing Service, Fisheries Industries Division, Fisheries Department, Food and Agriculture Organization of the United Nations, Rome, Italy.

[4]Value per kilogram data used in this paper represent the total value of frozen shrimp imports divided by the total volume of frozen shrimp imports as reported by the exporting or importing country. Frozen shrimp vary greatly in price depending on the size of the shrimp and product form. For example, in the United States, frozen shrimp data may be reported as shrimp frozen other preparations, peeled frozen, or shell-on frozen in sizes ranging from 15/20 to over 70 count. The purpose of this discussion is to describe the world trade in frozen shrimp, and data were not intended to be developed by product form or count size. Sources are original data from: Japan, Customs Clearance Statistics, Ministry of Finance; United States, National Marine Fisheries Service, United States Department of Commerce; European Union, Fishery Information and Statistics Unit, FAO.

[5]The 16 countries referred to as Asia are: Bangladesh, China Mainland, China Taiwan, Hong Kong, India, Indonesia, Korea Republic, Malaysia, Myanmar, Pakistan, Philippines, Singapore, Thailand, Turkey, United Arab Emirates, and Viet Nam.

[6]The principal focus for seafood HACCP programs is to ensure the safety of seafood. For countries and plants lacking in technical and sanitary standards, the HACCP concept can only be used to improve seafood safety and quality once the plants are capable of achieving adequate sanitary standards. The goal in Bangladesh is to upgrade processed seafood to minimum safety and quality standards, and then to use HACCP to maintain safety and quality of the product.

[7]Data derived from the United States FDA import information program. Data for January to August, 1997, from the Import Detention System (IDS). Data for September, 1997, and later are from the new Operational and Administrative System of Import Support (OASIS). Some reasons for detentions duplicated since some shipments were detained for more than one reason.

[8]Eighteen plants reported volume and value data for 1997. The average volume of the 18 was used as the volume of the remaining plant to determine the percentage of total industry value for the 19 plants.

[9]Based on personal visits to Bangladesh by Carlos A. Lima dos Santos, December 1997 and February 1998.

**References**

Ahmed, Md. Kador. 1998. Unpublished Paper. Present Status of Seafood Quality Control and HACCP Programme in Bangladesh. Industry Seminar on Export Promotion of Value-Added Products and Their Sustainable Development. Chittagong, Bangladesh: Fish Inspection and Quality Control, Department of Fisheries.

Ali, Md. Liaquat. April 15, 1998. Personal Communication via Fax. Director General. Department of Fisheries. Matsya Bhaban, Ramna, Dhaka, Bangladesh.

CFC/INFOFISH. 1997. Progress Report (1 January - 30 June 1997). CFC/INFOFISH Project on Export Promotion of Value-Added Fishery Products and their Sustainable Development. Kuala Lumpur, Malaysia: INFOFISH.

Codex, Codex Alimentarius Commission. 1995. *Codex Alimentarius General Principles of Food Hygiene.* Alinorm 95/13. Rome, Italy: Food and Agriculture Organization of the United Nations.

Codex, Codex Alimentarius Commission. *1978 Codex Alimentarius, Recommended International Code of Practice For Shrimps and Prawns.* Volume B. Second Edition. CAC/RCP-1978. Rome, Italy: Food and Agriculture Organization of the United Nations.

EEC Commission Decision 97/513/EC of 30 July, 1997. 1997. Concerning Certain Protective Measures with Regard to Certain Fishery Products Originating in Bangladesh. *Official Journal of the European Communities.* No. L. 214/46.

EEC Commission Decision 98/147/EC of 13 February 1998. 1998. Laying down Special Conditions Governing Imports of Fishery and Aquaculture Products Originating in Bangladesh. *Official Journal of the European Communities.* No. L. 46/13-17.

FAO. 1995. Preparation of a HACCP-based Fish Quality Assurance Programme for Bangladesh. Rome, Italy: FAO, Technical Co-operation Programme, Fish Utilization and Marketing Service.

FAO/GLOBEFISH. January, 1998. *Highlights.* Rome, Italy: Food and Agriculture Organization of the United Nations.

INFOFISH. 1996. Export Promotion of Value-Added Fishery Products and their Sustainable Development. Annual Work Programme and Budget. Kuala Lampur, Malaysia: INFOFISH.

Konuma, Hiroyuki. 1998. FAO Representative's Annual Report on Bangladesh for 1997. Dhaka, Bangladesh: Food and Agriculture Organization of the United Nations.

National Fisheries Education and Research Foundation, Inc. Undated, About 1991. Economic Impacts of HACCP Models for Breaded, Cooked and Raw Shrimp and Raw Fish. Unpublished document prepared by Kearney/Centaur Division, A. T. Kearney. Alexandria, Virginia. 101 Pages.

Toyofuku, Hojima. 1998. Personal Communication (via email). Director, Institute of Public Health, Ministry of Health and Welfare. Tokyo, Japan.

Chapter 24

# The Economics of HACCP Application in Argentine Fish Products

Aurora Zugarramurdi, M. A. Parin, L. Gadaleta and Hector M. Lupin[1]

## Introduction

In the fish processing industry, two factors have to be simultaneously evaluated: the safety and the quality of fish and fishery products. As a rule, regulations in most countries require that provisions in the Good Manufacturing Practices (GMP) and Sanitation Standard Operating Procedures (SSOPs) manuals must be fulfilled first. This is a necessary condition but not sufficient to assure the safety of seafood; therefore, the application of a preventive program, HACCP, is needed. With the systematical and objective approach of this program, food risks are in fact reduced (Huss 1995).

The concept of quality can be applied to a product when it is innocuous, nutritive and appetizing, and if it satisfies the expectations of the market. The additional consideration of the process technology and competitiveness of the product require the application of economic engineering techniques to the Total Quality systems (Zugarramurdi et al. 1995). The evaluation of the quality cost and its relationship to the level of product quality constitutes a useful tool to make decisions intended to implement systems of quality assurance (Hosking 1984; Porter et al. 1992). Their analysis permits the establishment of the basis to evaluate the effectiveness of the proposed system and to optimize the technological process.

## Objectives

The main aims of this work are:
- to analyze economic, quality and technological parameters; and
- to study the relationships of these parameters to quality costs, total production costs, final product quality, and selling price.

## Materials and Methods

To calculate the investment and production costs, the methodology developed by Zugarramurdi et al. was used (Zugarramurdi 1981; Zugarramurdi and Parin 1988; Parin and Zugarramurdi 1994; Zugarramurdi et al. 1995).

The study of the quality costs was performed using the Prevention, Appraisal, Failure (PAF) model proposed by Feigenbaum (Feigenbaum 1974). The model allows the study of the relation between the three main costs involved in quality:

- *Prevention costs ($C_P$)*. The costs associated with any intended action to investigate, prevent, or reduce defects and failures in all company units. All expenses associated with a preventive action is considered an investment since this is recovered upon achieving a decrease in defective products.
- *Appraisal costs ($C_A$)*. The costs that derive from sampling, inspection, and test actions performed to evaluate if the level of predetermined quality is maintained.
- *Failure costs ($C_F$)*. The costs related to the defects detected in the plant (internal failures), or after the product is delivered (external failures). In fact, these costs are losses, being evaluated against the optimum parameters achieved with ideal conditions.

In Figure 1 a qualitative graph of the PAF model is shown (BSI 1981; BSI 1990). For food processing plants, the values found for low quality

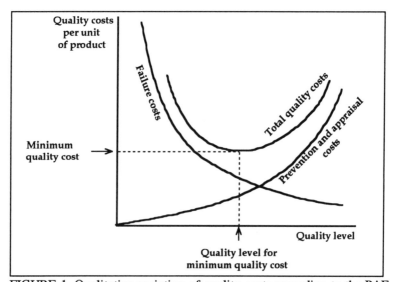

FIGURE 1. Qualitative variation of quality costs according to the PAF model.

products are 5, 25, and 70, expressed as a percentage of the $C_P$, $C_A$, and $C_F$, with respect to the total quality costs (Morgan 1984). If the actual relationship between the components of the costs of quality is the one exposed before, there are many failures and no preventive actions. If the goal is to produce at a better quality level, these figures indicate that an increase in the preventive efforts will reduce the defects and simultaneously increase the quality (Sterling 1985; Sandholm 1987).

## Results and Discussion

Quality costs are analyzed for the Argentine export fishing industry of the following manually elaborated products: frozen blocks of hake fillets (*Merluccius hubbsi*) and salted anchovy (*Engraulis anchoita*). In addition, the results obtained at Cuban plants for frozen lobster (*Panulirus argus*) are presented.

Variables and parameters for the evaluation of quality costs were elaborated from data obtained in previous projects and from experience in actual fish processing plants (Zugarramurdi et al. 1995). Starting from the results obtained from previous work (Montaner et al. 1994a; Montaner et al. 1994b; Montaner et al. 1994c; Crupkin et al. 1996; Montecchia et al. 1997), the influence of quality inputs in relation to product quality was analyzed. A linear correlation between initial raw material quality and final product quality was observed.

Raw material quality and training of workers mainly affects the yield and productivity of fishing plants (Amaria 1974; Kelsen et al. 1981; Zugarramurdi et al. 1995). This is observed in Figure 2. For salting plants, in the heading and gutting step, yield and productivity increase 28 and 40

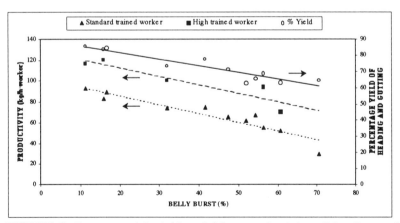

FIGURE 2. Raw material yield and labor productivity of heading and gutting operation as a function of initial raw material quality and worker's training level.

percent, respectively, with training when raw material is of good quality.

The required fixed investment for fishing plants was calculated as a function of capacity (tons of product per day) (Montaner et al. 1995). Food processing industries are usually variable costs intensive. The variable production costs structure for a level of standard quality was taken from Zugarramurdi et al. (1995) and Montaner and Zugarramurdi (1995).

The items considered for calculating the costs of resulting and controllable quality are presented in Table 1.

The $P_1$ includes the expenses for designing the quality plan, the audit system, designing and developing the quality measurements, and the control equipment. As a rule, they are estimated as a percentage of the fixed investment ($I_F$). One percent was used since the plants complied with GMP and SSOPs (NOAA 1993).

The $P_2$ includes design and permanent training programs in quality for suppliers and workers. Consequently, uniformity in size and quality of each lot is achieved. This was calculated as a function of the savings in sampling in the reception step (QMP 1989).

Cleaning is one of the most important activities of failure prevention in food plants (Huss 1994; Zugarramurdi et al. 1995). $P_3$ includes input costs (detergent; disinfectants) and labor to accomplish the cleaning under established procedures. It is a function of the direct labor cost (Dunsmore 1983).

In order to evaluate $P_4$, costs of adjustments and repairs that assure the right operation of the equipment were considered. They were estimated as a percentage of the maintenance cost of the plant. The maintenance cost was calculated as two to six percent of $I_F$ (Jelen and Black 1983; Zugarramurdi et al. 1995).

**TABLE 1**

List of Categories and Elements of Quality Costs

| Controllable Costs | Resulting Costs |
|---|---|
| *Prevention Costs* | *Internal Failure Costs* |
| P1. Design, Development and Implementation of a Quality Assurance Program | F1. Reprocessing or Spoilage |
| P2. Quality Training Programs for Suppliers and Workers | F2. Low Labor Productivity |
| P3. Hygiene and Sanitation | F3. Low Process Yield |
| P4. Preventive Maintenance | |
| *Appraisal Costs* | *External Failure Costs* |
| E1. Inputs and Plant Inspection | F4. Claims and Products Rejected and Returned |
| E2. Sampling and Laboratory Analysis | |
| E3. In-Process Control; Supervision | |

$E_1$ and $E_2$ were estimated as a percentage of the direct labor cost. $E_1$ includes the following costs: inspection of the plant, control of raw material (salt, fish), and inspection of the process. For $E_2$ calculation, tests to evaluate the quality of the finished products or products in process were considered. Both $E_1$ and $E_2$ diminish when training activities are accomplished ($P_2$). $E_3$ grows by the extra supervision required to reach high levels of quality of product (Zugarramurdi et al. 1995).

$F_1$, $F_2$ and $F_3$ have an impact on the loss in yield due to waste and low efficiency of labor rather then on the productivity and the profitability of the company (Valdimarson 1992; Bonnell 1994). They are estimated as losses or decreases in sale price (less quality) and inefficiencies in labor (less productivity) and in raw material (less yield).

External failures occur when a quality assurance plan does not exist or the plan is not applied in the correct way. $F_4$ is calculated as an exponential function of $C_A$ and is based on the number of claims and on the product's sale price (Saita 1991).

The integration of the concepts exposed for different levels of quality of product permitted the calculation of the total quality cost per unit of product (TQC). It can be expressed as a function of product quality level (q):

(1) $TQC(q) = C_P(q) + C_A(q) + C_F(q)$

In Figure 3 the controllable costs for plants of anchovy salting (Argentina), frozen hake (Argentina), and frozen lobster (Cuba) are presented where the values of quality were normalized and the costs were made dimensionless. Furthermore, the correlation obtained for the plants as a whole is shown as a first approximation to a general quality cost model for fishing plants ($R^2 = 0.85$).

The exponential nature of the curves of both the resulting and the controllable costs are shown in Figures 4 and 5. The increase in the $C_P$ and $C_A$ is correlated with a decrease in the internal and external failure cost; therefore, there is a point at which the quality cost per unit of product is at a minimum.

For salting plants (Figure 4), it is observed that:
- $C_P + C_A$ increases from 30 to 88 percent of the TQC when the quality level is increased from good to very good.
- $C_F$ reduces from 99 to 70 percent of the TQC when quality level is increased from regular to good.
- The minimum TQC is reached at 80 percent of the maximum product quality.

For freezing plants (Figure 5), it is observed that:
- $C_P + C_A$ increases from 35.5 to 64 percent of the TQC when there is an increase in the quality level from good to very good.

- C_F decreases from 97 to 64.5 percent of the TQC when there is an increase in the quality level from regular to good.
- The minimum TQC is reached at 80 percent of the maximum product quality.

Given the existence of a minimum quality cost, it is necessary to reach an economic balance to find the level of product quality for which the resources assigned to prevent and evaluate failures compensate the benefits due to failure decrease.

At the same time, it is important to calculate the average total production cost per unit of product (ATC) since, as observed in Figures 4 and 5, it continues reducing beyond the point of minimum quality cost per unit. This is due to the decrease of production costs by the increase in plant productivity associated with high levels of quality. Total quality costs descend from 40 to 21 percent of total production cost when the quality level is increased from bad to very good.

However, if the relationship between the final product quality and its selling price (P) is considered, a maximum benefit point is obtained that is not necessarily coincidental with the previous minimum points. Consequently, an interval of quality level exists within which the company will decide the optimum operation point according to the characteristics of demand function associated with the product.

Without the recognition of the existence of failure costs, the optimization of quality costs is impossible. In those cases, the usual

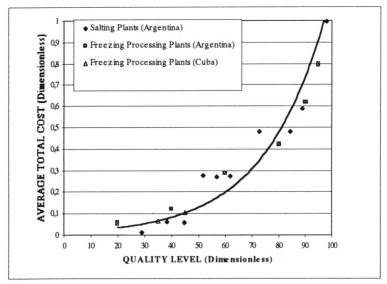

FIGURE 3. Controllable costs as a function of product quality level for fish processing plants.

actions of the companies are the reduction of prevention and appraisal costs that lead to lower quality levels.

The company could decide to increase the prevention and appraisal costs above the minimum point based on considerations such as volume of sales, safety, prestige (brand image), and company reputation.

## Conclusions

The quality costs in installed plants of the fishing sector are analyzed. Quality costs are divided into controllable (prevention and evaluation) and resulting costs (external and internal failures). The controllable costs can be taken to represent the costs of implementing HACCP, and the reduction

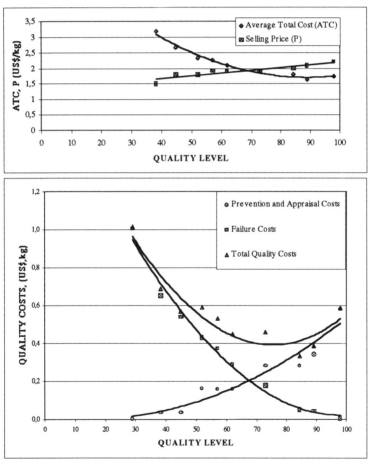

FIGURE 4. Average total cost, selling price, and quality costs as a function of product quality level for salting plants.

in resulting costs are the benefits from HACCP. It is observed that the more representative the controllable costs to reach a good level of product quality are inspection of raw material, training of labor, and production control.

Results show that due to the poor quality of inputs, failure costs are over 95 percent of total quality cost. For a very good quality level, failure costs descend below 20 percent of total quality cost. At the same time, total quality cost descends from 40 to 21 percent of total production cost when the level of product quality is increased from poor to very good. Also, different minimum points for total quality cost and total production cost and a maximum benefit point are observed, indicating an advisable

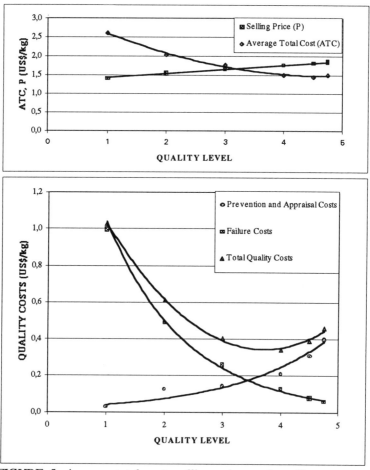

FIGURE 5. Average total cost, selling price, and quality costs as a function of product quality level for freezing processing plants.

working zone within 80 to 90 percent of the optimum quality level.

**Notes**

[1]A. Zugarramurdi is a Technological Research Member and Director of the Southern Regional Center (CEMSUR-CITEP), National Institute of Industrial Technology (INTI), Mar del Plata (Argentina). M.A. Parin and L. Gadaleta are Research and Development Staff (CONICET), the Southern Regional Center (CEMSUR-CITEP). H.M. Lupin is working at Fishery Industries Division, FAO Fisheries Department, Rome (Italy). The authors express their thanks for the financial support by CONICET, CIC, and FAO/DANIDA. Much appreciated was the assistance of the fishing industry associations of Mar del Plata (Argentina), which allowed the authors to make the necessary industrial assays.

**References**

Amaria P.J., 1974. Productivity Studies in Fish Processing: Some Factors Affecting Work Performance. Atlantic Fisheries Technological Conference, Quebec City.
British Standards Institution BS 6143 (BSI), 1981. Guide to the Determination and Use of Quality Related Costs.
British Standards Institution BS 6143 (BSI), 1990. Guide to the Economics of Quality, Part 2: Prevention, Appraisal and Failure.
Bonnell A.D., 1994. Quality Assurance in Seafood Processing: A Practical Guide. Chapman & Hall.
Crupkin, M., Montecchia, C., Montaner, M.I., Parin, M.A., Gadaleta, L.B., Beas, V. and Zugarramurdi, A., 1996. Influence of the Reproductive Cycle of Argentinian Hake on Fillet Yield and Operating Costs. Journal of Aquatic Food Product Technology, 5(3), 29-39.
Dunsmore D.G., Tomlinson P. and Ashley R.J., 1983. Cleaning Practices in the Australian Seafood Processing Industry. Food Technology in Australia, 35(12), 566-570.
Feigenbaum A.V., 1974. Total Quality Control. New York: McGraw Hill.
Hosking G., 1984. Quality Cost Measurement in the Food Industry. Food Technology in Australia, 36(4), 165-167.
Huss H.H., 1994. Assurance of Seafood Quality. FAO. Fisheries Technical Paper No. 334, 169pp.
Huss H.H , 1995. Quality and Quality Changes in Fresh Fish. FAO. Fisheries Technical Paper No. 348, 195pp.
Jelen F.C. and Black J., 1983. Cost and Optimization Engineering. New York: McGraw Hill, 538.
Kelsen S.E. Mujica, Belloni R. and Malan C., 1981. Manual sobre Métodos e Investigaciones Económicas en la Industria Pesquera. Instituto Nacional de Pesca, Uruguay, Informe Técnico, 24.
Montaner M.I., Parin M.A., Zugarramurdi A. and Lupín H.M., 1994a. Requerimiento de Insumos de la Industria Pesquera. I. Materia prima. Revista de Tecnología e Higiene de los Alimentos: Alimentaria (253), 19-24.
Montaner M.I., Parin M.A., Zugarramurdi A. and Lupín H.M., 1994b. Requerimiento de Insumos de la Industria Pesquera. II. Mano de obra. Revista de Tecnología e Higiene de los Alimentos: Alimentaria (254), 81-85.

Montaner M.I., Parin M.A., Zugarramurdi A. and Lupìn H.M., 1994c. Requerimiento de Insumos de la Industria Pesquera. III. Servicios Auxiliares. Revista de Tecnología e Higiene de los Alimentos: Alimentaria (255), 27-29.

Montaner, M.I., Gadaleta, L, Parin, M.A. and Zugarramurdi, A., 1995. Estimation of Investment Costs in Fish Processing Plants. International Journal of Production Economics, 40, 153-161.

Montaner, M.I. and Zugarramurdi, A., 1995. Influence of Anchovy Quality on Yield and Productivity in Salting Plants. Journal of Food Quality, 18, 69-82.

Montecchia, C.L., Roura, S.I., Roldán, H., Perez Borla, O. and Crupkin, M., 1997. Biochemical and Physicochemical Properties of Actomyosin from Frozen Pre-and Post-spawned Hake. Journal of Food Science, 62 (3).

Morgan Anderson R. 1984. Controlling Food Plant Quality Costs. Food Technology in USA. April, 111-112.

NOAA, National Oceanic and Atmospheric Administration, 1993. U.S. Department of Commerce, National Marine Fisheries Services. Office of Trade and Industry Services. The Report of the Model Seafood Surveillance Project. Draft.Washington, DC.

Parin M.A. and Zugarramurdi A., 1994. Investment and Production Costs Analysis in Food Processing Plants. International Journal of Production Economics, (34), 83-89.

Porter, L.J and Rayner, P., 1992. Quality Costing for Total Quality Management. International Journal of Production Economics, 27, 69-81.

QMP, 1989. Canadian Quality Management Programme.

Saita, M., 1991. Economia della Qualit. Strategia e Costi. ISEDI Petrini Editore, Torino, Italia, 306 p.

Sandholm L., 1987. Reducing Quality Costs to Improve Export Earnings. International Trade Forum, Oct.-Dec., 26-31.

Sterling R., 1985. Relationship Between Manufacturing and Quality Costs. Food Technology in USA, September, 54-55.

Valdimarson G., 1992. Developments in Fish Processing - Technological Aspects of Quality. In Quality Assurance in the Fish Industry, ed. H.H. Huss et al., 169-183, Elsevier Science Publications.

Zugarramurdi A., 1981. Estimación de Costos en la Industria Pesquera. La Industria Carnica Latinoamericana, 8(40), 16-22.

Zugarramurdi A. and Parin M.A., 1988. Economic Comparison of Manual and Mechanical Hake Filleting. Eng. Costs & Prod. Econ., (13), 89-95.

Zugarramurdi A., Parin M.A and Lupin H.M., 1995. Economic Engineering applied to the fishery industry. FAO Fish. Tech. Pap., (351), 295.